PROBLEMS OF CALIBRATION OF ABSOLUTE MAGNITUDES AND TEMPERATURE OF STARS

INTERNATIONAL ASTRONOMICAL UNION
UNION ASTRONOMIQUE INTERNATIONALE

SYMPOSIUM No. 54

HELD IN GENEVA, SWITZERLAND, SEPTEMBER 12–15, 1972

PROBLEMS OF CALIBRATION OF ABSOLUTE MAGNITUDES AND TEMPERATURE OF STARS

EDITED BY

B. HAUCK

*Institut d'astronomie de l'Université de Lausanne
et Observatoire de Genève, Szitzerland*

AND

B. E. WESTERLUND

ESO, Santiago, Chile

D. REIDEL PUBLISHING COMPANY

DORDRECHT-HOLLAND / BOSTON-U.S.A.

1973

Published on behalf of
the International Astronomical Union
by
D. Reidel Publishing Company, P. O. Box 17, Dordrecht, Holland

Sold and distributed in the U.S.A., Canada, and Mexico
by D. Reidel Publishing Company, Inc.
306 Dartmouth Street, Boston,
Mass. 02116, U.S.A.

Library of Congress Catalog Card Number 73–83562

ISBN-13:978-90-277-0372-9   e-ISBN-13:978-94-010-2645-1
DOI: 10.1007/978-94-010-2645-1

TABLE OF CONTENTS

PART IV / ABSOLUTE MAGNITUDE DETERMINATIONS FROM HYDROGEN-LINE PHOTOMETRY

PART V / GROUND BASED AND EXTRATERRESTRIAL OBSERVATIONS OF STELLAR FLUX

PART VI / THE USE OF MODEL ATMOSPHERES FOR TEMPERATURE

PART VII / STELLAR TEMPERATURE SCALE AND BOLOMETRIC CORRECTIONS

PART VIII / CHOICE OF STANDARD STARS

PREFACE

In connection with arranging IAU Symposium No. 50 on 'Spectral Classification and Multicolour Photometry', sponsored by Commissions 45 and 25, it was decided to exclude all calibration problems. Instead it was agreed that we should attempt to arrange a separate symposium, dealing with the fundamental problems of the calibration of absolute magnitudes and temperatures of stars.

The Executive Committee of the IAU accepted our proposal, and IAU Symposium No. 54 was held in Geneva on September 12–15, 1972, sponsored by the following IAU Commissions: 24, 25, 29, 33, 35, 37, 44 and 45. It was attended by about 90 scientists representing 16 countries.

The Symposium was divided into eight sessions. Each session started with a review paper by an invited speaker; this was followed by a general discussion including a few contributed papers.

The contents of the present volume follow closely the programmes of the individual sessions of the Symposium. Most of the recorded discussions have been kept, and only in a few cases have the order of questions and comments been altered so as to obtain more homogeneity in the presentation.

Financial assistance was provided by the Executive Committee of the International Astronomical Union for travel grants to certain participants. Some travel grants were also supplied by the 'Fonds national suisse de la recherche scientifique'. We are grateful to the Universities of Geneva and Lausanne for additional support such as secretarial help and other facilities. Mrs M. Zuger and Miss B. Zwenzner are gratefully acknowledged for their assistance at all stages of preparation of the present volume.

SCIENTIFIC ORGANIZING COMMITTEE

C. Jaschek (Chairman), A. D. Code, D. L. Crawford, Y. Fujita, M. Golay,
W. J. Luyten, V. L. Straizys, H. C. Thomas, B. E. Westerlund

LOCAL ORGANIZING COMMITTEE

M. Golay (Chairman), B. Hauck, F. Rufener, U. Steinlin, R. West

PART I

ABSOLUTE MAGNITUDES
FROM TRIGONOMETRIC AND
STATISTICAL PARALLAXES

ABSOLUTE MAGNITUDES FROM TRIGONOMETRIC
AND STATISTICAL PARALLAXES

P. VAN DE KAMP

Sproul Observatory, Swarthmore College, Swarthmore, Pa., U.S.A.

Abstract. The attainable accuracy of photographic trigonometric parallaxes may be improved by the use of modern measuring engines, larger number of exposures, facilitated by increased speed of emulsions, more refined knowledge of reduction to absolute parallax. Systematic errors remain an obstacle both in trigonometric and statistical parallaxes.

With an attainable goal of $\pm 0''002$ for the probable error of a parallax it appears difficult to obtain absolute magnitudes with a probable error less than 0.2 mag. for stars beyond 20 pc.

I believe that the best contribution I can make to the subject of absolute magnitudes is a presentation of the attainable accuracy in trigonometric parallaxes and a brief reference to the same problem in statistical parallaxes.

The evaluation of the accuracy of determining trigonometric annual parallaxes has been frequently studied and reported. Among the exhaustive studies, I mention those by Land (1948), Strand (1963), and Lippincott (1970). It is not easy to improve on these investigations but perhaps some general remarks may be in order today.

Greater accuracy may be reached in several ways. There is (1) the obvious choice of observing times, (2) the potential increase in accuracy due to the vastly increased speed of photographic emulsions, (3) the introduction of new measuring techniques, (4) better knowledge of reduction from relative to absolute parallax, (5) elimination of orbital motion in the case of series of observations covering a long time interval, both for known visual binaries and perturbations, (6) recognition and possible elimination of systematic errors, which vary with the interval of observation, (7) as far as absolute magnitudes are concerned: recognition of duplicity in order to obtain proper absolute magnitudes of the components.

1. Observing Times

Ideally, or rather traditionally, long focus photographic parallax observations are made at dawn and at dusk, for the purpose of obtaining the largest possible range in parallax factors in right ascension. The extreme range thus obtainable varies with the position of the star in the sky, the observer's latitude, and the pattern of clear nights as it varies with the time of the year. A large range in parallax factors, up to a maximum of 2.0, may be obtained for stars in certain right ascensions, while in other parts of the sky, at best, one has to be satisfied with smaller range, say only 1.5, and this may not even be possible if weather conditions are unfavorable. As a rule a smaller range in right ascension is accompanied by a larger range of parallax factors in declination. At the Sproul Observatory, as a matter of policy, all stars are measured in declination also, not primarily for parallax, but for the obvious purpose of observing orbital motion

B. Hauck and B. E. Westerlund (eds.), Problems of Calibration of Absolute Magnitudes and Temperature of Stars, 3–10.
All Rights Reserved. Copyright © 1973 by the IAU.

(or perturbation). While on the average the weight of parallax determinations in declination is only 15% of that in right ascension, this procedure statistically, at least, has contributed to an evaluation of systematic accuracy. From a study of 124 stars, Lippincott has found an average difference, in the sense R. A. minus Decl., of $-0\overset{''}{.}0018 \pm 0\overset{''}{.}0011$ (p.e.). While this is no proof that either or both coordinates are free from systematic errors, the close agreement may be taken to exclude large systematic errors, at least *on the average*, for the whole sky.

In the early days of long-focus photographic parallax determinations, measurements were frequently made in celestial longitude, thus yielding a somewhat larger range in parallax factors. I do not believe that a return to this approach should be made. In all observational work it remains important to measure in basic, natural coordinates, as they result from the unavoidable observational approach and to study the errors, both accidental and systematic, inherent in this approach. Right Ascension and Declination are these natural coordinates, affected in a straight forward manner by such errors as are due to tracking, or guiding, in right ascension, while refraction and dispersion play a primary role in declination. For observations *off* the meridian, refraction and dispersion enter into right ascension, proportional to the hour angle and with opposite signs, for eastern or western hour angles. Schlesinger and others pointed this out and argued for as small a range in hour angle as possible, preferably hovering around the meridian. Personally, I take a very dim view of certain recent experiments with extremely large ranges in hour angle. This, in my opinion, is asking for trouble, no matter how elaborate the system of parameters and computer facilities may be. I continue to believe that prevention is better than the cure.

Ideally, parallax factors (in right ascension) of $+1.0$ and -1.0 are the astrometrist's dream. However, as pointed out earlier this ideal usually remains a castle in the air. This is no reason for despair; after all, it is the *range* in parallax factor which counts.

There is the related problem of matching the number of evening and morning observations. While ideally the number of plates with positive and negative parallax factors should be evenly matched, in practice one may settle or have to settle for considerable imbalance without much harm. Assuming a distribution of material which permits elimination (i.e., determination) of proper motion, the weight of a parallax determination depends on the total weight of the material and the range in parallax factors. To illustrate this, let us assume equal negative (evening) parallax factors $(-P)$ and equal (morning) parallax factors $(+P)$, hence a range $2P$ between morning and evening. Assume fractions f and $1 - f$ of the material with opposite parallax factors. The weight of the parallax determination is given by

$$nP^2 \left[1 - (1 - 2f)^2\right] = nP^2 \left[4f - 4f^2\right],$$

where n is the total weight of the material (see following table on next page).

An imbalance in the ratio 2 to 1 leads to a weight reduction by a factor 0.89, a ratio 3 to 1 to a reduction of 0.75. Such imbalance, if accompanied by a *good range* in parallax factors, may often be quite an acceptable compromise, if not the only obtainable goal.

f	$1-f$	Weight of parallax	Relative weight
0.5	0.5	nP^2	1.00
0.6	0.4	$0.96\,nP^2$	0.96
0.7	0.3	$0.84\,nP^2$	0.84
0.8	0.2	$0.64\,nP^2$	0.64
0.9	0.1	$0.36\,nP^2$	0.36

2. Photographic Emulsion; Law of Diminishing Returns

An important historical factor in obtaining higher accuracy is the tremendous increase in sensitivity of photographic emulsions, conservatively speaking, fifty fold over the past fifty years. Where 'in the old days' we used to struggle to get 20 plates for a parallax determination, with the same observational effort we can now obtain 200 plates or more. The resulting increase in accuracy of a parallax determination is obvious if we may assume that increase of the number of exposures yields a corresponding increase in accuracy, i.e., that the law of diminishing returns does not enter to an appreciable extent. There is some evidence for small inhibiting effects of this law in the form of night errors which discourage an unduly enlarged observational effort for any one star on any one night. I feel that 16 to 20 exposures distributed over four plates still appear a desirable maximum goal for any one night.

3. Measurements on Modern Machines

Measurements on modern machines have proven that an impersonal evaluation by photoelectric scanning of photographic star positions together with superior precision screws, such as in the Grant two-screw machine, or the sophisticated optical approach, which is the basis of the U.S. Naval Observatory machine, lead to a substantially improved positional accuracy. A good indication of the superior accuracy obtained with the USNO machine is a comparison of measurements made by John L. Hershey of plates on CC 1213 taken with the Sproul 24 in. refractor over the interval 1959 to 1969.

Machine	No. of reference stars	P.E. 1
Manual Swarthmore	3	$\pm\,0^{\mathrm{mm}}.00139$
USNO	3	$\pm\,0^{\mathrm{mm}}.00098$
USNO	8	$\pm\,0^{\mathrm{mm}}.00096$
USNO	11	$\pm\,0^{\mathrm{mm}}.00091$

Provisional results obtained with the 2-screw Grant machine now in operation at the Sproul Observatory indicate a similar favorable factor of weight increase. We list the following comparison; the positions refer to 3 to 5 reference stars.

	P.E. 1
One-screw machines built in 1916, 1943, 1963	$\pm 0^{mm}.0020$
Grant 2-screw machine	$\pm 0^{mm}.0012$

The decrease in probable error from the bisection to the scanning method is striking. If these results are confirmed and maintained we reach the important conclusion that automatic or semi-automatic positioning by modern measuring engines leads, conservatively speaking, to a *doubling of weight* compared with earlier manual bisections. Here then is a significant accomplishment toward increasing the accuracy of long-focus trigonometric parallaxes. It raises the question whether all existing early plate material should be measured, to double, in a manner of speaking, that material, or rather, for the first time to exploit this material to the fullest; to extract from it all that can be obtained in the way of positional accuracy.

I view with reservation certain current attempts to materially increase parallax accuracy by using a large number of reference stars. The weight of a position referred to n reference stars is proportional $n/(n+1)$ for the ideal case of equal dependences, or weights, of the reference stars. The increase in weight with increasing number of reference stars is seen from the following tabulation:

No. of reference stars	Weight	Weight relative to 3 reference stars
3	0.75	1.00
4	0.80	1.07
5	0.833	1.11
6	0.857	1.14
7	0.875	1.17
8	0.889	1.19
–	–	–
11	0.917	1.22
–	–	–
∞	1.00	1.33

Recent results obtained at Sproul for CC 1213 yield relative weights of 1.04 and 1.11 for 8 and 11 star-, relative to the 3 star-configuration, actually below the values 1.19 and 1.22, which hold for equal dependences; all this for the case of linear plate constants. Introduction of other parameters, such as magnitude, color and non-linear terms, may be a reason for using more reference stars. Experience will tell.

4. Reduction from Relative to Absolute Parallax

This correction did not have to be taken too seriously in the days of conventional parallax determinations based on some twenty plates and yielding probable errors of about $\pm 0\rlap{.}''010$. But currently, when such determinations are sometimes based on hundreds, or even thousands of plates, and with the vastly improved measuring techniques, probable errors of $\pm 0\rlap{.}''003$ and less are becoming commonplace. Where in the past statistical corrections, based on magnitude and galactic latitude were adequate, there is now need for refining the evaluations of the parallaxes of the reference stars. In my opinion magnitude, spectra, coupled with some knowledge of the proper motions of the reference stars, are the best parameters for estimating the parallaxes of the reference stars and, hence, the reduction to absolute in the form of the dependence weighted mean of their individual parallaxes. An increase in number of reference stars has been proposed as a means of improving the reduction to absolute. In principle, this is true; in practice, little improvement would be reached.

5. Elimination of Orbital Motion (Perturbation)

Early parallax determinations based on a series of plates extending over a few years only, could hardly be affected by orbital motion. Series spread over several decades are needed in the determination of mass-ratio and perturbations. The conditional equation for proper motion and (relative) parallax:

$$X = c + \mu t + \pi P$$

is now replaced by

$$X = c + \mu t + \pi P + \alpha Q$$

or its equivalent, where α is the semi-major axis of the orbit of the observed position (primary, secondary, or photocenter). In case of known or recognized orbital motion, allowance can be made and the parallax determination will thus be not affected by orbital motion. It is likely however that for a number of-alleged-single stars orbital motion may go undetected and thus affect the parallax depending on the relative size of π and α.

6. Recognition of Systematic Errors Depending on the Interval of Observation

While the afore mentioned orbital effects would be different for different stars, there also may be systematic effects depending on the time interval which have nothing to do with the possible orbital behaviour of the individual star, but which depend on the specific time interval, decade 1960-70 for example, and/or on the right ascension and declination of the star.

There is evidence for such effects in the recent study by Miss Lippincott (1970).

These effects may be related to adjustments of the optics of the telescope, or possible subtle changes in observing procedures.

7. Recognition of Duplicity in Order to Obtain Proper Absolute Magnitudes

The attainable accuracy in parallaxes has some bearing on the determination and interpretation of absolute magnitudes. Is the observed star single or an unrecognized binary, not detected by any other means? An abnormally bright absolute magnitude may often be regarded with reservation or suspicion, and may be indicative of duplicity.

All in all, we may look forward toward appreciably improved accuracy of trigonometric parallaxes. Increases in weights up to a factor of 10 or more are obtained, or within reach, for large numbers of stars. The corresponding decrease in probable error, therefore, permits us with the same accuracy as before to explore stellar distances by the long-focus photographic method at least three times farther than we have done so far. But there are still other problems. Increased accidental accuracy does not at all necessarily mean increased systematic accuracy. The bug bear of systematic errors remains with us, although it may eventually become easier to pinpoint these errors because of increased accidental accuracy.

A similar pessimistic attitude may be maintained toward such ambitious projects as determining the trigonometric parallax of a star cluster such as the Hyades. The parallax of the Hyades is close to $+0\overset{''}{.}025$ and is rather well known from studies of the perspective effect in the proper motions of the cluster members. To improve this value we should reach a probable error of well below $\pm 0\overset{''}{.}001$ for the trigonometric parallax of the cluster. This expected value, would be based on many determinations for individual members of the cluster. With an investment of several thousand plates such a small error may well be reached, formally speaking, but who is going to assess the systematic error of such a result?

A few final remarks concerning the use of parallaxes. The principal significance of determining trigonometric parallaxes is their use for determining absolute magnitudes (M) and masses $(\mathcal{M})$:

$$M = m + 5 + 5 \log p \quad \text{or} \quad \log L = 0.4\,(m - M_\odot) - 2 - 2 \log p$$

$$\mathcal{M}_1 + \mathcal{M}_2 = \frac{a^3}{P^2} \cdot \frac{1}{p^3} \quad \text{or} \quad \log(\mathcal{M}_1 + \mathcal{M}_2) = \log \frac{a^3}{P^2} - 3 \log p.$$

The error situation is as follows:

$$dM = 2.2 \frac{dp}{p} \qquad \frac{dL}{L} = -2 \frac{dp}{p}$$

$$\frac{d(\mathcal{M}_1 + \mathcal{M}_2)}{\mathcal{M}_1 + \mathcal{M}_2} = -3 \frac{dp}{p} \qquad d \log({}_1\mathcal{M} + \mathcal{M}_2) = -1.3 \frac{dp}{p}$$

The fractional errors in the luminosity and in the total mass of a binary are proportional to the fractional error of the parallax; the same holds for the masses of the components, ignoring the error in the mass ratio. To maintain the same relative accuracy at greater distances, the weight of the parallax determination should increase with the square of the distance. We remind ourselves of the well known factor 3, which makes it difficult to obtain precise masses even for the nearest visual binaries.

The following table summarizes the ratios between the fractional errors of the various quantities and the fractional error in parallax:

Absolute Magnitude	$+ 2.2$	Luminosity	$- 2$
Mass	$- 3$	log Mass	$- 1.3$

$$\frac{\text{Absolute Magnitude}}{\text{log Mass}} \quad -\frac{5}{3}$$

$$\frac{\text{Luminosity}}{\text{Mass}} \quad +\frac{2}{3}$$

$$\frac{\text{Luminosity}}{\text{Absolute Magnitude}} \quad -0.9$$

$$\frac{\text{Mass}}{\text{log Mass}} \quad +2.3$$

An example of the error situation is given below:

Parallax	Probable Error	Probable Error in abs. mag.	Fractional Error in total mass
0″100	± 0″005	0.11	0.15
	0.002	0.04	0.06
	0.001	0.02	0.03
0.050	± 0.005	0.22	0.30
	0.002	0.09	0.12
	0.001	0.04	0.06
0.020	± 0.005	0.5	0.75
	0.002	0.2	0.30
	0.001	0.1	0.15

The tabulated figures speak for themselves. Assuming a probable error of $\pm 0″002$ for parallax as a reasonable and obtainable goal, it becomes difficult to obtain stellar masses with a relative probable error smaller than 10% for stars beyond 20 pc. As to absolute magnitudes it appears difficult to obtain them with a probable error less than 0.2 mag. beyond 20 pc. And we have not yet taken into account, since we cannot, systematic errors!

On the subject of absolute magnitudes from statistical parallaxes, my colleague Luyten will deal with this problem particularly as it refers to the fainter end of the luminosity curve.

I have not been active in this field for several decades and feel that I cannot do much more than refer to the following survey papers:

(1) 'Statistical Parallaxes', by A. N. Vyssotsky, *The Astronomical Journal* **63**, 166, presented at the Conference on the Cosmic Distance Scale, held in Charlottesville, Virginia, April 5–7, 1956.

Vyssotsky stresses the limitations of the two principal methods used for deriving mean parallaxes: parallactic motions and dispersion in peculiar motions, i.e. v- and τ-components.

As to parallactic motions, they are sensitive to systematic errors in proper motions and errors in the precession constants. Moreover, knowledge of radial velocities is required.

As to peculiar motions, high accuracy is required as well as knowledge of radial velocities.

An additional complication is, of course, the dispersion in absolute magnitudes which affects the conversion from mean parallax to mean distance.

(2) 'Statistical Parallaxes', by Fricke (1964), presented at *Aspects of Stellar Evolution*, Flagstaff, Arizona, June 22–24, 1964.

Fricke points out the existence of errors up to $0^{m}2$ and more depending on the fundamental system used.

References

Fricke, W.: 1964, *Vistas in Astronomy* **8**, 211.
Land, G.: 1948, *Trans. Astron. Obs.*, Yale University, Vol. 15, Part 3.
Lippincott, S. L.: 1970, *Publ. Leander McCormick Obs.* **16**, 99.
Strand, K. Aa.: 1963, *Stars and Stellar Systems* **3**, 55.
Vyssotsky, A. N.: 1956, *Astron. J.* **63**, 166.

THE DETERMINATION OF ABSOLUTE MAGNITUDES
FROM PROPER MOTIONS

W. J. LUYTEN

The Observatory of the University of Minnesota, Minn., U.S.A.

Abstract. A brief review is given of the several ways in which stellar motions may be used statistically. In the case where the motions of all stars shown on plates of certain regions are measured these are generally used to determine kinematical relations in the Galaxy and only secondarily for the estimation of luminosities; when motions are measured for special groups of stars the solar motion is first determined and then used for the calculation of distances and luminosities.

When dealing with the statistical analysis of stellar motions we should never lose sight of the fact that on them depends, in final analysis, the derivation of the scale of the universe.

In actually utilizing proper motions for the estimation of stellar luminosities we must distinguish several different situations. The most general case is that where the motions of *all* stars shown on one or more plates covering certain areas in the sky have been measured. This is the situation dealt with by Vasilevskis and Klemola in the Lick Observatory program of determination of absolute proper motions relative to galaxies. Here, of course, the prime purpose is not so much the determination of stellar luminosities but rather an analysis of the kinematics of the Galaxy, solar motion, galactic rotation, etc., with, as one of the most important by-products, the derivation of the corrections to be applied to convert relative to absolute proper motions. In general, in these cases, there is no dearth of material, thousands of stars are normally available, and the procedures employed are straightforward. But, as usual, there are pitfalls, such as, e.g. whether to include or reject, or give only partial weight to any, or all stars of large proper motion, since even one star with a very large motion can dominate the solution for a fairly substantial area in the sky. Another one is that, if one wishes to use such data and analyses for the derivation of luminosities, one must make a guess as to the value of the solar velocity – usually from the declination of the apex determined – for, with the very faint stars considered in these investigations, there never are radial velocities available.

Perhaps the most satisfactory situation in the statistical utilization of proper motions occurs in the application to moving clusters. The classical case is that of the Hyades cluster where we have available substantial numbers of very accurate absolute proper motions as well as radial velocities, and the solar motion as such is not involved. It has now become almost possible to determine accurate individual distances and luminosities for the stars in the cluster, but this is obviously a very special case.

In another, somewhat more general case, we deal with groups of stars which, at best, appear to belong to a fairly homogeneous class, astrophysically speaking, either from their spectra, or from other properties, such as Cepheids, Long-period variables, N or R stars, Planetary Nebulae etc. Here the usual approach is, first, to try and deter-

B. Hauck and B. E. Westerlund (eds.), Problems of Calibration of Absolute Magnitudes and Temperature of Stars, 11–17.

mine the solar apex from the motions alone, then select a value for the solar velocity and proceed from there. Sometimes there are radial velocities available, which can give at least an approximate solution for the value of the solar velocity, but if there are very few, or no radial velocities available, one must, again, do the best one can and estimate the solar velocity from the declination of the apex found. But, one must be very careful and watchful for surprises such as discovering that what at first sight appeared to be a homogeneous group, later turns out to be a mixture of two or more subgroups with different kinematical properties – witness what happened in the case of the W Virginis type of Cepheids.

Also, one must be very careful that pre-conceived astrophysical beliefs do not force the proper motions into a straight-jacket. I am thinking of such situations as with the Faint Blue Stars where the spectroscopists identified large numbers of White Dwarfs among them, even though many of these had been known before, or were found later to possess virtually no proper motions. This was especially glaring since most of these Faint Blue Stars were in high galactic latitudes where we would expect the stars to have rather larger tangential velocities than usual, and this is, indeed, borne out by or proper motion surveys. When a group of tenth-magnitude blue stars near the North Galactic Pole, selected only from their colours, all turn out to have proper motions of $0''1$ annually, or less, they are not likely to be, *all* of them, White Dwarfs with absolute magnitudes around $+11$.

In the spectroscopic estimation of absolute magnitude there are many hidden unknowns and uncertainties. We know there are many large uncertainties in results derived from proper motions but I think it is fair to say that most of these are very visible and on the surface since they result from the kinematic distribution – and hence one can make fairly reliable estimates of them.

Summarizing I would say that in those cases, where one hopes one is dealing with a single astrophysical group, presumably with a well-defined mean luminosity, and not too large a dispersion around it, about the only way to determine this mean luminosity from the proper motions is to make a solution for solar motion and infer the solar velocity from the declination of the apex found. But, if the stars are not sufficiently well distributed to yield a reliable solar apex, and one must use parallactic motions for single areas, from an assumed apex, then, obviously, the resultant uncertainties in the luminosities become very much larger.

The situation with which I personally am mostly familiar is that of the determination of luminosities for stars which had been pre-selected according to size of proper motion. This problem is basic to the derivation of the function describing the distribution of stellar luminosities and involves the estimation af absolute magnitudes in individual cases as well as the determination of mean absolute magnitudes and their dispersion for different groups of stars of known proper motion and apparent magnitude.

There are two main approaches toward a solution: the use of mean-parallax formula in individual cases, which was pioneered by Kapteyn, and, again, the determination of the solar motion, and subsequent evaluation of the parallactic motion.

Kapteyn's formula was a simple one:

$$\log p = a + bm + c \log \mu$$

and one determines the constants from data for the stars of known, and reliable trigonometric parallaxes, When one is dealing with the proper motions of the very faint stars involved in the derivation of the luminosity function there are actually few, if any stars with known trigonometric parallaxes, and one must extrapolate from data for brighter stars. It is for this reason that I have tried a different formula, viz.

$$M = a + bH \quad \text{where} \quad H = m + 5 + 5 \log \mu,$$

the so-called 'reduced proper motion', first used by Hertzsprung. This being a two-constant formula it cannot, of course, be as accurate as Kapteyn's which uses three constants, but because of the large extrapolation and inherent uncertainties involved I have felt that, because it is simpler, it has an advantage in usage.

The real difficulty in the application of either formula is that there are *no* parallaxes or radial velocities for these very faint stars – fainter than the eighteenth apparent magnitude, and hence it is not now possible to make a reliable determination of the real dispersion in the absolute magnitudes derived, as due to the dispersion in tangential velocities and the only thing one can do is, obviously, to extrapolate from the known data for brighter stars.

Since there have been several recent attempts to criticize my derivation of the luminosity function from the data in the Bruce and Palomar Proper Motion Surveys in 1938 and 1968, I may perhaps digress here a little and deal with those criticisms.

Wanner has, for the third time now, claimed that the mean absolute magnitudes I derived for the proper-motion stars used were too bright and repeatedly has derived new reduction formulae which would indicate that my absolute magnitudes are too bright by about $1^{m}2$ in the region where it counts most, between $M = +13$ and $M = +16$, near the maximum of my luminosity function. Yet when he is finally through with his own derivation of the luminosity function, he ends up with a maximum for this function which is more than *three* magnitudes *brighter* than mine (+12.3 as against +15.7).

On the other hand, Murray and Sanduleak have, on the basis of an analysis of the proper motions of all of 21 stars near the North Galactic Pole, reached the conclusion that the frequencies in the region of the maximum of my luminosity function must be multiplied by a factor of *five* – then they also obtain agreement with Oort's mass-density per cubic parsec. Now the 21 stars they used were found on objective-prism plates, and classified as M. Since they are of the apparent magnitude 15 vis, the argument is used that they cannot be giants, hence they must be M-dwarfs, with a mean absolute magnitude of around +11. Now of the 21 stars they used, seven had been found before, in my proper motion survey, to have proper motions averaging about 0".10 annually. Seven more were found later to have small proper motions, averaging around 0".06 annually, and the last seven appear to have proper motions so small that I would call them marginal. Yet, according to the spectroscopists they must all be M-dwarfs and should have parallaxes around $0''.02$ on the average, which would imply that their average tangential velocity must be around 14 km s^{-1} – and this at the North Galactic Pole.

Other recent determinations of the luminosity function find, in one case, a constant frequency from $M = +8$ to $M = +16$, and, in another case, a continuously increasing frequency from supergiants all the way to, I believe, micrometeorites.

Since all these recent analyses appear to pull my luminosity function in several different directions at once, and since, as of now, there is no material inexistence even faintly comparable to that from the Palomar Survey – which most of these researches ignore – I see no reason to change the present luminosity function. I want to emphasize again that it is necessarily a very preliminary one. When the Palomar Survey is at least half completed we shall have a lot more, and better data on the frequency of proper-motion stars. But beyond that we shall need at least hundred trigonometric parallaxes for very faint stars down to at least the twentieth photographic magnitude in order to determine both the mean absolute magnitude and the dispersion in tangential velocities for these stars. Until we have these data I believe it is useless to talk about a determination of the luminosity function other than an extreme preliminary and cursory evaluation of it, and, equally, there is no point in trying to find any second order undulation in the shape of the curve: a smooth curve is all that the present data warrant.

Finally, there is the problem that sometimes comes up with an attempt to find a rough luminosity for a single object of a new astrophysical type, perhaps, when no spectroscopic data are available and only a very rough preliminary proper motion is known. If it is a star found from its large proper motion then, perhaps $p = \mu/12$ is as good a guess as any, but if the proper motion is determined after the object had been found then $p = \mu/10$ is perhaps better. The direction of the motion is also a good indication, for one must remember that no self-respecting high-velocity star goes north, or toward the solar apex.

References

Luyten, W. J.: 1939, *Publ. Obs. Minn.* **2**, No. 7.
Luyten, W. J.: 1968, *Monthly Notices Roy. Astron. Soc.* **139**, 221.
Murray, B. C. and Sanduleak, N.: 1972, *Monthly Notices Roy. Astron. Soc.* (in press).
Vasilevski, A. E. and Klemola, A. R.: 1971, *Publ. Lick Obs.* **22**, No. 2.
Wanner, J. F.: 1972, *Monthly Notices Roy. Astron. Soc.* **155**, 463, 468.

DISCUSSION

Strand: Since Dr Luyten has referred so frequently to Naval Observatory parallaxes, I like to show a diagram showing the present status of this program, as far as data now completed, and now in press in the Observatory's publications. In this M_v vs $B-V$ diagram it is noted that there is a clear separation between the main sequence and the degenerate branch, with no white dwarfs beyond $B-V$ larger than approximately one. It is noted *for* the stars of immediate colour that those with *small* parallaxes lie approximately on the average one magnitude *below* the main sequence. These are high velocity sub-dwarfs.

Pecker: Do you have some T-Tauri and UV Ceti or at least stars with IR excesses in your survey? Where do they fall on your diagram?

Strand: There are no T-Tauri stars on our program at this time.

Murray: I would like to congratulate Strand on the results he has shown for his 61-in. reflector. However, he is fortunate in having a dedicated telescope for parallax work. It is desirable for others to observe faint stars. This means using large telescopes and taking what time one can get. Hence, as in our programme with the Isaac Newton Telescope, it is essential to use the time available, whatever the hour angles involved.

Jaschek: Is any effort being made to take spectra of the stars you are observing for parallax?

Strand: No. Only Greenstein is observing the brighter ones, many of them being faint.

Luyten: In connection with determining parallaxes in declination I should like to point out that not only are the amounts much smaller that in RA but furthermore, while one can get the extremes of the parallax ellipse in right ascension it is in the nature of things impossible to get one of the extremes in declination, because then the star is in conjunction with the sun and inobservable. Only if the star is very close to the pole could one observe it, but then it will be below the pole and in that case other troubles appear.

Van Altena: In an attempt to increase the accuracy of trigonometric parallaxes several changes in the observation and reduction methods have been made at Lick and Yerkes Observatories. Observations are made at large hour angles to increase the parallax factor. Traditional observations near the meridian give average parallax factors of 0.6–0.7, while at Yerkes an average value of 0.95 is obtained. The effects of differential colour refraction can be accurately determined observationally and theoretically and agree very well. The use of a large number of reference stars (~ 24) allows one to: (1) solve explicitly for the magnitude equation; (2) compensate for quadratic effects in the coordinates; (3) minimize the importance of a single reference star in the position of the parallax star; (4) obtain the correction to absolute parallax from the proper motion dispersion and the expected velocity dispersion. The results of these modified methods give increased accuracy of parallaxes and parallaxes in agreement with their expected position in the colour-magnitude diagram.

Van de Kamp: I would like to ask Dr Strand at what hour angles he makes his observations?

Strand: Less than 15 min.

Van de Kamp: It is an interesting idea to use a large number of reference stars in order to obtain the correction to absolute parallax. Also, by using a large number you obtain 'luxury parallaxes'; most of us determine 'economy parallaxes'. I would like to ask Van Altena why only quadratic terms are considered; why not also other powers? Also, if you observe at large hour angles, you are playing with fire.

Van Altena: Automatic measuring machines make it possible to measure many reference stars nearly as easy as a few reference stars. In my opinion, the advantage of many reference stars outweigh the slight economy of a few. Plots of the residuals versus the coordinates occasionally show quadratic terms, but not higher power terms. Since we have many reference stars distributed nearly uniformly over the inner 150 mm of the plate, it should be possible to eliminate any systematic errors introduced by observing at large hour angles. (Differential colour refraction is eliminated from the measured coordinates before further proceeding.) The lack of any systematic trends in the residuals supports this contention.

Murray: Concerning the question of selection of reference stars and the reduction to absolute, at Herstmonceux we normally use six reference stars. In the first 20 fields, which have recently been submitted for publication, we found two reference stars, or about 2% which showed significant parallax.

Strand: Dr van de Kamp mentioned that the *p.e.* of unit weight decreases from $1.4\,\mu$ to $1\,\mu$ when the Sproul plates measured manually were measured with the U.S. Naval Observatory machine. A recent Sproul Series measured with this machine actually decreased to $0.8\,\mu$; of the same size as with the material from the 61-in. astronomic reflector.

Pecker: The number of people working on parallaxes is decreasing, whereas the needs are still large. Could a list of essential programs of parallaxes be done for such instruments as the Nice 76-cm refractor, where some scientists, although not experienced, express a definite interest for that kind of observation? These programs could be done according to the interest of the stars from a purely astrophysical point of view. What is your opinion?

Van de Kamp: Yes. The need for more parallax determinations is as great or greater than ever before. Astronomers at active parallax observatories would be glad to assist and give advice in the formulation of observing programs.

Pecker: Are the masses of the perturbators in the perturbed cases you have shown (ROSS 614, etc.) known? Wat are they, if the reply is yes?

Van de Kamp: Yes and no. For ROSS 614, which is now a resolved binary the latest study (Lippincott and Herskey) yields a mass of 0.06 solar mass for the companion. For other perturbations, where the companions have not yet been seen the estimated masses range from approximately 0.008 to close to one solar mass, except for the possible planetary companion or companions of Barnard's Star.

Jaschek: Is any effort being made to re-observe nearby stars with large parallax errors?

Van de Kamp: Yes.

Blaauw: Already more than 20 yr ago we noted that there are large systematic differences between parallax catalogues, f.i. 0″.005 between Allegheny and Cape. Have these systematic differences been re-analyzed, and has more recent work thrown some light on where in between these catalogues (or outside this range) the true parallax system lies?

Van de Kamp: We need more parallax Observatories, particularly in the Southern Hemisphere. How can we improve on systematic differences between different observatories, such as Allegheny and Cape? Do the telescopes remain the same and what about decreased activity at some observatories?

Gliese: I refer to a paper 'Errors in Trigonometric Parallaxes' which I have presented at Herstmonceux 1971 and which is now published in *Quart. J. Astron. Soc.* **13**. The inclusion of the modern observations does not change the systematic difference $\pi_{\text{Cape,Yale}} - \pi_{\text{Alleghyeny}} = +0″.005$. Moreover, I have combined (1) a comparison of trigonometric parallaxes with the moving cluster parallaxes of the Hyades by Upton, (2) a very preliminary comparison between first 61-in. results and the Jenkins' system, and (3) a comparison between luminosity calibrations based on large Allegheny parallaxes and calibrations based on small Allegheny parallaxes. The resulting 'true system' of trigonometric parallaxes is: $\pi_{\text{Allegheny}} + 0″.002 \pm 0″.0014$ (s.d.) which corresponds to $\pi_{\text{Cape,Yale}} - 0″.003$.

Murray: Just to keep the record straight, according to my information parallax working is still going on at the Cape Observatory.

Maeder: In connection with the problem of parallaxes, I would like to make a brief comment on the accuracy of apparent magnitudes, which may in fact also limit the accuracy of the absolute M_v. For example, in the Catalogue of Bright Stars, careful comparisons have shown that more than 40% of the stars have appreciable *systematic* errors of about 0^m10, and we shall return in a following session (with Dr Rufener) to this point.

Jaschek: Is somebody making a critical revision of the parallax catalogue in order to eliminate determinations with small weights, which are still quoted, although improved values exist?

Gliese: One should reject all stars with large errors, whatsoever the value of the parallax is.

Lippincott: I wish to say a few more words on the accuracy of trigonometric parallax determination. At the Sproul Observatory, the long range astrometric program initiated by Peter van de Kamp, now yields series of plates on many nearby stars covering an interval of over thirty years. Those series may be considered to be made up to 5 or 6 individual parallax series of 6 or 5 yr each. Thus considered, there appears a far larger range in parallax values than would be generated by purely accidental errors for some series. This has occurred despite the extreme effort to obtain a homogenous material. It seems likely that these systematic errors are due to a lack of proper collimation over a period of time. Therefore it is not surprising to find that there are significant systematic differences between parallaxes determined at different observatories. Also since one and the same telescope performs differently over different time intervals it makes it even more difficult to evaluate the differences between observatories.

The second point I would like to stress which is reassuring, is the determination at the Sproul Observatory of the difference between parallax in right ascension and declination for the average of 124 stars already quoted by Dr van de Kamp as $-0″.0018 \pm 0″.0011$ (*p.e.*) Here the systematic errors are for the most part independent.

Gliese: On the basis of reliable trigonometric parallaxes the following mean relations for main-sequence stars from dFS to dM have been derived and published 1971 in *Veröff, astr. Rechen-Inst. Heidelberg, No. 24*:

M_v, MK type	M_v, $B-V$
M_v, Mt. Wilson type	M_v, $U-B$ (from F8 to K7)
M_v, Kuiper type	M_v, $(U-B)$ Cape (from F8 to K7)
	M_v, $R-I$.

For late type main-sequence stars the latter relation, obviously, is the most reliable one. But since the number of stars with U, B, V photometry is still larger than the number of stars with $R - I$ data, I have tried to improve the (M_v, $B - V$) relation by a simple formula:

$$M_v = M(B - V) + b \cdot \delta(U - B)$$

where $\delta(U - B)$ is the UV – excess. In the region $B - V > +0.45$ the derived values of b are always positive. In H. L. Johnson's colour system b is about $+7$ for F dwarfs; it decreases to $+4$ in the K dwarf region and it is $+8$ for $B - V > +1.36$. Eggen's system give a similar run of b but somewhat smaller values ($+2 < b < +6$). The correlation coefficient r between ($M_{\text{trigonometric}}$ minus M_{B-V}) and $\delta(U - B)$ decreases from $+0.69$ (dF) to $+0.51$ (dK) and increases to $+0.89$ for M dwarfs. These variations of b and r probably are caused by the effects of the observational errors in $\delta(U - B)$ which are largest in the K dwarf region and which diminish b and r.

I summarize: Obviously there is a correlation between the ultraviolet excess of late-type main-sequence stars and their luminosity. But the evaluation of additive corrections as a function of $\delta(U - B)$, depends on the colour system and seems to be so sensitive to observational errors that an effective application of corrections is not yet recommended.

Strand: We at the Naval Observatory are now determining ($V - I$) for all red dwarfs on the parallax program.

NOTE ON ABSOLUTE MAGNITUDES IN THE RED AND
INFRARED OF LATE M-TYPE DWARFS

S. L. LIPPINCOTT

Sproul Observatory, Swarthmore College, U.S.A.

Abstract. The importance of red and infrared magnitudes for late dwarf M-type stars is stressed. The absolute magnitudes for the nearest stars with $(R-I) \geqslant 1.20$ is re-assessed.

The Sproul Observatory astrometric program initiated in 1937 by Peter van de Kamp is primarily devoted to nearby stars which are to a great extent faint red dwarf stars. One of the main objectives is to ascertain whether these stars have unseen companions revealed by the discovery of variable proper motion. A by-product of this program is the accurate determination of parallaxes which leads to absolute magnitudes. Magnitudes in the red and infrared may also be important in pointing to unresolved binaries from red or infrared excess.

For late M dwarfs there is little difference in the $(B-V)$ with increasing M_v whereas the $(R-I)$ changes significantly with M_R for the very late type stars. Therefore, red and infrared colors and magnitudes can be important as standards. Kron and his associates' (Kron *et al.*, 1957) monumental photometric red and infrared work on 282 stars with known parallaxes includes many red dwarfs.

In this note I am particularly concerned with the study of the dispersion of color and magnitude for the very late M-dwarfs where $(R-I)$, Kron's system, is $+1.2$ mag. or greater. This constitutes stars with spectral types M4 and later with M_v greater than $+12$. Kron (Kron *et al.*, 1957) has pointed out the intrinsic scatter, from $(B-V)$ vs $(R-I)$ plots, up to several tenths magnitudes far exceeding the probable errors assigned to the colors.

Turning to a plot of M_R against $(R-I)$ for $(R-I) \geqslant 1.20$, (figure 1), the author has re-assessed the parallax determinations of each star to achieve the best possible M_R. The probable error of each M_R as determined from the probable error of the parallax is indicated. Since these are nearby stars with $\pi \gtrsim +0\rlap{.}''10$ the p.e. in M_R is generally smaller than $\pm 0\rlap{.}^{m}15$. The positions in the diagram show a scatter of 1 mag. in M_R and 0.2 to 0.3 in $(R-I)$. Kron and associates (Kron *et al.*, 1957) found systematic differences that depend on space velocity. This appears true in the diagram for stars with $(R-I) < 1.3$, in the sense that high velocity stars are too faint for the $(R-I)$. The relatively few stars available, which are redder, do not show this systematic behavior. Components of double stars have been singled out to see if both components deviate from the average in the same way. In general this seems not to be the case. BD$+19°5116$ A, which is not red enough to appear in the accompanying diagram, is close to the average of other stars with similar color but the B component is outstanding. One obvious explanation is that BD$+19°5116$ B is a close double of near equal magnitudes. The

B. Hauck and B. E. Westerlund (eds.), Problems of Calibration of Absolute Magnitudes and Temperature of Stars, 18–20.
All Rights Reserved. Copyright © 1973 by the IAU.

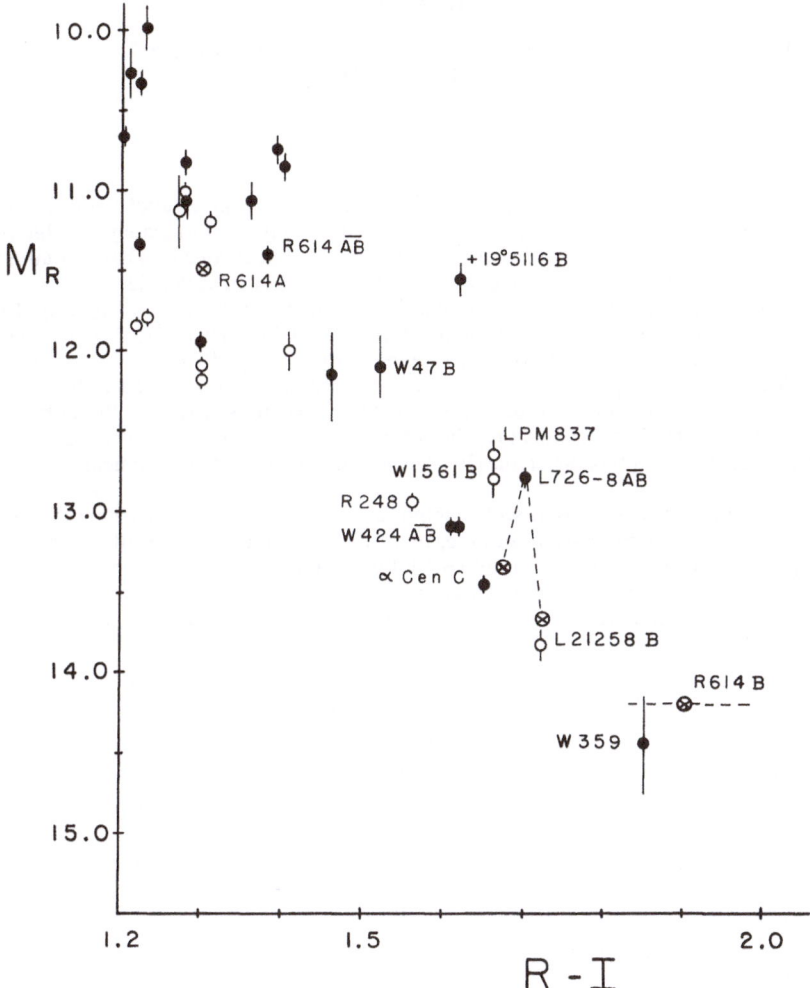

Fig. 1. $M_R - (R - I)$ relation for nearby late M type dwarfs. The vertical lines represent the probable errors of the M_R. The open circles indicate high velocity stars. The open circles with crosses inside represent estimates.

separate components of L 726–8 differing by 0.5 visual magnitude are estimated to lie close to α Cen C and Lal 21258 B, respectively.

The infrared magnitudes of Ross 614 A and B have been inferred by the shift in the blended image of the system toward their center of mass on Sproul astrometric infrared sensitive plates compared to the position of the blended image in the visual region. This shift yields a value for the Δ m in the infrared compared to that in the visual region (Lippincott and Hershey, 1972).

References

Kron, G. E., Gascoigne, S. C. B., and White, H. S.: 1957, *Astron. J.* **62**, 205.
Lippincott, S. L. and Hershey, J. L.: 1972, *Astron. J.* **77**, 259.

DISCUSSION

Murray: I feel that there really is a problem of the red dwarfs. Any kinematic investigation of these stars must be based on a selection by photometric criteria only. Unfortunately suitable plates for proper motion work, reaching $m_v \sim 17$ are not easily available, but anyone who has a suitable material should be encouraged to determine proper motions of all red stars in each field.

McCarthy: Concerning the need for more data for red dwarf stars discussed by Miss Lippincott and also by Murray I wish to point out the importance of red dwarfs in the Pleiades cluster. Ambartsumian has predicted that some 1000 stars fainter than $m_v = 13$ will prove to be cluster members at the faint end of the main sequence. Thus far some 200 flare stars have been discovered at the Byurakan, Asiago and Tonantzintla observatories. Most recently (August 1972) another 200 flare stars have been reported in a Buyrakan reprint. Many of those flare stars are identical with the faint red stars selected in objective prism studies by Treanor and myself and in preliminary blink estimates by Luyten.

The determination of $R - I$ colours will be important for these stars. In December 1971 I obtained 24 plates centered on the Pleiades in infrared, red, yellow and blue wave lengths with the Palomar 48″ (120 cm) Schmidt telescope. At Castel Gandolfo we have begun the determination of pseudo-magnitudes and colours for these stars. I welcome any suggestions and aids in determining a faint red and infrared photoelectric sequence as a basis for photoelectric interpolation.

MAGNITUDE ABSOLUE MOYENNE DES RR LYRAE

A. HECK

Institut d'Astrophysique de l'Université de Liège, Belgique

Résumé. Par une méthode basée sur le principe du maximum de vraisemblance, nous obtenons, pour un échantillon de 102 RR Lyrae, $\bar{M}_{pg} = 0.6 \pm 0.2$ et $\sigma_{M_{pg}} = 0.5 \pm 0.2$. Les résultats de Clube et Jones (1971) ne sont pas confirmés. L'étude des corrélations entre $\bar{M}$, la période, l'indice de métallicité ΔS et l'amplitude de variation en magnitude montre que, sur la base des données actuelles, aucune corrélation significative ne peut être établie.

Abstract. By a method based on the principle of maximum likelihood, we obtained for a sample of 102 RR Lyrae stars $\bar{M}_{pg} = 0.6 \pm 0.2$ and $\sigma_{M_{pg}} = 0.5 \pm 0.2$. We could not confirm the results of Clube and Jones (1971). The study of the correlations between $\bar{M}$, the period, the abundance parameter ΔS and the light variation amplitude showed that on ground of the present data, no significant correlation can be established.

Le travail dont il est question ici consiste en l'application de la méthode basée sur le principe du maximum de vraisemblance, mise au point par Jung (1968, 1970), à un échantillon de RR Lyrae.

L'échantillon que nous avons considéré est constitué des RR Lyrae tabulées par Van Herk (1965) dont les mouvements propres et la vitesse radiale sont connus. Nous avons éliminé les étoiles à grande erreur sur les mouvements propres ($> ''014$) et deux étoiles (RV Cap et AS Vir) dont les vitesses spatiales sont très importantes. Il est en effet peu probable que ces étoiles appartiennent à une population parente. L'échantillon résultant se composait de 102 étoiles.

Les erreurs probables sur les mouvements propres ont été prises en considération. Le biais introduit a été calculé à l'aide d'expériences numériques qui ont en outre permis de déterminer la précision des résultats.

Les magnitudes sont des magnitudes photographiques et elles ont été corrigées de l'absorption par la loi:

$$A_{pg} = 0.19 [1 - \exp(-0.01r|\sin b|)]/|\sin b|.$$

Compte tenu du biais, le résultat final est le suivant:

$$\bar{M}_{pg} = 0.6 \pm 0.2$$
$$\sigma_{M_{pg}} = 0.5 \pm 0.2.$$

Un problème intéressant soulevé par Clube et Jones était de savoir si les mouvements propres sont affectés d'erreurs systématiques provenant d'erreurs sur le mouvement solaire, particulièrement dans la direction de la rotation galactique, des étoiles de référence. La comparaison des mouvements propres des étoiles communes à notre échantillon et à l'AGK3 nous a amenés à conclure à une erreur systématique de $0''006$. La correction des mouvements propres sur base de ce résultat ne conduit cependant qu'à des modifications négligeables.

Nous avons également appliqué la méthode du maximum de vraisemblance à

B. Hauck and B. E. Westerlund (eds.), Problems of Calibration of Absolute Magnitudes and Temperature of Stars, 21–22.

l'échantillon utilisé par Clube et Jones (1971) en vue de confirmer ou d'infirmer leurs valeurs $\bar{M}_v = 1.30$ et $\sigma_{M_v} = 0.5$. Mais, même en tenant compte d'une erreur systématique sur les mouvements propres (0˝015), nous ne pouvons pas confirmer ces résultats. Les valeurs obtenues (après correction du biais et de l'erreur systématique) sont $\bar{M}_{pg} = 0.6$ et $\sigma_{M_{pg}} = 0.8$. Nous rappelons que, selon Van Herk (1965), il faut ajouter la correction $(B - V)_o = 0.19$ aux magnitudes visuelles pour les transformer en photographiques, ce qui accroît encore le désaccord.

Enfin, nous avons recherché les relations susceptibles d'exister entre $\bar{M}$, la période, l'indice de métallicité ΔS et l'amplitude de variation en magnitude. Les résultats montrent que si certaines relations paraissent être réelles en considérant les valeurs moyennes, les grandes dispersions nous amènent plutôt à considérer de telles relations avec prudence. Il serait vivement souhaitable de reprendre ces calculs avec un échantillon plus important! Pour plus de détails, nous renvoyons à Heck (1972).

Bibliographie

Clube, S. V. M. et Jones, D. H. P.: 1971, *Monthly Notices Roy. Astron. Soc.* **151**, 231.
Heck, A.: 1972, *Astron. Astrophys.* **21**, 231.
Herk, G. van: 1965, *Bull. Astron. Inst. Neth.* **18**, 71.
Jung, J.: 1968, *Bull. Astron. Paris Ser. III*, **3**, 461.
Jung, J.: 1970, *Astron. Astrophys.* **4**, 53.

A PROGRAM TO DETERMINE AN ACCURATE
TRIGONOMETRIC PARALLAX FOR THE HYADES CLUSTER

W. F. VAN ALTENA

Yerkes Observatory, University of Chicago, Williams Bay, Wisc., U.S.A.

Abstract. A list is presented of very probable Hyades cluster members that are suitable for an accurate determination of the cluster's trigonometric parallax.

In 1968 I prepared a list of very probable Hyades cluster members which were suitable for parallax determination in an effort to improve the distance to the Hyades cluster and the population I distance scale. This list (Table 1) contains 20 very probable Hyades members lying within 3° of the cluster center. The criteria for membership are: (1) photoelectric photometry is available which indicates either main sequence or white dwarf membership; and (2) proper motions are available from at least three sources yielding errors of $\leqslant \pm 3°$ (m.e.) and $\pm 0''006 \ \mathrm{yr}^{-1}$ (m.e.), These conditions are relaxed

TABLE I
Very probable Hyades members suitable for π determination

No.	N	μ	$\sigma\mu$	$\Delta\mu$	θ	$\sigma\theta$	$\Delta\theta$	V	B−V	U−B	Remarks Priority
					Main sequence						
276	5	0''136	±5	+20	100°	±2°	−1	10.49	1.24	1.18	7
294	4	0.126	6	+12	97	3	−1	10.90	1.30	1.30	8
310	4	0.119	5	+3	104	3	−1	9.99	1.06	0.95	5
363	3	0.112	5	−3	106	3	+1	9.01	0.94	0.66 double?	14
459	4	0.114	6	+3	97	3	−2	9.52	0.93	0.70	4
472	4	0.118	6	+8	101	3	+3	9.03	0.84	0.51	2
475	4	0.105	6	−7	103	3	−1	11.13	1.37	1.24	10
500	4	0.101	5	−10	99	3	−4	10.71	1.36	1.18 double?	15
502	5	0.109	5	−2	102	2	−2	11.96	1.44	1.24	11
529	5	0.111	6	+1	102	3	+1	12.44	1.47	1.02 (1)	13
548	4	0.104	5	−7	101	3	−3	10.32	1.17	1.08	6
560	4	0.115	5	+4	104	3	0	9.10	0.85	0.58	3
587	4	0.096	3	−14	101	2	−1	8.93	0.84	0.49	1
638	5	0.106	6	−2	97	3	−5	12.17	1.47	1.00 (1)	12
645	5	0.116	6	+8	102	3	0	11.04	1.31	1.23	9
747	4	0.092	4	−14	99	2	−4	9.69	0.91	0.68	16
			0''000 ±1			−1°±1°	−0.04±.06				
					White dwarfs						
292	4	0.124	7	+8	102	3	0	14.29	−0.02	−0.84	19
490	4	0.101	8	−15	103	4	−2	14.02	−0.09	−0.97	17
673	5	0.118	5	+9	110	2	+3	13.95	+0.33	−0.69 double	20
722	3	0.096	6	−9	96	2	−1	14.18	−0.03	−0.89	18
			−0''002 ±3			0°±2°					

B. Hauck and B. E. Westerlund (eds.), Problems of Calibration of Absolute Magnitudes and Temperature of Stars, 23–24.

in the case of the white dwarfs), and the position angle of the motion must lie within two mean errors of the convergent point.

Only the classical DA white dwarfs are included. As indicated in the table the average motion of these stars with respect to the Hyades motion is not statistically significant. The average difference of the 16 main sequence stars from the average Hyades main sequence is not statistically significant. The priorities are ordered by apparent magnitude, except for those stars either known to be double or suspected of being double, and number 747 which I feel has the weakest case for being a member. Cross identifications and finding charts may be found in Tables IIIA and IIIC and in Plate I of *Astron. J.* **74**, 2, 1969.

At the present time seven Observatories (Allegheny, Cape, Herstmonceux, Lick, McCormick, Van Vleck, and Yerkes) are co-operating in this joint program which should yield an average parallax for the Hyades with an error of $\sim \pm 0\rlap{.}''001$ (m.e.) for each observatory. Within the next few years it should be possible to determine an accurate parallax for the Hyades cluster from these data that will have approximately the same accuracy as that found from the convergent point method.

In addition to the primary aim of this project, it will also be possible to evaluate the systematic differences between observatories with modern data for the Hyades region of the sky.

DISCUSSION

Worley: Eggen commented that the double star orbits from which the masses of the Hyades binaries are determined result from observations made over a long period of time by observers of varying skill. How much this inhomogeneity affects the resultant masses is the question.

Together with a colleague G. G. Douglas, I have been examining the problem of the mass discrepancy of the Hyades binaries vs the 'normal' (Sun-Sirius) ML relation. Unless we are willing to accept a very peculiar systematic error, which affects the observations made in the region of the Hyades, but which does not appear in the rest of the sky, we cannot attribute the ML discrepancy to systematic effects in the observations.

The above investigation also pointed out that a more general study of systematic errors was needed, since the last such investigation was made many years ago. Such a study is now in progress.

Wesselink: The Leiden Observatory (Miss Kluyvers) is conducting a new study of proper motions convergent point, distance modulus of the Hyades. This study may give rise to an interesting new answer to the modulus problem.

Van Altena: A redetermination of the Hyades convergent point is being made from absolute proper motions with respect to faint galaxies by Robert Hanson at Lick Observatory.

Thomas: I have a question concerning the chemical composition of the Hyades derived with Watson opacities. Did I understand you correctly that these values stem from fitting the theoretical slopes of the mass-radius and mass-luminosity relation to the observations. Since chemical composition does not enter the homology relations, is this an effect of the non-homology of computed models?

Demarque: No. The theoretical slopes of the mass-radius and mass-luminosity relations are practically independent of chemical composition in the relevant mass range. The fitting was done by requiring consistency between the distance modulus used in deriving the masses of Hyades binaries and that obtained from the position of the main sequence.

Crawford: We find in our work that the Hyades is more discrepant than any other cluster. Any change in the distance modulus will not solve the problem. I will describe the discrepancy a bit tomorrow.

Pecker: Within clusters (Hyades, Praesepe...) is not the reddening correction to be questioned because of inter-cluster additional reddening? (a sort of circumstellar reddening?)

Eggen: There is no evidence for appreciable reddening either in the Hyades or in Praesepe.

PART II

ABSOLUTE MAGNITUDES FROM GALACTIC
CLUSTERS AND ASSOCIATIONS

ATTEMPT TO APPLY THE $U\ B\ V\ B_1\ B_2\ V_1\ G$ PHOTOMETRY TO THE DETERMINATION OF THE DISTANCE OF THE HYADES

M. GOLAY

Observatoire de Genève, Switzerland

Abstract. A photometric method, using the $U\ B\ V\ B_1\ B_2\ V_1\ G$ system, is suggested to determine the distance of clusters. This method based on the assumption that stars with same seven colours have also same absolute visual magnitude is applied, as an example, to Hyades and Praesepe.

1. Introduction

Several methods have been used to determine the distance of the Hyades. According to Lutz (1970) the distance moduli obtained are between 3.03 and 3.25. Hodge and Wallerstein (1966) have even suggested that 3.42 may be acceptable. We try here to tackle this problem by means of a method totally independant of the above mentioned ones. We now dispose of yet unpublished measurements of 3500 stars in the 7 colours $U\ B\ V\ B_1\ B_2\ V_1\ G$. Of these stars, more than 200 belong to the catalogue of nearby stars by Gliese (1969), more than 100 are presumed to be members of the Hyades, more than 100 to be members of Praesepe. The sample of stars for which we have accurate measurements of the 7 colours is large enough to test the following hypothesis: "The stars for which the 7 colours are almost identical also have almost identical values of θ_{eff}, M_v, χ (chemical composition characterised here by [Fe/H]]". The use of models which allow the theoretical computation of fluxes with line blocking shows that two stars for which the 7 corresponding colours do not differ by more than 0.01 mag. have effective temperatures which differ by less than 200 K, gravities which differ by a maximum value of 0.2 in $\log g$, chemical composition different by a maximum of 0.2 in [Fe/H]. Therefore unless proved otherwise by observation, two stars having their 7 respective colours identical within 0.01 mag. could be considered as being identical in effective temperature, chemical composition and absolute magnitude. If this is the case, we can deduce the absolute magnitude of the Hyades stars by looking for all the stars of known trigonometric parallax of which the 7 colours differ by less than 0.01 mag. from 7 corresponding colours of members of the Hyades.

2. Test of the Proposed Hypothesis

The program of observations in $U\ B\ V\ B_1\ B_2\ V_1\ G$ has not been established in view of testing this hypothesis. At the present, we are making use of the capital of measurements collected. Under these conditions, the number of cases in which at least two stars have their 7 colours identical within a margin of $\delta = 0.01$ mag. is limited. Nevertheless, it is possible to put together 163 groups. Each group contains at least

B. Hauck and B. E. Westerlund (eds.), Problems of Calibration of Absolute Magnitudes and Temperature of Stars, 27–30.

2 stars. Often a group contains 3 to 4 stars, 5 stars being the maximum number at present. It is interesting to notice that strongly reddened stars never partake in the same group (except for 2 stars of the same cluster), even when the tolerated difference of the individual colours is extended to $\delta = 0.02$ mag. Only those stars having difference of colour excess smaller than 0.01 mag. coexist in a same group. We can conclude that a strongly reddened star does not have the 7 colours of a cooler unreddened star (yet these may coincide in the case of three colours as the UBV diagram shows). We have shown (1973) (see 'Remarks on the Photometric Criteria of Choice of the Standards Stars' in the present Proceedings) that the presence of a companion affects the colours of the main component if the difference in magnitude between these is smaller than 5. There too, at least in our present collection of groups, the 7 colours of a binary rarely coincide with the colours of an isolated star, except when the difference in magnitude is > 5 or when both components are nearly identical (except the giants). As far as rotation is concerned, we notice that only those stars having $V \sin i < 150$ km s^{-1} (when this is known) are found in the groups. On the other hand, a great number of stars having a high value of $V \sin i$ belong to no group. In view of showing to what extent the members of a same group are identical, we have only considered 63 groups of which the member stars are unreddened, not binaries or having a companion weaker by 5 mag., and having measurements of weight $p > 3$. We have already published elsewhere (1969) the results concerning the identity of the spectral classes and luminosity classes of stars of a same group. We have shown that the observed dispersion reaches 0.1–0.2 of a spectral class and is within the limits of dispersion due to the heterogeneity of the spectral classifications. If we make use of a very homogeneous classification like that of Cowley *et al.* (1969), for A stars, dispersion rarely reaches $\frac{1}{10}$ of spectral class (15% of cases) and dispersion in luminosity class is negligible. In our groups, there is no coexistence of Am stars with normal stars when these have been the object of a careful classification. On the other hand, marginal Am stars may sometimes coexist with normal stars. In such cases, the stars have identical H and metallic lines and differ only by the Ca IIK line. Similarly, in groups where all the stars are Am, these may differ among each other by the Ca II K line, all other lines being identical. As for groups which contain a single Ap star, our information is still limited. Yet it seems that the stars of a same group may be all Ap, not all of these necessarily having the same particularity. Let us now examine the greatest differences in absolute magnitude ΔM_v among stars of a same group. We distinguish four cases according to the provenance of the absolute magnitude.

(a) Spectroscopic absolute magnitude.

(b) Absolute magnitude obtained from trigonometric parallaxes (quality classes *A* to *F*).

(c) Difference of absolute magnitudes deduced from the difference of apparent magnitudes of stars belonging to a same group and to a same cluster.

(d) Absolute magnitude obtained by means of the relations established by Olsen (1969) for the Copenhagen photometry.

We find the average values of the greatest differences to be:

Case (a) 0.95 mag.
Case (b) 0.33
Case (c) 0.18
Case (d) 0.45

In the case of chemical composition, the greatest differences among stars of a same group obtained with the aid of measurements by Peat and Pemberton (1969) are less than 0.07 for MgH and to 0.016 for NaD. As for the greatest differences in [Fe/H] computed by means of the relations established by Olsen, we find an average value of 0.07.

Thus, in spite of the heterogeneous nature of the sources for values of M_v, [Fe/H], Sp, etc., the hypothesis of identity in M_v, θ_{eff}, χ of stars having the same 7 colours could well be acceptable and merit further study.

3. Application to the Distance of the Hyades

In view of increasing the number of stars having their 7 respective colours within 0.01 mag. of those with accurately determined (classes A, B, C) trigonometric parallaxes, we have also considered the stars of Praesepe. This procedure thus necessitates the preliminary determination of the difference of distance moduli between Praesepe and the Hyades. We obtain this by adjusting the V, $[d]$ sequences of these two clusters. The parameter $[d]$ of the $U\,B\,V\,B_1\,B_2\,V_1\,G$ photometry is a measurement of the Balmer discontinuity. The properties of $[d]$ are very similar to those of c_1 of the $uvby$ systems. The parameter $[d]$ has an advantage over the parameter Q of the UBV photometry in being little dependant on chemical composition. In the interval F8–A7, the $B-V$ indices varies by 0.4 mag. whereas $[d]$ varies by 0.8 mag. This wide interval of variation leads to a lesser slope $\Delta V/\Delta\,[d]$ than $\Delta V/\Delta(B-V)$ which facilitates the separation of binaries, non members, fast rotating stars seen pole-on, reddened stars ($[d]$ is practically independant of extinction for small masses of interstellar matter). The adjustment of both sequences is done by translation along the axis of magnitudes.

We find

$$(m_v - M_v)_{\mathrm{Praesepe}} - (m_v - M_v)_{\mathrm{Hyades}} = 3.0 \pm 0.05.$$

In Table I, we give the list of groups of stars used. The two members of each group have corresponding colours which differ by less than 0.01 mag. In each group, one of the members has an absolute trigonometric magnitude taken from Gliese's catalogue (1969), the other member belongs either to the Hyades or to Praesepe.

The numbers of the Hyades stars are those of Van Bueren's (1952) catalogue and those of the Praesepe stars are those of the catalogue of Klein-Wassink (1927). The V magnitudes have been determined by Rufener and Maeder (1972). The star KW 203 has had its magnitude corrected to account for the presence of a companion

TABLE I

Groups of stars used

No.		M_v (Gliese) V (cluster)
Gr1	HD 17206 Praesepe KW 293	$M_v = 3,8$ E $V = 9.85$
Gr2	HD 48682 Hyades 113	$M_v = 4,4$ C $V = 7,26$
Gr3	HD 114710 Praesepe KW 162	$M_v = 4,66$ B $V = 10,74$
Gr4	HD 187013 Hyades 78	$M_v = 3,35$ C $V = 6,92$
Gr5	HD 203280 Praesepe KW 203	$M_v = 1,5$ C $V = 8,2$

which is nevertheless sufficiently weak not to influence the colours too much. By assigning to the Hyades and Praesepe stars the same absolute magnitudes as those of co-members of the same groups for which this is known, and by taking into account the difference of distance moduli between the Hyades and Praesepe clusters, we find a distance modulus for the Hyades of 3.25 mag. ± 0.3. This distance modulus is obtained on the basis of a very limited number of groups. We will soon be able to quadruple the number of groups by increasing the number of observations of stars of Gliese's catalogue (work in progress) and by introducing observations of other clusters having approximately the same V, $[d]$ diagram as that of the Hyades and Praesepe.

References

Bueren, H. G. van: 1952, *Bull. Astron. Inst. Neth.* **11**, 452.
Cowley, A., Cowley, Ch., Jaschek, M., and Jaschek, C.: 1969, *Astron. J.* **74**, 375.
Gliese, W.: 1969, *Veröffentl. Astron. Rechen-Inst. Heidelberg*, No. 22.
Golay, M.: 1973, this volume, p. 275.
Golay, M., Peytremann, C., and Maeder, A.: 1969, *Publ. Obs. Genève* **76**.
Hodge, P. W. and Wallerstein, G.: 1966, *Publ. Astron. Soc. Pacific* **78**, 411.
Klein-Wassink, W. J.: 1927, *Publ. Kapteyn Astron. Lab.*, No. 4.
Lutz, T. E.: 1970, *Astron. J.* **75**, 1007.
Olsen, E. H.: 1969, Thesis, Copenhagen University Observatory.
Peat, D. W. and Pemberton, A. C.: 1969, *Monthly Notices Roy. Astron. Soc.* **140**, 21.
Rufener, F. G. and Maeder, A.: 1972, in preparation.

A FAR-ULTRAVIOLET FLUX DIFFERENCE BETWEEN
HYADES AND PLEIADES STARS

R. E. SCHILD

Smithsonian Astrophysical Observatory, Cambridge, Mass. 02138, U.S.A.

Abstract. Four ultraviolet fluxes measured in U_2 (centered at λ_{eff} 2300 Å) by the Celescope experiment aboard OAO 2, reveal an important flux difference between Pleiades and Hyades stars. Available blanketed stellar models show that the difference is too large to be understood as a blocking effect for admissible metal overabundance in Hyades stars. Known rotation and presence of Ap and Am stars in the Hyades and Pleiades apparently cannot account for the discrepancy.

One of the important assumptions of the cluster-fitting method is that two main-sequence stars of the same spectral type or intrinsic colour have the same absolute magnitude. I wish to show that Pleiades and Hyades stars of the same spectral type or $b-y$ colour differ by a factor of 2 in their 2000 Å ultraviolet fluxes, and so presumably differ also in their absolute visual magnitudes.

The data for this paper were acquired as part of the Celescope sky-mapping experiment aboard the successful OAO 2 satellite. The Celescope package incorporated four ultraviolet-sensitive television scanners, or Uvicons, which, in conjunction with four filter sets, provided stellar fluxes at four wavelength bands between 1200 and 3000 Å. I should like to present results obtained with the U_2 filter, which, because of its high peak transmission and broad bandpass, yielded the greatest amount of data. The U_2 filter has a bandpass of 1000 Å centered at approximately 2300 Å. Reduction of the Celescope pictures has been completed under the direction of Dr R. J. Davis, Dr W. A. Deutschman and Mrs K. Haramundanis and will be discussed in the final printed catalog. The rms deviation of a single observation has been assessed, from repeated observations of various stars, to be $\sigma_{rms} = 0.15$ mag.

Strömgren four-colour and Hβ photometry from Crawford and Perry (1966) and from other unpublished data by Crawford was used to analyze the ultraviolet photometry. The Hyades were assumed to be unreddened, and the Pleiades were dereddened by using the calibrations and procedures outlined by Crawford and Perry (1966), with a value of 4.75 used for the ratio of $E(U_2-V)/E(B-V)$ (Haramundanis and Payne-Gaposchkin, 1972). Because reddening of the Pleiades stars is almost zero, essentially none of the conclusions of this paper is likely to be an artifact of the reddening corrections.

Figure 1 shows the comparison of $U_2 - V$ fluxes of the Hyades stars and all Pleiades members having $(b-y)_o \geqslant 0.1$. In this graph we see a clear separation between the two clusters.

Several points must be made concerning the validity of the difference between Pleiades and Hyades stars, as seen in Figure 1. Because the Pleiades and Hyades are close in the sky and because the Pleiades measurements were made between several sets of Hyades measurements, it is very unlikely that small orientation or time-

B. Hauck and B. E. Westerlund (eds.), Problems of Calibration of Absolute Magnitudes and Temperature of Stars. 31–35.

R. E. SCHILD

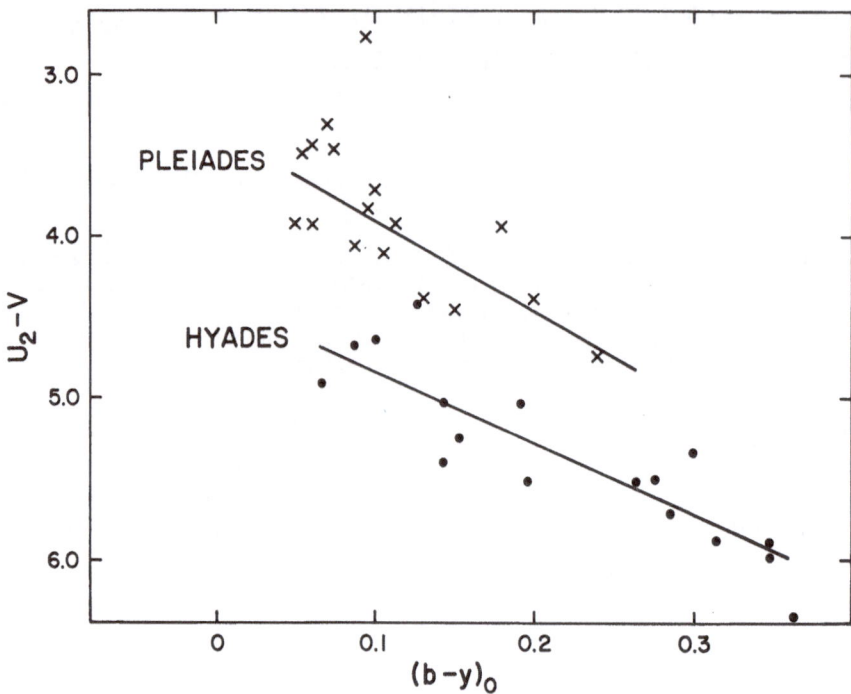

Fig. 1. Two-colour diagram showing $U_2 - V$ as a function of $(b-y)_0$, with reddening of Pleiades stars corrected for by use of a $\beta - (b-y)$ diagram. Data for hotter Pleiades stars are not shown, because the dereddening procedure can no longer be applied. Data for Pleiades stars redder than $b-y = 0.1$ may be affected by completeness, since observations were continued to a magnitude limit in U_2.

variable calibration effects caused the observed difference. The Pleiades data may be affected by completeness, especially for $b-y \geqslant 0.15$; however, the mean Pleiades relation in Figure 1 can be seen to be a smooth continuation of the $U_2 - V$ relation for hotter Pleiades stars and for hotter stars in the θ Car cluster.

Some differences between the Pleiades and the Hyades stars have previously been noted in the literature. Chaffee *et al.* (1971) pointed out that Hyades stars have 50% higher metal abundances than do Pleiades stars and the Sun. The Hyades are known to have many metallic-line stars, and Struve (1945) has shown the Pleiades to have significantly higher mean projected rotational velocities.

Before considering the effects of abundance and rotation on the ultraviolet fluxes, we might ask how field stars compare to the mean Pleiades and Hyades relations. In Figure 2, we have plotted $U_2 - V$ against spectral type, since four-colour photometry is not available for a significant number of field stars having Celescope photometry. We see from the figure that the A stars in the field have far-ultraviolet fluxes similar to the Hyades stars, whereas the F stars are, if anything, fainter in the ultraviolet.

Now we can consider the effects of abundances on the far-ultraviolet and visible fluxes. I have reviewed the models of Chaffee *et al.* (1971) to determine the effects of metal overabundance on the U_2 and V fluxes. In Figure 3, the far-ultraviolet fluxes for 25% and 50% metal overabundances are plotted; the 50% overabundance is the value adopted by Chaffee *et al.* For a 50% overabundance, these models predict that the ultraviolet fluxes in U_2 will be depressed by $0\overset{m}{.}21$ while the V magnitude will be increased by $0\overset{m}{.}025$. If we extrapolate these results, we would need a factor of 2.5 overabundance of metals in Hyades and field stars relative to the sun and Pleiades stars in order to account for the Hyades (and field star!) ultraviolet deficiencies. Such large metal overabundances seem precluded by direct abundance determinations. We note that even such a large amount of ultraviolet line blocking, interpreted as an overabundance effect, appears to cause no more than an $0\overset{m}{.}1$ increase in the V magnitude at constant T_{eff}. This is because the V magnitude is also strongly affected by blocking, and much of the radiation escapes in the infrared.

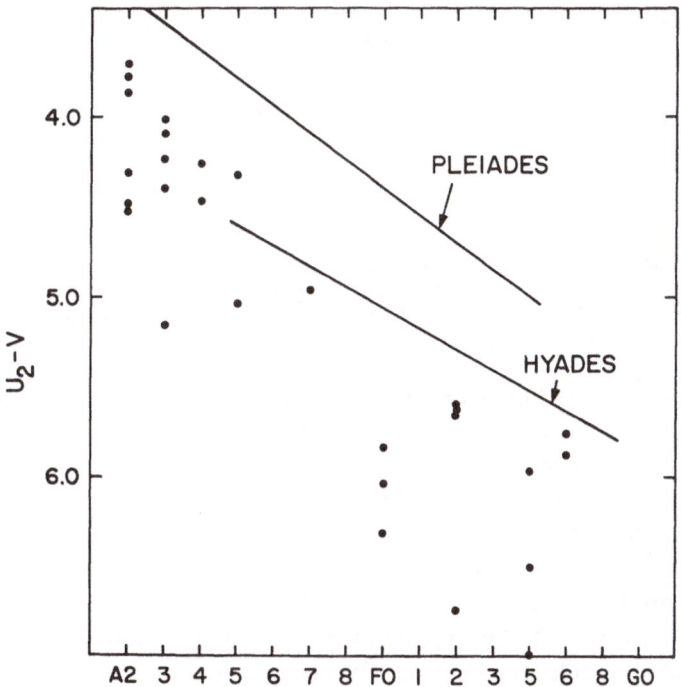

Fig. 2.   A comparison of the $U_2 - V$ fluxes of field stars, as a function of spectral type, with the mean Pleiades and Hyades relations. Not only are the field stars fainter at U_2, but the scatter is much greater than for stars in coeval groups.

We conclude that if the Hyades ultraviolet deficiency is an effect of line blocking due to metal overabundance, the visual magnitudes are not likely to be affected by more than $0\overset{m}{.}1$; however, a metal overabundance sufficient to depress the ultraviolet as observed should have been easily detected in direct abundance determinations from coudé spectra.

We next consider the possibility that differences in rotation cause the observed

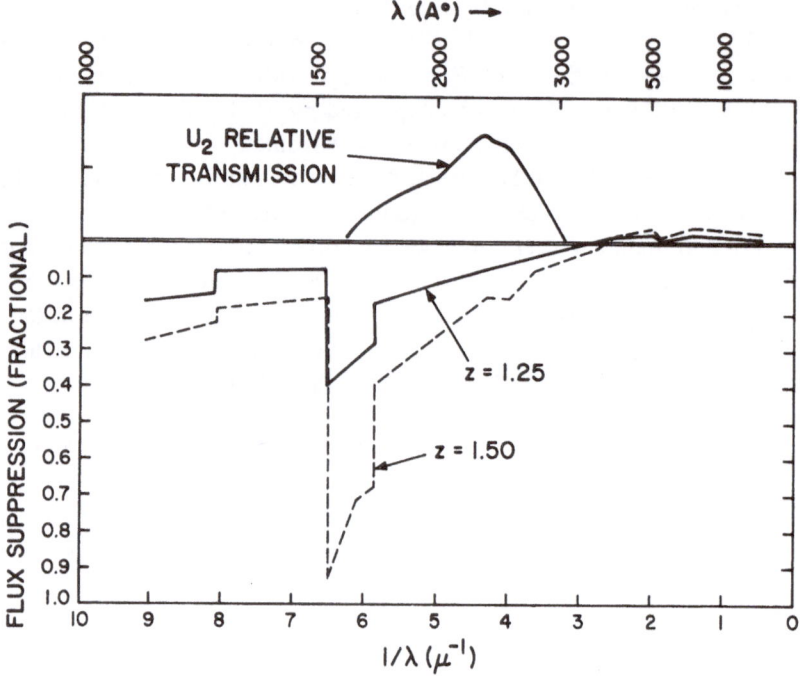

Fig. 3. Ultraviolet-flux suppression as a result of metal overabundance in a $T_{eff} = 7500$ K main-sequence star. The two curves show the suppression computed for 25% ($z = 1.25$) and 50% ($z = 1.50$) metal overabundances relative to the Sun. Note that the curves change sign in the visible and near infrared, showing that the flux escapes the star in these spectral regions.

differences in ultraviolet flux. Hardorp and Strittmatter (1968) have computed energy distributions of nonrotating and critically rotating stars. We note their result for a $T_{eff} = 8600$ K star: Compared to a nonrotating star, a pole-on star rotating at 99% of breakup velocity is only $0^m.06$ bluer in $U_2 - V$. Of course, both the pole-on star and the rapid rotator have sharp spectral lines. Relative to the sharp-lined stars, the critically rotating stars seen equator-on are $0^m.6$ *fainter* in $U_2 - V$. Thus, whereas rotation can produce large changes in the ultraviolet fluxes, the sense is wrong to account for the fact that the apparently more rapidly rotating Pleiades stars are brighter in the ultraviolet than are the Hyades and field stars.

It is well known that there are a relatively greater number of Am and Ap stars in the Hyades than in the Pleiades. Since the Am stars are binaries, could the presence of binary components cause the Pleiades-Hyades difference? For example, if secondary components contribute to the V magnitude, then the binaries will be fainter in $U_2 - V$.

We must conclude that binary secondaries do not cause the observed Pleiades-Hyades differences in ultraviolet flux, for the following reasons:

(i) From the amount of the effect, essentially all the binary systems would have to have equally luminous components.

(ii) The known spectroscopic binaries and Am stars are not displaced from the nonbinaries in Figure 1.

(iii) The nonbinary Hyades are redder in $U_2 - V$ than the Pleiades spectroscopic binaries.

We conclude that the ultraviolet faintness of the Hyades relative to the Pleiades measured in Project Celescope does not have a simple explanation in abundance, rotation, or spectrum peculiarity. As we have seen, for the observed ultraviolet faintness of the Hyades to be an abundance effect, the metal abundance would have to be so high as to have been detected on coudé spectra. Suppression of the ultraviolet by rotation effects is in the wrong sense for the known difference in projected rotational velocities between the Pleiades and the Hyades. And the identification of the stars in Figure 1 known to be spectroscopically peculiar shows them not to be responsible for the effect. Until the origin of this ultraviolet difference is understood, the method of cluster fitting based on the Hyades must be applied with caution.

Acknowledgements

I thank the members of the Project Celescope staff, especially R. Davis, W. Deutschman, and K. Haramundanis, for making the *Celescope Catalog of Ultraviolet Observations* available to me in advance of publication. It is a special privilege to acknowledge fruitful scientific discussions with Dr C. Payne-Gaposchkin. Mr Carl Woebcke assisted in the data reductions.

References

Chaffee, F., Carbon, D., and Strom, S.: 1971, *Astrophys. J.* **166**, 593.
Crawford, D. and Perry, C.: 1966, *Astron. J.* **71**, 206.
Haramundanis, K. and Payne-Gaposchkin, C.: 1972, Abstract to be presented at AAS Meeting, East Lansing, Michigan, 15–18 August.
Hardorp, J. and Strittmatter, P.: 1968, *Astrophys. J.* **151**, 1057.
Struve, O.: 1945, *Pop. Astron.* **53**, 259.

DISCUSSION

Jaschek: Apparently, from the two communications, one has to conclude that the difference between theory and observation in the case of the Hyades can only be explained in two ways:
(1) by changes in the bolometric corrections,
(2) by having a large number of undetected binaries which affect the total magnitudes of the stars.
Could any one of the speakers comment upon this?

Schild: My answer can be brief; my observations, since they are reported as colour-colour diagrams, are independent of bolometric corrections. Similarly, undetected binary secondaries are likely to be so much fainter, and redder, that it is hard to imagine their contributing significantly to the far ultraviolet fluxes.

Pecker: The Pleiades vs Hyades diagram $U_2 - V$ vs $b - y$ is essentially characteristic of atmospheric properties. If we exclude Jaschek's good suggestion for double-star phenomena affecting the measurements, we must consider that a large difference in abundances cannot be excluded (as said by Schild) on the basis of *differential* curves of growth, as the atmospheres themselves may have a definitely different structure (possibly connected to deeper convective regions which, according to Demarque's comment, are coming in the picture). Differential methods have to be strongly criticized in such problems.

PHOTO-ELECTRIC MEASURES OF Hα AND Hβ IN THE SCORPIO-CENTAURUS ASSOCIATION

D. H. P. JONES

Royal Greenwich Observatory, England

and

D. W. CARRICK

*Mount Stromlo and Siding Spring Observatories, Research School of Physical Sciences,
Australian National University, Australia*

Abstract. Photo-electric measures of Hα and Hβ have been made in an attempt to confirm the scatter in the Luminosity-Balmer line relation predicted by Hardorp and Strittmatter as an effect of Stellar rotation.

There is a well known correlation between luminosity and Balmer line strength for early type stars. In this connection Jones (1971) derived new luminosities for stars in the Scorpio-Centaurus association and compared them with luminosities derived from a number of Balmer line indices. In a recent paper, Hardorp and Strittmatter (1968b) have suggested that comparisons of the kind performed by Jones should suffer from an irreducible scatter because of the effects of Stellar rotation. A cosmic scatter may also arise from differences in temperature and gravity at constant luminosity, or from emission lines arising in a shell surrounding the star. The present investigation was undertaken in an attempt to evaluate the irreducible scatter in the luminosity-Balmer line relation after all identifiable sources of dispersion had been removed.

The two-channel scanner at the Cassegrain focus of the Mt. Stromlo 50 in. telescope was used to observe Hα and Hβ in stars in the Scorpio-Centaurus association listed by Jones (1971). The scanner was set to an exit slit width of 30.6 Å, and programmed to read at six wavelengths. These instrumental readings $R(\lambda)$ were combined into two indices.

$$I\beta = \frac{2 \times R(4861)}{R(4961) + R(4761)}$$

$$I\alpha = \frac{2 \times R(6563)}{r(6642) + R(6487)}.$$

The $I\alpha$ index is the reciprocal of the $R\alpha$ defined by Andrews (1968) who used a slightly different bandwidth, 36 Å. Even when the bandwidth difference is ignored, the correspondence between them is

$$R\alpha^{-1} - I\alpha = + 0.004 \pm 0.008 \text{ s.e. per star}$$

for 25 absorption line stars. The correspondence for emission line stars is not so good but this almost certainly arises from intrinsic variability.

B. Hauck and B. E. Westerlund (eds.), Problems of Calibration of Absolute Magnitudes and Temperature of Stars, 36–40.

In general Hα and Hβ are formed by the same processes in stellar photospheres and may be expected to correlate strongly. On the other hand if the Balmer lines appear in emission in an optically thin region surrounding a star their decrement is very different with Hα much the stronger. Several stars have $I\alpha > 1$. and are indisputably emission line objects: HD 105435, 112091, 120324, 142983 and 148184. In Figure 1 the other stars show a strong correlation between $I\alpha$ and $I\beta$ with only one deviant star, HD 142184. This is also a known emission star (de Vancouleurs, 1957).

In Figure 2, M_v from Jones (1971) is plotted against $I\beta$. The emission line stars are arrowed and the curved line refers only to the remainder. Apart from errors of observation the residual scatter is a combination of variation in $I\beta$ with temperature,

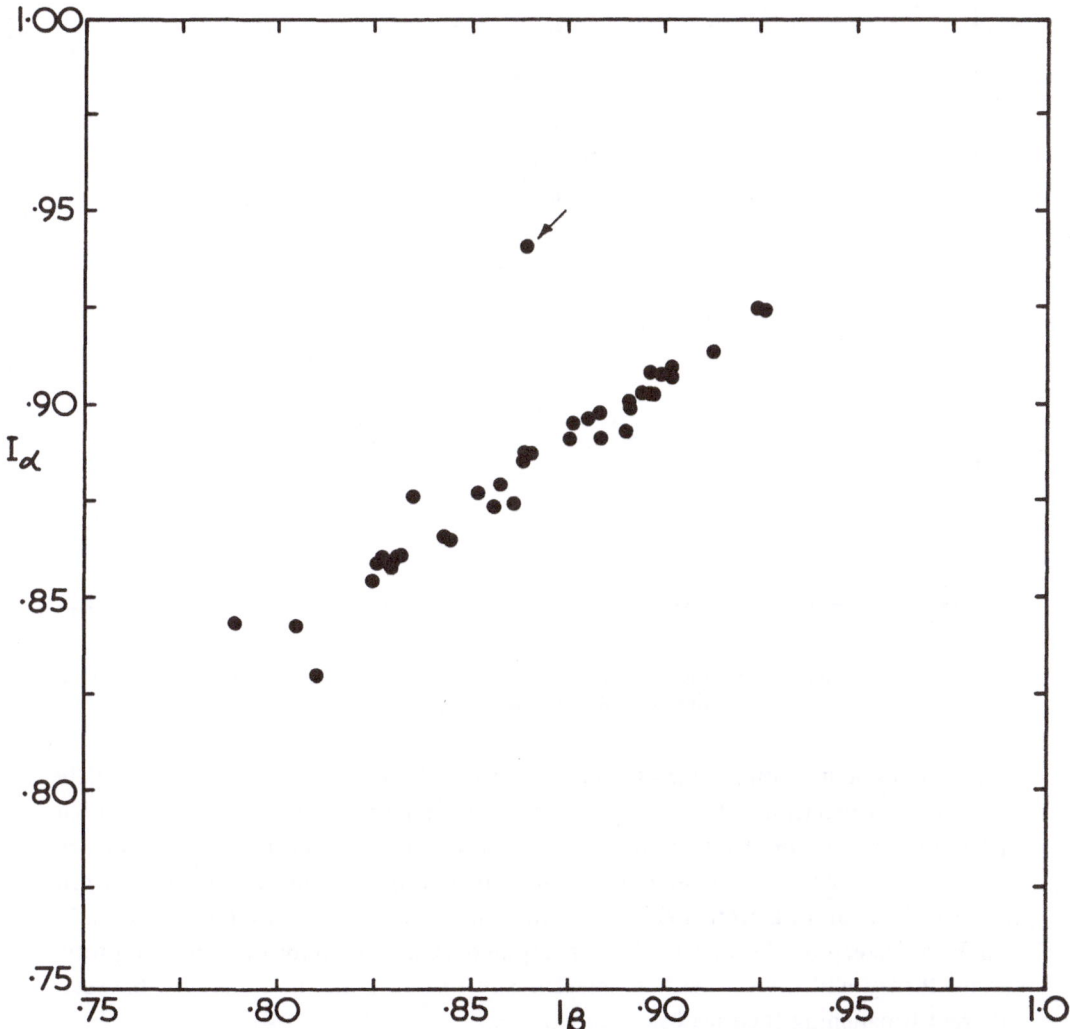

Fig. 1.   Correlation of $I\alpha$ with $I\beta$, stars with $I\alpha > 1$. (strong Hα emission omitted). The arrowed star
is HD 142 184 which is also a known emission line star (de Vaucouleurs, 1957).

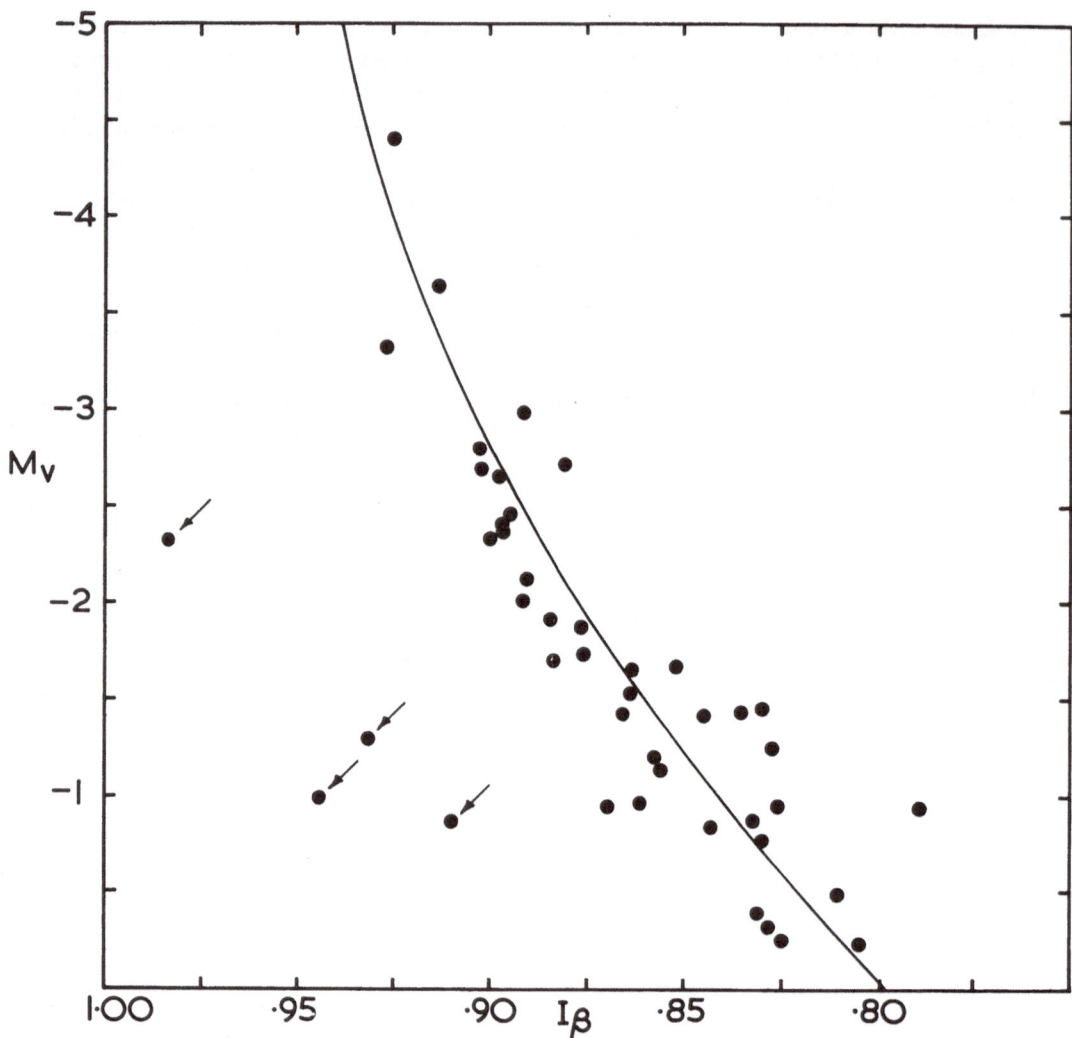

Fig. 2. M_v from Jones (1971) plotted against $I\beta$. The arrowed stars have emission lines. The curved
line is drawn to fit only the others.

variation in gravity with luminosity and any effects of Stellar rotation. It is practically
impossible to determine the dependence of $I\beta$ on temperature because of the high
internal correlation of $I\beta$ on M_v and, in turn, of M_v on temperature. The dependence
must be determined theoretically and we are calculating the required quantities from
the van Citters and Morton (1970) model atmospheres in co-operation with Drs B.A.
and V. A. Peterson. To date we have computed $I\alpha$ and $I\beta$ only for the asymptotic
Griem theory and do not wish to present any results until the difference between
different broadening theories has been explored.

   In Figure 3 the deviations $\Delta I\beta$ from the line in Figure 2 are plotted against $\Delta (U-B)_o$,
the deviation of each star in Figure 10 of Jones (1971) from the lower envelope in

that figure defined as

$$(U - B)_0 = -0.672 + 0.1215M_v.$$

Figure 3 exhibits a definite correlation but some such correlation is expected from the scatter in M_v which affects both $\Delta I\beta$ and $\Delta(U-B)_o$. The expected rms scatter of ±0.16 in M_v should produce the scatter shown by vector **A** in Figure 3. The approximate effects of rotation are shown by the vectors **B** and **C**. If a non-rotating star corresponding to Hardorp and Strittmatter's (1968a, b) model II is placed at the origin then the corresponding pole-on star at break up velocity will be carried along the vector **B** while the corresponding equator-on star is carried along the vector **C**.

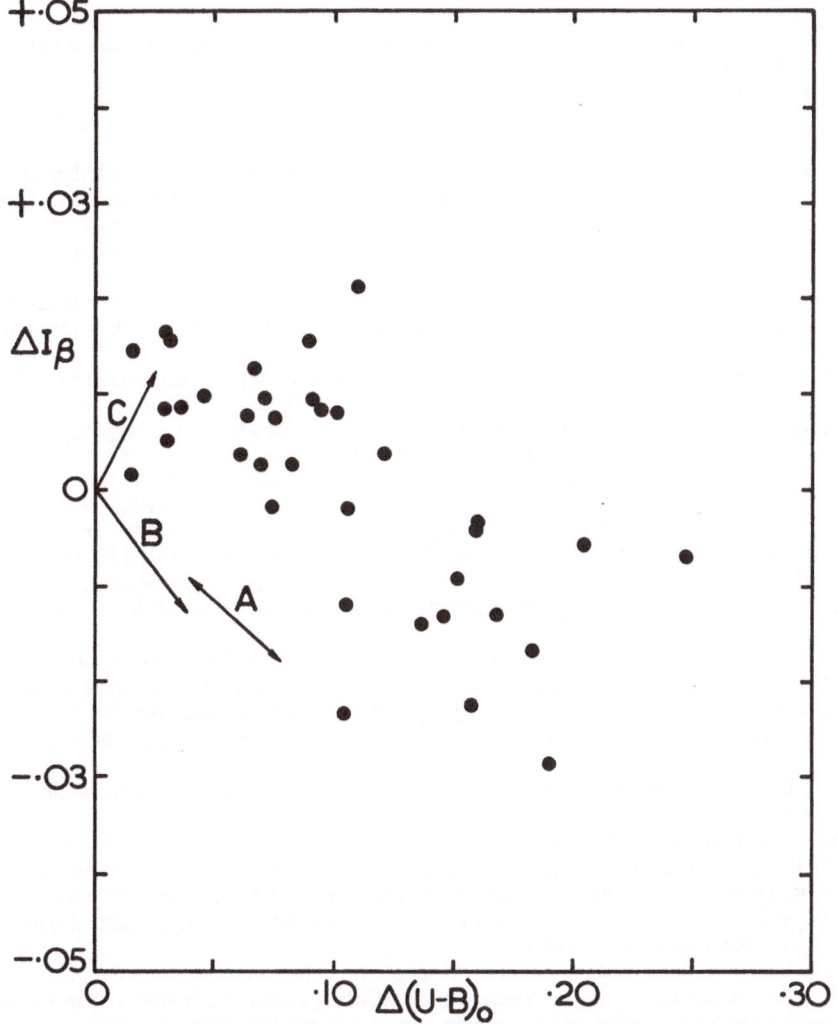

Fig. 3. $\Delta I\beta$ deviation in Figure 2 plotted against $\Delta(U-B)_o$ as defined in the text. Vector A represents the effect of the expected errors in M_v (two standard errors in Length). The vectors **B** and C represent the effect of rotation at break up velocity (**B** pole-on and C equator-on).

The work completed so far is thus qualitative confirmation of Hardorp and Stritt-matter predictions. It is hoped to make this confirmation quantitative in the near future.

References

Andrews, P. J.: 1968, *Mem. Roy. Astron. Soc.* **72**, 35.
Citters, G. W. van and Morton, D. C.: 1970, *Astrophys. J.* **161**, 695.
Hardorp, J. and Strittmatter, P. A.: 1968a, *Astrophys. J.* **151**, 1057.
Hardorp, J. and Strittmatter, P. A.: 1968b, *Astrophys. J.* **153**, 465.
Jones, D. H. P.: 1971, *Monthly Notices Roy. Astron. Soc.* **152**, 231.
Vaucouleurs, A. de: 1957, *Monthly Notices Roy. Astron. Soc.* **117**, 449.

DISCUSSION

McCarthy: (1) Can you tell us the spectral range of the Scorpio-Centaurus stars observed here? (2) How many B8 stars were included?

Jones: From B0 to B8, with at least one of the latter.

Newell: Have you considered the possibility of a range in mass existing at each (g, θ_{eff})-pair, and whether or not such an effect can contribute to the scatter that you observe?

Jones: While a unique mass-luminosity law exists, gravity and luminosity will be linked at constant θ. When the star leaves the main-sequence there will be a change in luminosity which will break the perfect correlation with gravity. In Figure 3 this will appear as an error in luminosity, and will move the stars in a parallel direction to the vector **A**. Motion in a perpendicular direction cannot be explained by this effect.

Garrison: I will discuss this work tomorrow in more detail during the session on calibration of the MK system, but I would like to take a minute discuss one of the results which is relevant to the distance of the Hyades.

There is a problem which arises if the distance modulus of the Hyades is changed from 3.0 to 3.2 or 3.4. I have constructed a composite HR diagram, using MK classifications by either Morgan or myself or both of us, of 3 moving clusters with distances determined by the convergent point method. These are the Hyades, α Persei, and the Inner Region of Upper Scorpius. (See Figure 1, page 120.) The point relevant to today's discussion is that, assuming a distance modulus of 3.0 for the Hyades, I get a distance modulus of 6.2 for Upper Scorpius, which is exactly what Bertiau obtained in 1958. However, Derek Jones has recently redetermined the Scorpius distance modulus and has obtained 5.8. If the Hyades is moved further away, then Upper Scorpius must be moved by about the same amount. This will be in the wrong direction, however, and the discrepancy between the cluster fitting and Derek Jones' distance then becomes more than half a magnitude, which is not easy to explain away.

Jones: Within the errors of observation my convergent point for Scorpio-Centaurus agrees with that of Bertiau. In any case the stars are nearly 90° from the vertex on the average and luminosities are insensitive to changes in the convergent point. The only critical quantities are the size of the proper motions on the FK4 system and the correction to Newcomb's precession and motion of the equinox.

Thomas: Could there be differences in the mean rotation of the clusters, which would alter your conclusions?

Garrison: There seems to be a considerable difference of opinion, as illustrated by todays' discussion, concering the effect of rotation on photometry. It is, in any case, unlikely that small differences in mean rotation will make a difference. Extremely rapidly rotating stars have been left out of the formation of the main sequence, mostly because it is difficult to determine a unique spectral type for such stars which exhibit a range in excitation.

Newell: Have you considered the possible effects of abundance differences on the positions of the main sequences in the HR-diagram? There are two distinct effects. I am referring to the effect of abundance on the *intrinsic* location of the main sequence in the HR-diagram.

Garrison: It is possible, by spectral classification, to detect metal abundance differences as peculiarities, even though the MK system is defined in only two dimensions. In the case of the Hyades stars and the α Persei stars, there seem to be no significant differences.

ROTATIONAL EFFECTS ON THE PHOTOMETRIC DETERMINATIONS OF ABSOLUTE MAGNITUDES FOR A- AND F-TYPE STARS

A. MAEDER

Observatoire de Genève, Switzerland

Abstract. The effects of rotation on the determinations of absolute magnitudes by means of the c_1 vs $(b-y)$ and d vs $(B_2 - V_1)$ diagrams are discussed.

These effects mainly concern field stars, for which absolute magnitudes are determined by means of the c_1 vs $(b-y)$ diagram (Strömgren, 1963) or the d vs $(B_2 - V_1)$ diagram (Hauck, 1966). We consider here the case of uniform rotation, which has received some observational supports for upper main-sequence stars (e.g. Maeder and Peytremann, 1970, 1972). Figure 1 shows the rotational tracks in the c_1 vs $(b-y)$

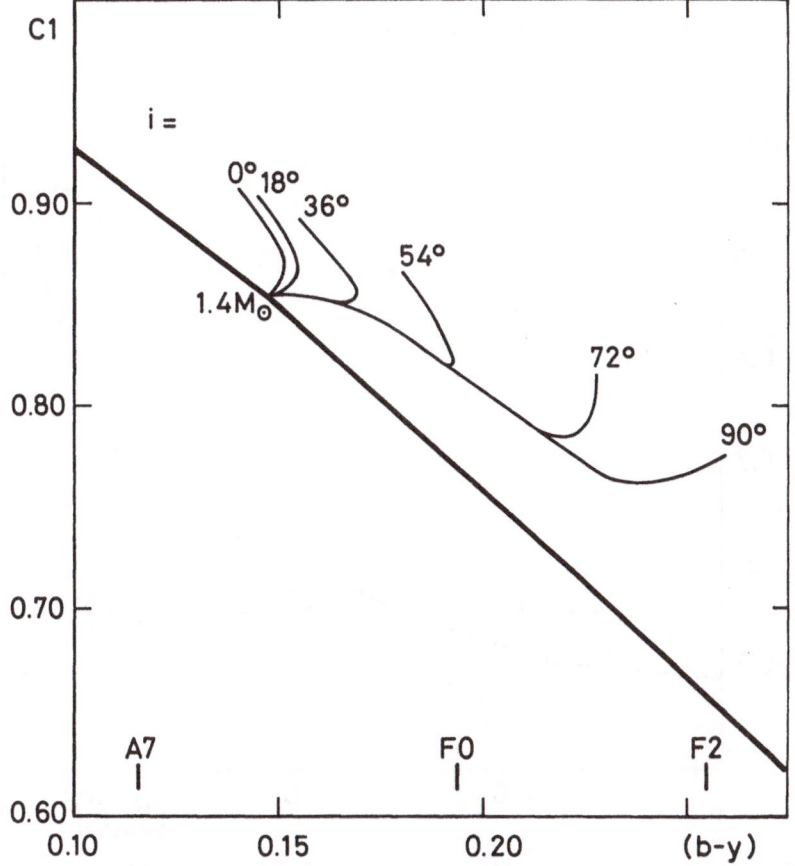

Fig. 1.   Rotational tracks in the c_1 vs $(b-y)$ diagram.

B. Hauck and B. E. Westerlund (eds.), Problems of Calibration of Absolute Magnitudes and Temperature of Stars, 41–43.

diagram, as they are predicted by the models of uniformly rotating stars with hydrogen- and metallic-line blanketing (Maeder and Peytremann, 1972). The determinations of M_v make use of the ratios $\Delta M_v/\Delta c_1$ or $\Delta M_v/\Delta d$, which have been calibrated by Strömgren (1963) and Hauck (1972). Table I gives the rotational shifts ΔM_v, Δc_1, Δd, as they are predicted by the models; ω is the ratio of the angular velocity of the star to the value at the break-up point, v_R is the equatorial velocity and i the aspect angle. One sees that the mentioned ratios are far from being conserved by rotation. The last two columns of Table I give the difference δM_v between the real M_v and the value found by the application of the methods using the c_1 vs $(b-y)$ and the d vs $(B_2 - V_1)$ diagrams. Figure 2 illustrates these differences in the case of the d vs $(B_2 - V_1)$ diagram. Both methods lead to an overestimate of brightness for rapidly rotating stars, which are seen equator-on and to an underestimate for those, which are seen pole-on. For rapidly rotating stars later than F0, attention has to be given to the fact, that all observed effects appear in the same sense, but twice as large (Maeder and Peytremann, 1972) as those predicted by the models. Figure 2 does not give a mean for correcting the effects of rotation, because this would require the knowledge of the angle i; it only gives an indication on the sense and size of the effects of rapid rotation in M_v determinations of this kind.

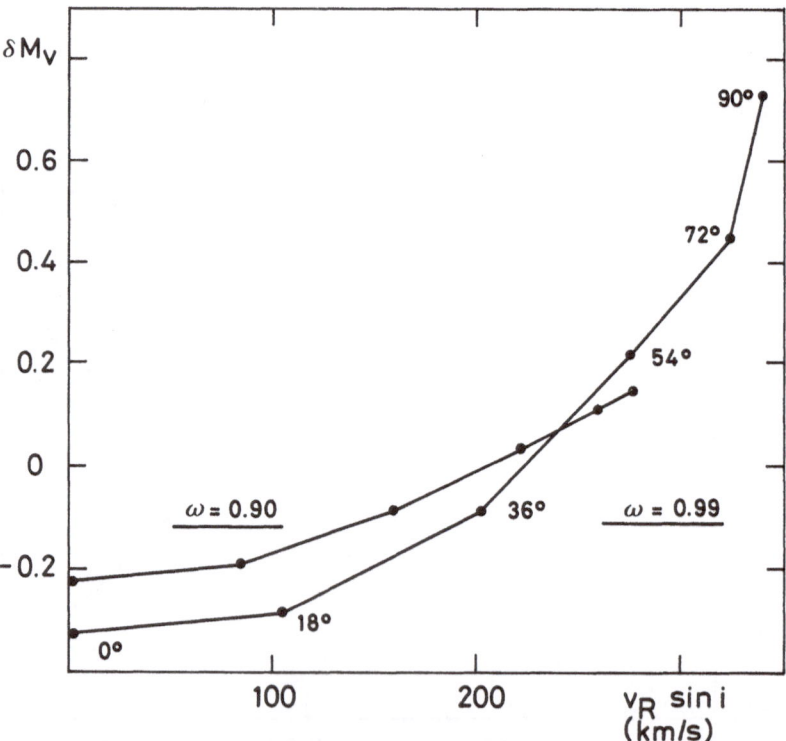

Fig. 2.   Difference between the real M_v and the value of M_v determined by means of the d vs $(B_2 - V_1)$ diagram in function of $v_R \sin i$ and i for $\omega = 0.9$ and 0.99.

TABLE I

Results of the 1.4 $M_{\odot}$ model

ω	v_R (km s^{-1})	i	ΔM_v	Δc_1	Δd	δM_v $(c_1/b-y)$	δM_v (d/B_2-V_1)
0.90	272	0°	0.33	0.026	0.016	−0.12	−0.23
		54°	0.26	0.045	0.048	+0.10	+0.03
		90°	0.33	0.064	0.076	+0.22	+0.13
0.99	341	0°	0.46	0.043	0.024	−0.12	−0.32
		54°	0.19	0.072	0.067	+0.39	+0.21
		90°	0.33	0.126	0.140	+0.83	+0.72

References

Hauck, B.: 1966, *Publ. Obs. Genève* **72**, 181.
Hauck, B.: 1972, this volume, p. 117.
Maeder, A. and Peytremann, E.: 1970, *Astron. Astrophys.* **7**, 120.
Maeder, A. and Peytremann, E.: 1972, *Astron. Astrophys.*, **21**, 279.
Strömgren, B.: 1963, *Stars and Stellar Systems* **3**, 123.

DISCUSSION

Pecker: I doubt if even a good index can give valuable information on the rotation; the reason is that the interpretation is strongly depending upon the *assumed* distribution of temperature with latitude, and upon, therefore, the internal structure etc.

Crawford: Have you compared your model predictions with observations? I do not like Praesepe for these comparisons. The Pleiades or α Per cluster is better, for the A stars are not evolved. One can more easily separate parameters therefore.

Maeder: The comparisons between the observations and H-lines blanketed atmospheric models of rotating stars have been made in *Astron. Astrophys.* **7**, 120 (1970), and for the hydrogen-and metallic-lines atmospheric models in *Astron. Astrophys.*, in press (1972). Rotational effects may only be put in evidence by discussion of *differential* effects. So the fact that the A-type stars of Praesepe are evolved has no importance, because it suffices to compare the slow and the fast rotating stars without considering the position of the ZAMS. Praesepe has the advantage, compared to the Pleiades and the α Persei cluster, that no differential reddening corrections have to be applied.

Kodaira: What coefficient of the ω^2-term have you adopted? In the last years the coefficient, adopted by Strittmatter-Hardorp, was revised to a smaller one. The theoretical model is still not acurate enough, one should make more effort to detect the effect observationally. My attempt to find the effect in Balmer-jump excess for a given $(b-y)_0$, resulted in a contradiction of the theoretical prediction (*Astrophys. J.* **159**, 931). The deviations found by me are rather along the line of 'pole-on' effect (equivalent to enlarging log$g$), in agreement with the Mt. Stromlo data presented here today by Jones. The possible error in M_v, indicated by I_β of the Mt. Stromlo data, would be still explained by a rotational effect, or an unknown effect equivalent to enlarging log$g$.

Maeder: I have adopted the coefficients given by Faulkner *et al.* (1968, *Astrophys. J.* **151**, 203), which are also in agreement with those given by Sackmann and Anand (1970, *Astrophys. J.* **162**, 105).

Thomas: In the paper by Roxbourgh *et al.* the coefficient for ω^2 corresponded to a maximum decrease in luminosity of 25% but all later authors obtained values of about 7% to 8%, which is the value also used by Dr Maeder in his talk.

Code: I should like to comment with respect to rotation that comparison of stars in the ultraviolet from the Balmer jump to about 1500 Å, based on OAO-2 spectrophotometry show no dependence on $V\sin i$ even for large rotational velocities. This photometry is accurate to about $0\overset{m}{.}02$.

PART III

CALIBRATION OF
SPECTROSCOPIC PARALLAXES

THE CALIBRATION OF LUMINOSITY CRITERIA

A. BLAAUW

European Southern Observatory, Hamburg, F.R.G.

Abstract. Attention is drawn to important systematic effects in the calibration procedure due to the accidental errors in the measured luminosity criteria. The present state of the luminosity calibrations in the MK system is reviewed with reference to recent work based on proper motions and radial velocities, and on trigonometric parallaxes, resulting in evidence for corrections of about -0.4 mag. for the K0–K5, III stars. Brief reference is made to the developments with regard to the $M_v(K)$ system.

The present report summarizes the principal elements of the review presented at the symposium.

1. Avoidance of Using a 'Biassed' Calibration Curve

The problem of the calibration of luminosity criteria is essentially that of establishing the relation between the measured quantity I (say, an intensity ratio in the spectrum) and the absolute magnitude, M. It is complicated by (a) the observational errors in I; (b) the circumstance that for a given I, the luminosity may depend on still other factors such as unresolvable duplicity, or stellar rotation and the angle of orientation of the rotational axis, or chemical abundance. Factors like these may be described to cause a 'cosmic' scatter with respect to a standard relation between I and M. They will be discussed in more detail elsewhere during this symposium. For the present introduction I shall ignore these 'cosmic errors' in M and assume that for errorfree values of I a strict relation between I and M exists. Attention will be drawn here to the importance of the random errors in I as a cause of *systematic* errors in the derived M, if no proper precautions are taken in deriving the calibration relation. The matter has been discussed previously (Blaauw, 1963), but a re-presentation, with a somewhat simpler approach, seems useful.

The systematic effect referred to is a function of the frequency distribution of the absolute magnitudes, $\varphi(M)$; for reasons of simplicity we shall assume it to be gaussian with dispersion σ. We further, also for the sake of simplicity, shall assume the relation between M and true I to be linear, $M = \alpha I + \beta$, and the mean error of I to be μ_i. The frequency distribution of I, $F(I)$, is then also gaussian, with dispersion $\sigma/\alpha = \sigma_i$. Figures 1a, b, c show, respectively, a section of the distribution function $\varphi(M)$, the corresponding distribution $F(I)$, and the relationship between I and M.

Suppose we select from the sample a subgroup with observed values of I in the interval ΔI around I_0' (I_0' chosen arbitrarily). For these, a mean absolute magnitude is determined, either by trigonometric or by secular parallaxes: M'. We shall assume this value of M' to be error-free. The dashed relation in Figure 1c, is obtained by plotting the value M' obtained in this way for the selected subgroup at I_0' as well as the value M'' obtained similarly for a subgroup at I_0''. I_0'' is also arbitrarily chosen. Now, this dashed relation is *not* to be identified with the strict relation $M = \alpha I + \beta$, which ought to be used as a calibration curve for converting observed I into M.

B. Hauck and B. E. Westerlund (eds.), Problems of Calibration of Absolute Magnitudes and Temperature of Stars, 47–56.

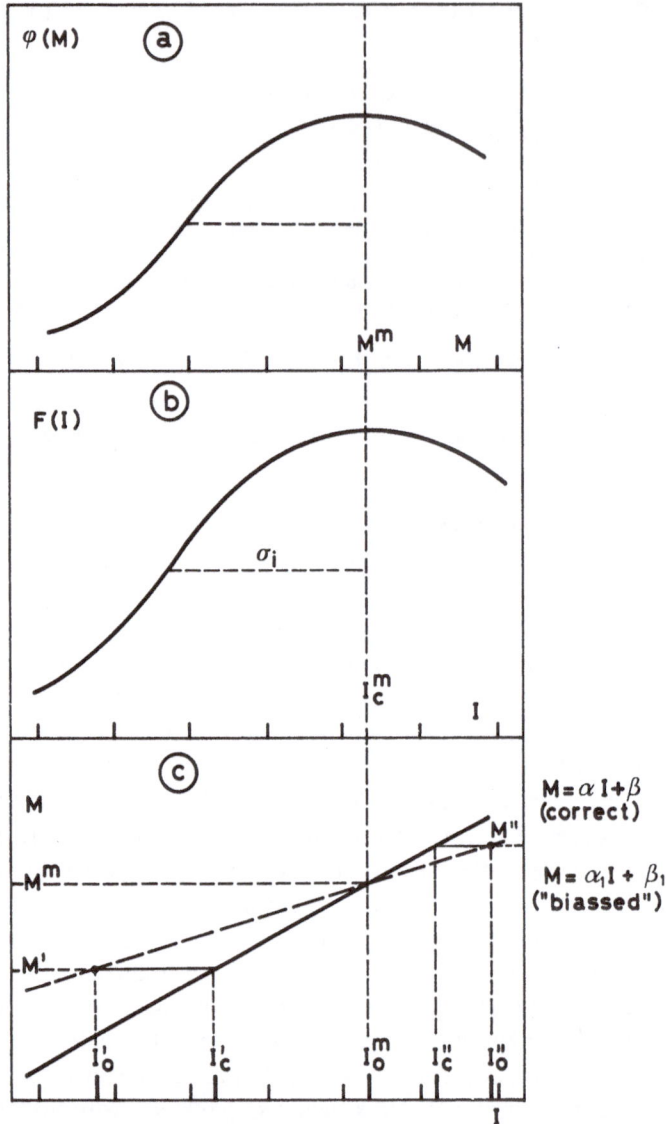

Fig. 1.   (a) The frequency distribution of absolute magnitudes, assumed to be gaussian. (b) The corresponding frequency distribution for the luminosity criterion I, assuming a linear relation between M and I. (c) The biassed calibration relation $M = \alpha_1 I + \beta_1$ (dashed line) and the correct relation. $M = \alpha I + \beta$ (drawn line).

Use of this biassed (sometimes called 'partial') dashed line $M = \alpha_1 I + \beta_1$ has, in the past, led to spurious conclusions, as we shall mention below.

For determining the proper relation $M = \alpha I + \beta$, we proceed by selecting the values M' or M'' obtained before and asking: which are the proper values of I belonging to them? We shall denote these by I'_c and I''_c, sothat $M' = \alpha I'_c + \beta$ etc. I'_c differs from

I_0' because, in the subsample $\Delta I_0'$ we have collected error-affected values of I_0 which happen to have arrived in this interval as a result of the measuring errors in I. The essential point, then, is, that the mean of the true values of I in this subsample is normally *not* identical to I_0' but differs from it systematically. In the present case, more objects with true value $I > I_0'$ have crept in than objects with $I < I_0'$, due to the slope of the function $F(I)$. A positive correction therefore has to be applied to I_0' to obtain the true value of I corresponding to M'. Similarly a negative correction has to be applied to I_0''. No correction need be applied to the value I_0^m corresponding to M^m, the absolute magnitude for which $\varphi(M)$ is at maximum, for reasons of symmetry. We have, $M^m = \alpha I^m + \beta = \alpha I_0^m + \beta$. It can be shown easily that

$$\frac{I_c' - I_0^m}{I_0' - I_0^m} = \frac{I_c'' - I_0^m}{I_0'' - I_0^m} = \frac{\sigma_i^2 + \mu_i^2}{\sigma_i^2}.$$

The thus corrected points (M', I_c') and (M'', I_c'') define the correct calibration relation $M = \alpha I + \beta$; it contains the point $(M^m, I_c^m) \equiv (M^m, I_0^m)$.

Clearly, the change in slope from the true relation to the biassed one is such, that using the latter for converting observed values of I into M leads to an underestimate of the spread in the absolute magnitudes. Typical examples of this effect, which can be very serious, have been mentioned in the article quoted (Blaauw, 1963).

Once the correct relation $(M = \alpha I + \beta)$ is used for calibrating a given set of values of I into M, the resulting distribution of M is, of course, affected by the random errors of I in such a way as to broaden the (gaussian) distribution of M. This effect must be considered quite separately from that described before. E.g. the determination of the proper calibration relation may be based on a quite different sample of stars than those for which the calibration is used, the latter for instance with mean error μ_j of I. The 'broadened' dispersion of M then is larger than the true one in the proportion $(\sigma_i^2 + \mu_j^2)^{1/2}/\sigma_i$. If $\mu_i \equiv \mu_j$, then the resulting effect of using the biassed calibration curve is still a narrowing, in the ratio $\sigma_i/(\sigma_i^2 + \mu_i^2)^{1/2}$. Neglecting the narrowing effect of the use of the biassed calibration curve may, for example, lead to a serious underestimate of the width of the giant branch in the HR diagram.

2. The Calibration of the MK Luminosity Classes

The MK spectral and luminosity classification system remains a most useful frame of reference for classification systems in general, although for certain domains of the array it is gradually being replaced by other, quantitative, systems. For the O, B and A stars this is the case with respect to the intermediate-band u, v, b, y system plus Hβ photometry, especially making use of the quantities $c_1 = (u-v)-(v-b)$ and $m_1 = (v-b)-(b-y)$ and thereby adding the third dimension: metal abundance. For the late type stars the luminosity estimates through the measures of the Ca^+ reversals gradually supersede the visual luminosity estimates. Several refined narrow and intermediate band systems, discussed elsewhere at this symposium, represent further improvements.

A. PRINCIPAL METHODS OF CALIBRATION

For a discussion of the status of the calibration of the MK system, reference to Figure 2 is useful. It indicates, by means of the differently hatched regions, the applicability of the three principal basic methods for the calibration: trigonometric parallaxes, secular parallaxes (more general: the use of proper motions plus radial velocities), and the zero-age main sequence fitting procedure.

The most fundamental method, trigonometric parallaxes, applies to the main sequence stars F5 through M, and also to a certain extent to the giants of classes

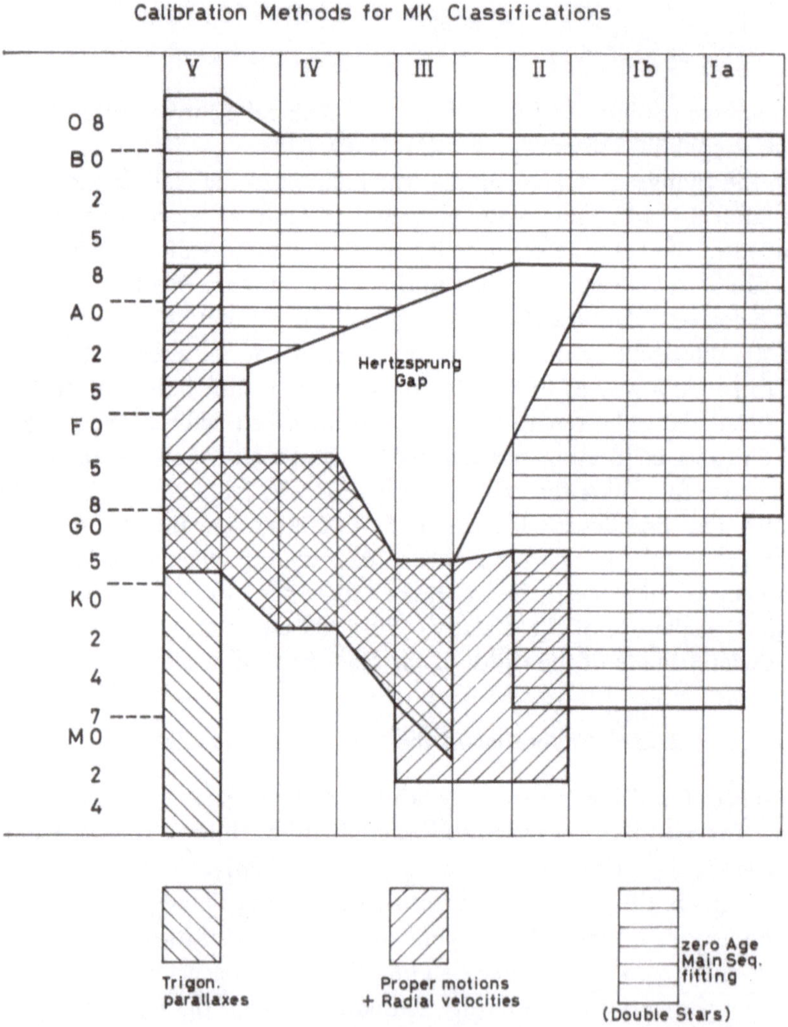

Calibration Methods for MK Classifications

Fig. 2.   The applicability of the three methods for luminosity calibration (trigonometric parallaxes, proper motions plus radial velocities, and fitting of the zero age main sequence) for the various domains of the HR diagram in the MK classification system.

G5 III to late K III, and to the intermediate classes IV. The secular parallax method overlaps with it for main sequence classes F5 through G5, for classes F5 IV through K0 IV, and for G5 III to late K III, but extends further along the main sequence upwards to class B8 V and along the giants down to the M III types, and to the giants G5 through K of class II. The zero age main sequence fitting procedure overlaps with the secular parallaxes upward from main sequence types A5, and somewhat in the domain of the G, K II stars.

Obviously this description is rather schematic, but it helps telling which method(s) will be most useful for the improvement of the present calibrations. In all methods, the region of the Hertzsprung gap (see Figure 2), due to its very scarce population, remains poorly calibrated, but due to this very scarcity the need for this calibration remains limited.

In the following paragraphs we review some recent improvements of the calibrations as compared to about ten years ago (Blaauw, 1963). See also Schmidt-Kaler (1965). We do not include in this discussion certain differential, though very interesting effects, like stellar rotation and abundance effects.

For a discussion of the accuracies within the classification system MK itself we refer to analysis by Jaschek and Jaschek (1971); It appears that the average uncertainty of a single classification (due to a variety of sources of error) is less than 0.6 luminosity class, and 0.6 or more in the spectral type.

B. RECENT IMPROVEMENTS THROUGH THE USE OF SECULAR PARALLAXES

The most comprehensive recent discussion is by Jung (1970). The procedure using proper motions and radial velocities, and referred to as a maximum likelihood method, aims at finding such a distribution of distances of stars of a given sample (spectral and luminosity type, apparent magnitude) as to give the 'best' fitting of proper motions (converted to tangential velocities) to radial velocities. It is essentially identical to the method applied long ago to more limited material and with more restricted computational facilities (for instance, Strömberg, 1933, 1936); however, the use of modern computers allows more diversified solutions. The method obviously is of greatest importance where trigonometric parallaxes fail or are of limited use: i.e. for stars of types A, F V and G, K, M III.

Basic material in Jung's analysis are the stars in the Bright Stars Catalogue (Hoffleit, 1964) for which proper motions and radial velocities are almost complete. This choice implies that the results apply virtually only to Population I and Disc population, the percentage of low metal abundance stars being very low.

The principal results by Jung are summarized in Tables I and II for classes V and III, respectively, in the columns headed 'P.M. + Rad. Veloc.'. The column BAD (1963) reproduces results given by the author (Blaauw, 1963) in *Basic Astronomical Data*. The numbers of stars used by Jung are indicated under *n*.

For classes V, B8–G8, the new results are systematically about 0.2 mag. brighter – but this is also the uncertainty inherent to the system due to the imperfections of the method. For classes III, G5–M4, the improvements are more striking: for K0–M

TABLE I
MK Class V; M_v per apparent mag.

	B.A.D. (1963)	P.M. + Rad. veloc.			Trigon. parall.		
		Jung (1970)	J − B	n	Jung (1971)	J − B	n
B 8	−0.5	−0.7	−0.2	140			
9	0.0	0.0	0.0				
A 0	+0.5	+0.5	0.0	134	+0.2	−0.4	34
1	+0.8	+0.6	−0.2				
2	+1.2	+0.7	−0.5	106	+0.6	−0.6	24
3	+1.5	+0.8	−0.7	81	+1.4	−0.1	22
5	+1.8:	+1.5	−0.3:	52	+1.2	−0.7:	17
7	+2.0:	+1.6	−0.4:				
F 0	+2.4:	+2.3	−0.1:	60	+2.5	−0.1:	26
2	+2.8:	+2.5	−0.3:				
5	+3.2	+3.0	−0.2	80	+3.3	+0.1	40
6	+3.5	+3.2	−0.3				
8	+4.0	+3.5	−0.5	60	+3.4	−0.6	35
G 0	+4.4	+4.1	−0.3	57	+4.3	−0.2	29
2	+4.7	+4.3	−0.4				
5	+5.1	+4.9	−0.2	29	+4.8	−0.5	32
8	+5.5	+5.5	0.0				

TABLE II
MK Class III; M_v per apparent mag.

	B.A.D. (1963)	P.M. + Rad. veloc.			Trigon. parall.		
		Jung (1970)	J − B	n	Jung (1971)	J − B	n
G 5	+0.4:	+0.3	−0.1:	51	+0.2	−0.2	70
8	+0.4	+0.2	−0.2	170			
K 0	+0.8	+0.1	−0.7	176	+0.4	−0.4	48
1	+0.8	+0.5	−0.3	98	+0.4	−0.4	19
2	+0.8	+0.2	−0.6	95	+0.6	−0.2	54
3	+0.1	−0.3	−0.4	88	−0.2	−0.3	48
4	−0.1	−0.8	−0.7	133	−0.3	−0.1	41
5	−0.3	−1.0	−0.7				
M 0	−0.4:	−1.2	−0.8:	66			
2	−0.4:	−1.5	−1.1:	28			
4	−0.5:	−1.7:	−1.2:				

the visual luminosities are about 0.5 mag. brighter, which should be a significant improvement in view of the uncertainty of ±0.2 mag.

Ljunggren and Oja (1965, 1966) arrived at similar corrections for G8–K5, III stars (see Table III), from an analysis of proper motions and radial velocities in the context of a calibration study of the Uppsala photometric system (see also below).

C. RECENT IMPROVEMENTS THROUGH THE USE OF TRIGONOMETRIC PARALLAXES

Results of a recent comprehensive study of the use of trigonometric parallaxes by Jung (1971) are also given in Tables I and II, last columns. Again, the stars in the *Bright Stars Catalogue* are used, with parallaxes from the catalogue of Jenkins (1954). A special effort was made to take into account the systematic effects possibly entering into the results as a consequence of the random errors in the parallaxes in combination with the special choice of the sample: objects with (observed) parallaxes exceeding a certain numerical limit. Also investigated was the influence of systematic corrections to the Jenkins parallaxes.

For classes V, F2–G8, the new results are systematically about 0.3 mag. brighter than the 1963 'BAD' values: the differences probably are significant. For the A V stars (Table I) a similar correction is indicated. Here the numbers of stars in the sample are necessarily very limited. For classes III, G8–K5 (Table II), again a systematic negative correction of several tenths of a magnitude is found which appears significant. It depends somewhat on whether the Jenkins parallaxes are systematically corrected or not by $-0\overset{''}{.}035$ as proposed by Schilt (1954). For these spectral classes Ljunggren and Oja (1966) arrive at a somewhat larger negative correction, about -0.6 mag. or the average (Table III).

Summarizing the results from secular and trigonometric parallaxes, we conclude that for luminosity class V, types A through G, the earlier calibrations were about 0.25 mag. too faint; that for luminosity class III types K0–K5 these early results were about 0.4 mag. too faint; and that for the G III stars and the M III starss maller, respectively larger negative corrections are indicated.

TABLE III

MK Class III; M_v per apparent mag.

	B.A.D. (1963)	Ljunggren + Oja (1966)			
		p.m. + rad.-vel.	LO − B	trig. parall.	t.p. − B
G 8	+0.4	+1.0	+0.6	+0.6	+0.2
K 0	+0.8	+0.2	−0.6	+0.1	−0.7
1	+0.8			+0.8	0.0
2	+0.8	+0.3	−0.5	−0.1	−0.9
3	+0.1			−0.2	−0.3
4	−0.1	−0.5	−0.4	−0.9	−0.8
5	−0.3			−1.4	−1.1

All values in Tables I to III refer to a selection of stars 'per apparent magnitude', i.e. containing the bias of intrinsically brighter stars of a certain subtype having been selected over a larger volume of space than the intrinsically faint stars. As has been pointed out before (Blaauw, 1963) these values are about half a magnitude brighter than values referring to a selection per volume of space.

Corrections to the upper part of the calibration table in the 'BAD' volume are not studied here; these would be due largely to the application of the zero age main sequence fitting procedure. In this domain of the spectral and luminosity classes (O, B, A) the practical value of the luminosities in the MK system is now rapidly being superseded by the quantitative photometric methods (see the contribution of Crawford at this symposium).

3. The $M_v(K)$ System

No detailed discussion of this system, introduced by Wilson and Bappu (1957) will be presented here. It clearly is going to be a most important source for absolute magnitudes in the range of spectral types G5 and later for all luminosity classes, and thereby also an important basis for calibrations of the various visual (MK) and photometric luminosity criteria. See, for instance, work by Häggkvist and Oja (1970) on the calibration of narrow band photometric criteria for F8–M4 stars, especially of luminosity class III. This, however, will require further evaluation of the influence of chemical abundance effects on which the results so far are not unambiguous. For a recent review we refer to Wilson (1970); more recent papers dealing with the sensitivity of the method to chemical abundance are by Kjaergaard (1970) and Hansen (1972), and by Yoss and Lutz (1971), whereas Wilson and Woolley (1970) discussed the relation with stellar ages.

References

Blaauw, A.: 1963, *Stars and Stellar Systems* 3, 383.
Häggkvist, L. and Oja, T.: 1970, *Astron. Astrophys. Suppl.* 1, 199.
Hansen, L.: 1972, (private communication).
Hoffleit, D.: 1964, *Catalogue of Bright Stars.*
Jaschek, C. and Jaschek, M.: 1971, in Ch. Fehrenbach and B. E. Westerlund (eds.), 'Spectral Classification and Multicolour Photometry', *IAU Symp.* 50, 43.
Jenkins, L. F.: 1954, *Catalogue of Trigparallaxes.*
Jung, J.: 1970, *Astron. Astrophys.* 4, 53.
Jung, J.: 1971, *Astron. Astrophys.* 11, 351.
Kjaergaard, P.: 1970, *Astron. Astrophys.* 5, 165.
Ljunggren, B. and Oja, T.: 1965, *Uppsala Astron. Obs. Medd.*, 150.
Ljunggren, B. and Oja, T.: 1966, in K. Lodén, L. O. Lodén and U. Sinnerstad (eds.), 'Spectral Classification and Multicolour Photometry', *IAU Symp.* 24, 317.
Schilt, J.: 1954, *Astron. J.* 59, 55.
Schmidt-Kaler, Th.: 1965, in Voigt, H. H. (ed.), *Landolt-Börnstein, Numerical Data and Functional Relationship in Science and Technology*, Vol. 1, Group IV, p. 299.
Strömberg, G.: 1933, *Astrophys. J.* 78, 178.
Strömberg, G.: 1936, *Astrophys. J.* 84, 412.
Wilson, O. C.: 1970, *Publ. Astron. Soc. Pacific* 82, 865.
Wilson, O. C. and Bappu, M. K. V.: 1957, *Astrophys. J.* 125, 661.

Wilson, O. C. and Woolley, v. d. R.: 1970, *Monthly Notices Roy. Astron. Soc.* **148**, 463.
Yoss, K. M. and Lutz, T. E.: 1971, *Mem. Roy. Astron. Soc.* **75**, No. 2.

DISCUSSION

Schmidt-Kaler: Before opening these papers for discussion I would like to make two remarks on some work done at Bochum concerning problems of intrinsic colours and absolute visual magnitudes of early type stars and supergiants.

1. The Upper Part of the Zero Age Main Sequence

Blaauw (1963) has constructed the upper ZAMS by fitting successively four open clusters and one association to the unevolved Hyades main sequence. It seems desirable (1) to bridge the gap by fitting an early-type cluster directly to the Pleiades or even to the Hyades, and (2) to do so for several clusters in order to check on possible systematic differences. In his doctoral thesis Vogt (1971) did this for the double cluster in Perseus. All stars in the $36' \times 60'$ field were measured down to $V = 18^{\rm m}0$ (altogether 6742 stars) as well as 4717 stars in eight symmetrically placed comparison fields of equal total area. The large numbers of stars and a careful study of the reddening and its variations in the field made it possible to determine the ridge line of the main sequence by statistical substraction down to $M_v = +3.6$ resp. $+4.0$. The result is that h Per gives a perfect fit within $\pm 0^{\rm m}1$ maximal deviation to Blaauw's ZAMS in the interval $M_v = +3.4 \ldots -1.3$ (corresponding to $(B-V)_o = +0.38 \ldots -0.21$) while the unevolved main sequence of χ Per in the interval $M_v = +2 \ldots -2.5$ resp. $(B-V)_o = +0.10 \ldots -0.27$ displays systematic deviations from Blaauw's ZAMS up to $0^{\rm m}3$. The cluster χ Per appears to be considerably younger than h Per; the distributions of the dwarf emission B-stars and of the supergiants are centered on it, the only O-star in the area belongs to it. A revised discussion of Vogt's photometry is in preparation.

2. The Absolute Magnitudes of OB-Stars and Supergiants

Recently, Stothers (1972) and Walborn (1972) presented recalibrations of the absolute magnitudes of supergiants and OB-stars, respectively. Although, of course, some improvement is possible we wish to point out that – with the present material – no significant differences between the new and the old calibrations (Blaauw, 1963; Schmidt-Kaler, 1965) are evident.

It is well known that the scatter of the absolute visual magnitudes for a given MK-type in this region of the Hertzsprung-Russell-diagram is about $\sigma_M = \pm 0^{\rm m}6$. For example, the mean standard deviation for supergiants from Stothers' (1972) work is $\pm 0^{\rm m}54$; a large part of this scatter is due to the fact that the MK-system puts the stars in discrete boxes.

Assuming that the distance determinations for the recalibrations and the first approximations of the absolute magnitudes as given in the Landolt-Börnstein Tables are free of errors we will certainly underestimate the variance of the differences of the calibrations. Applying Student's test to the differences ΔM_v (Stothers minus Schmidt-Kaler) only one difference significant on the 2% error-level (for the A Ia stars $+0.5 \pm 0.4$), and one just marginally significant difference (for the OB Ib stars -0.2 ± 0.2 on the 5% error-level) is found. The selection of A Ia stars is, however, for the most part taken from associations with uncertainties in distance modulus. Even the well-studied association Per OB1 gives rise to doubt: Stothers' discussion led him to assume $(m-M)_o = 11.5$ while Walborn's ends up with 11.65 which would lead to $\Delta M_v = +0.4$. At the level of $0^{\rm m}2$, of course, systematic errors of various origin come into play.

Applying Student's test on the same assumptions to the differences ΔM_v (Walborn minus Schmidt-Kaler) only two just marginally significant differences are found: for the O–B3 Ia stars $\Delta M_v = -0.5 \pm 0.4$, for the B0–2.5 V stars $+0.5 \pm 0.4$, both on the 5% error-level, is observed. Walborn's revised MK classification appears more accurate and may lead to somewhat smaller scatter. But even with $\sigma_M = \pm 0.3$ no other even marginally significant differences are noted. On the other hand Walborn's selection of main sequence B stars contains a large proportion of almost unevolved stars, the ZAMS being just about $0^{\rm m}5$ fainter than the average class V B star.

From the present discussion it is evident that the basic need in recalibrating the high luminosity areas of the MK-system is for more calibration stars. These should be members of many different groups and clusters in order to minimize possible systematic errors in individual cases. One and the same procedure should be used to find reddenings and distances; the colours should be evaluated simultaneously. Finally, a homogeneous distribution of calibrating stars over the areas considered should be aimed at. In the last three years Drs Moffat, Vogt and myself observed photoelectrically (in *UBV*) at the Bochum Southern Station 80 southern open clusters which had thus far never been studied, and photographically 12 northern open clusters thus far unstudied. They contain quite a few OB- and Be-stars and supergiants of all spectral classes. Work on their MK classification has just begun.

References

Blaauw, A.: 1963, *Stars and Stellar Systems* **3**, 383.
Schmidt-Kaler, Th.: 1965, in H. H. Voigt (ed.), *Astronomie und Astrophysik, Landolt-Börnstein*, Springer-Verlag, Berlin.
Stothers, R.: 1972, *Publ. Astron. Soc. Pacific* **84**, 373.
Vogt, N.: 1971, *Astron. Astrophys.* **11**, 359.
Walborn, N. R.: 1972, *Astron. J.* **77**, 312.

Eggen: Calibration of MK luminosity classification – Why? Even for the bright stars classifiers have difficulty in agreeing on the classifications and since the use on faint stars is the real end product, it would appear to be dangerous to apply any calibrations – at least at the present time.

Blaauw: I see the following use of the MK-luminosity classifications – and hence of their calibration in these fields:

(a) For individual stars a MK spectrum may give more easily the luminosity estimate than photometry, especially if one thinks of the reddening problem.

(b) For bright stars, the MK system provides a frame of reference for newly developing classification systems. E.g., a narrow band photometry system, as long as it has been applied to a limited number of stars, can better be calibrated with respect to MK than by analysis of trigonometric parallaxes and proper motions. This latter, more fundamental way of calibrating has to be used of course, once the sample in the photometric system has become large enough.

(c) For faint stars, objective prism plates (as we may now be getting with the new large Schmidt telescopes equipped with objective prisms, on sites of superior seeing) give wholesale MK classification possibilities. Whereas their follow up with photometry will frequently be desirable, the MK types already allow: 1, selection of the objects; 2, identification of luminosities and check on later to be obtained photometric luminosities. A control of spectral features in addition to the information from photometry seems as a rule very useful.

Jaschek: Partially the answer was given already by Prof. Blaauw but I would like to add that spectral classification – and therefore its calibration – must go on because of the impossibility of several photometric systems to distinguish between reddened and unreddened objects of spectroscopically different appearance. Of course one cannot do spectroscopy or photometry alone, but must do both and use them together.

Keenan: Is it not historically true that the systematic correction to the original MW calibration of spectroscopic parallaxes was made first by Öpik, whose earliest Tartu paper preceeded (I think) the work of Russell and Moore?

Blaauw: I am not very familiar with the early historical developments; Dr Keenan probably is right.

ABSOLUTE MAGNITUDES AND COLOURS OF
FIELD GIANT BRANCH STARS

H. NECKEL and P. KLAWITTER

Hamburg Observatory, F.R.G.

Abstract. Concerning the absolute magnitudes of the *giant* stars contained in the Mt. Wilson catalogue of Adams *et al.*, the following facts are pointed out:

(1) There is a linear relation between the Mt. Wilson magnitudes and the absolute magnitudes derived on account of the Wilson-Bappu effect (Figure 1). The scale factor is 2.3.

(2) The accuracy of the properly scaled Mt. Wilson magnitudes follows from the scatter in relation (1) and the accuracy of the Wilson-Bappu magnitudes (± 0.6). It is $\pm 0^{m}.6$ (mean error).

(3) The assured reliability of the corrected Mt. Wilson magnitudes allows a subdivision into bright, intermediate and faint giants (Figure 2). The 3 subgroups exhibit different 2-colour relations (Figures 3, 4, 5).

(4) The large number of giant stars present in the Mt. Wilson catalogue allows a trustworthy derivation of the *mean* absolute magnitudes of giant branch stars (Figure 6) and an estimate of its dispersion as well (Figure 7).

Our remarks concern the spectroscopic absolute magnitudes of giant stars given in the Mt. Wilson catalogue published by Adams *et al.* (1935). It seems to us that the reliability of these magnitudes has been considerably underestimated in the past. (See e.g. the remarks of Blaauw (1963) and of Wilson and Bappu (1957).)

Figure 1 shows the relation between these Mt. Wilson magnitudes (M_{MW}) and those (M_{WB}) published by Wilson (1959) and Wilson and Bappu (1957). Obviously, there is a common, linear relation for all stars of luminosity classes I to III, with a relatively low scatter. Stars of luminosity class IV are to be excluded because of a separate calibration procedure in the Mt. Wilson work.

The Mt. Wilson magnitudes may be brought into the 'Wilson-Bappu system' using the relation

$$M_v = M_{MW}^{corr} = 2.3 \, M_{MW} - 0.1 \,. \tag{1}$$

The multiplying factor is close to the rule of thumb value of 2 given by Keenan (1963) for the supergiants, but obviously it should be applied also in the case of the giants. The reason for the scale distortion of the original Mt. Wilson magnitudes has been discussed by Prof. Blaauw this morning.

The mean error of one corrected Mt. Wilson magnitude M_{MW}^{corr} follows from the scatter to be seen in Figure 1 and the mean error of the Wilson-Bappu magnitudes. It turns out to be $\pm 0^{m}.6$, which is just the same accuracy as it is obtainable from the Wilson-Bappu effect (1957).

Figure 2 is a part of the Mt. Wilson HR-diagram. The numerals indicate the numbers of stars present in the Mt. Wilson catalogue for each spectral subclass and each

B. Hauck and B. E. Westerlund (eds.), Problems of Calibration of Absolute Magnitudes and Temperature of Stars, 57–63.
All Rights Reserved. Copyright © 1973 by the IAU.

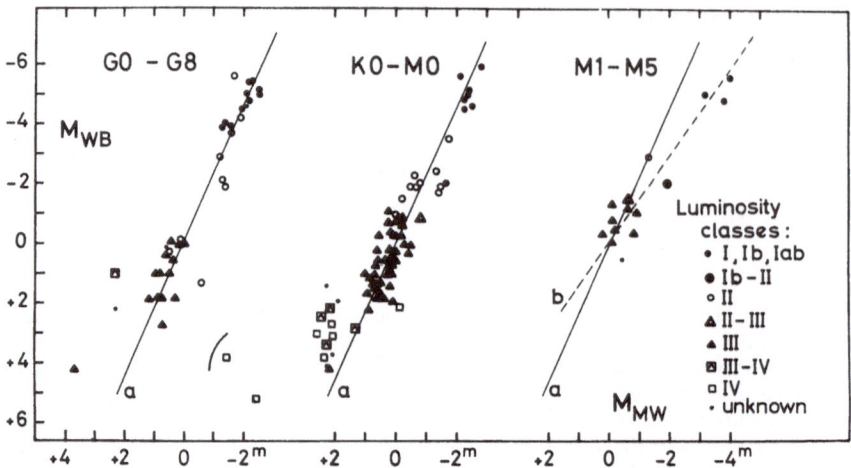

Fig. 1.   Relation between absolute magnitudes M_{MW} and M_{WB}.

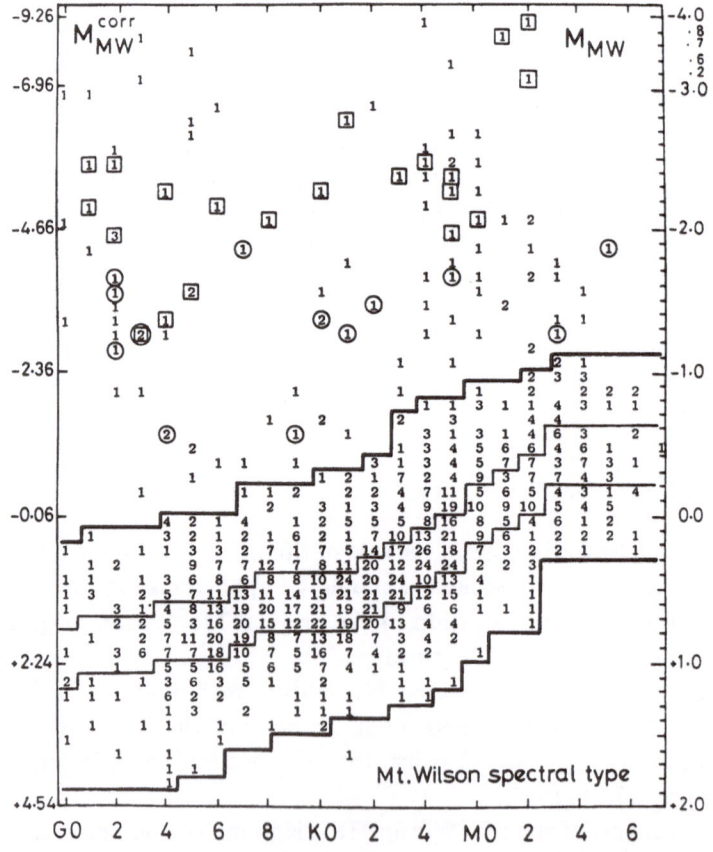

Fig. 2.   Upper right sector of the Mt. Wilson HR-diagram.

tenth of (uncorrected) magnitude. The left-hand scale gives the corrected Mt. Wilson magnitudes.

Numerals within squares indicate stars of luminosity class I, numerals within circles indicate stars of luminosity class II.

The stars between the heavy lines are considered to be *giant branch* stars, which is not necessarily identical with luminosity class III. The weak lines divide the giant branch into bright, intermediate and faint giants (α, $\alpha\beta$, β). This subdivision was

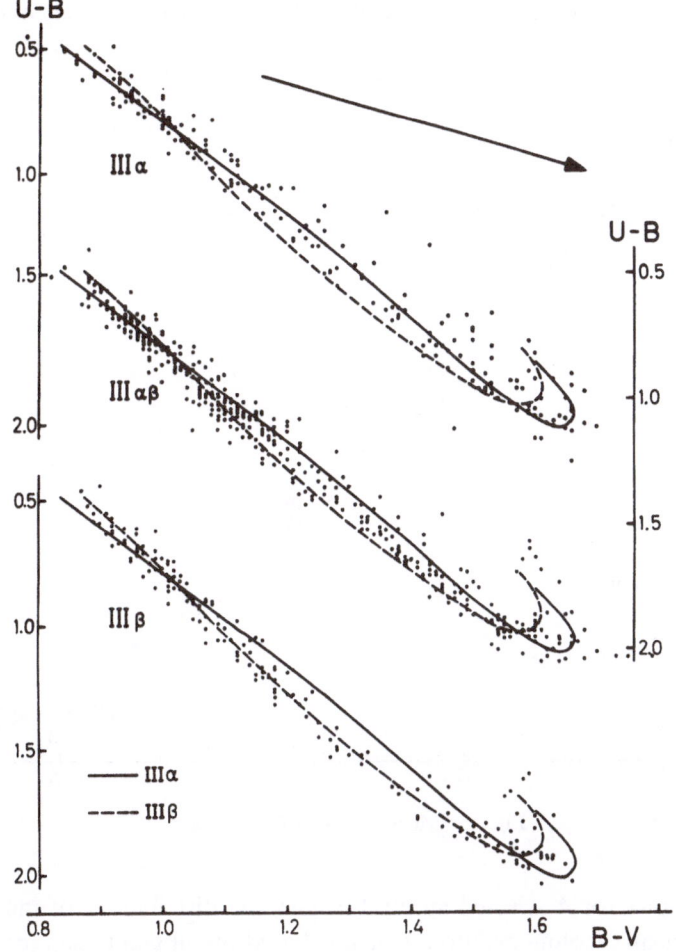

Fig. 3.   Two-colour diagrams for 3 luminosity groups of giant stars.

made purely arbitrarily, but intending to get about the same numbers of stars in each subbranch.

Figure 3 stresses the significance of this subdivision: Here we plotted the 2-colour

H. NECKEL AND P. KLAWITTER

diagrams separately for the 3 subgroups. The colours were taken from the photometric catalogue published by Blanco *et al.* (1970). The solid and the broken lines mark the mean relations for the upper and lower subgroup respectively. The arrow marks the reddening line. Apparently, the 3 subgroups occupy, on the average, different areas in the HR-diagram.

Figure 4 gives the *mean* colours for the different spectral classes. The crosses are based on all stars of the specific spectral class, which are included in both the Mt. Wilson catalogue and the photometric catalogue of Blanco *et al.* These values correspond to those given by FitzGerald (1970) for luminosity class III stars. Filled and open circles mark the mean colours for the bright (α) and faint (β) members of each spectral class.

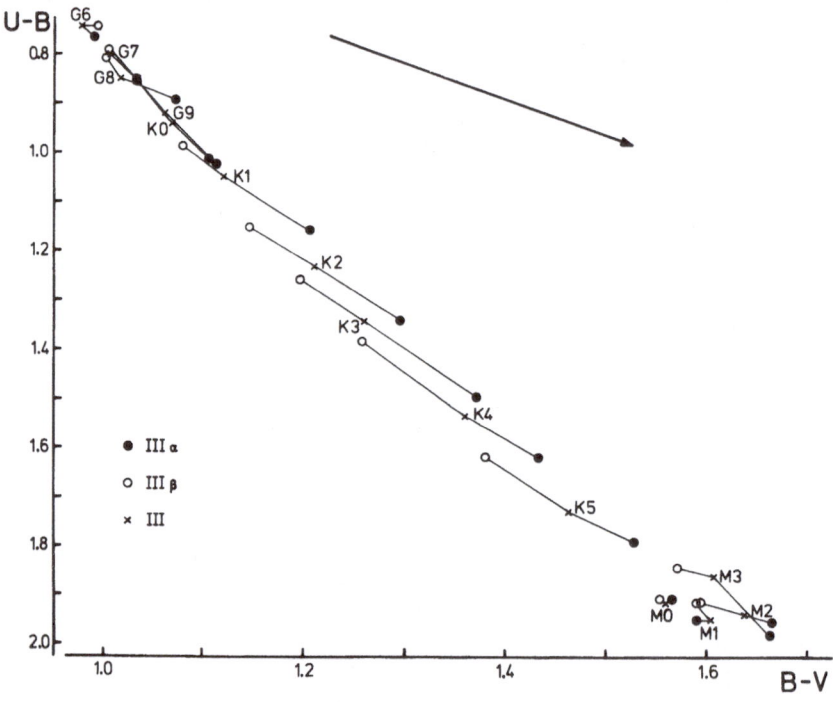

Fig. 4.   Mean colours of giant stars.

Figure 5 shows for 4 selected spectral classes the distribution of individual stars around that mean 2-colour-relation, which is based on all giant stars (connecting the crosses in Figure 4).

Figures 4 and 5 reveal that, on the average, the absolutely brighter giants are redder than the fainter objects, in both colours. It has been proven that this effect is not caused by interstellar reddening: There is not traceable dependency of the colour on the distance. Most of the stars are brighter than 6th magnitude and are within 100 pc.

(Remark during the discussion: There may be circumstellar reddening, which is dependent on absolute magnitude.)

Figure 6 shows the mean values of the corrected Mt. Wilson absolute magnitudes for the fields giants as a function of spectral type. A reduction to equal volume of space was not made yet, but the appropriate correction will be in the order of a tenth of a magnitude only. For comparison, the results or compilations of several other authors are given also.

In Figure 7 we have plotted the number of giants and supergiants present in the Mt. Wilson catalogue versus the distance $M_v - \bar{M}_v$ from the mean giant branch (step-

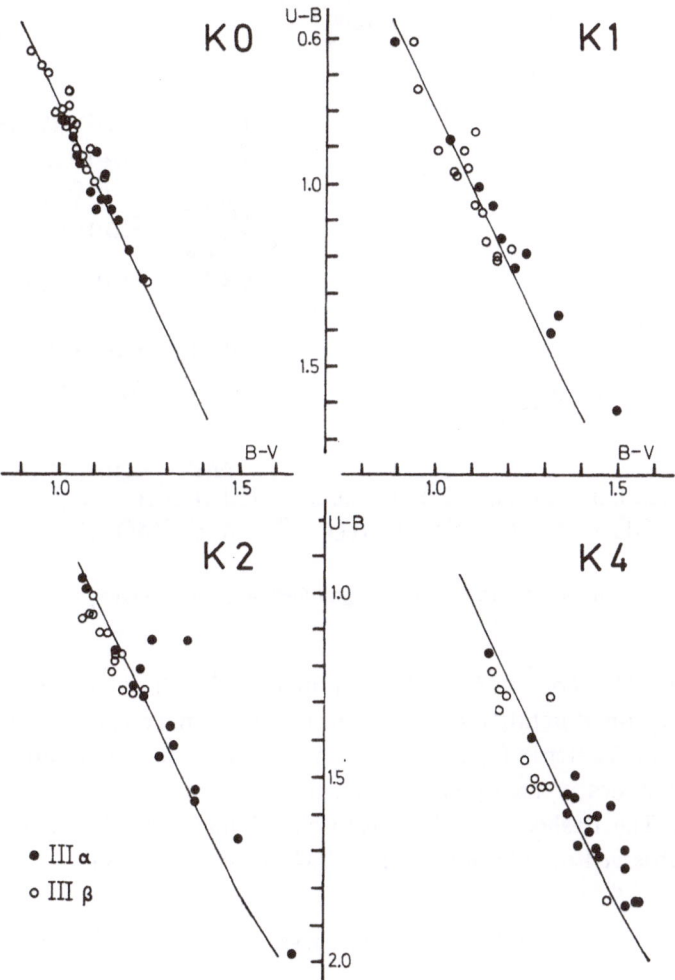

Fig. 5.   Two-colour diagrams for stars of 4 selected spectral classes.

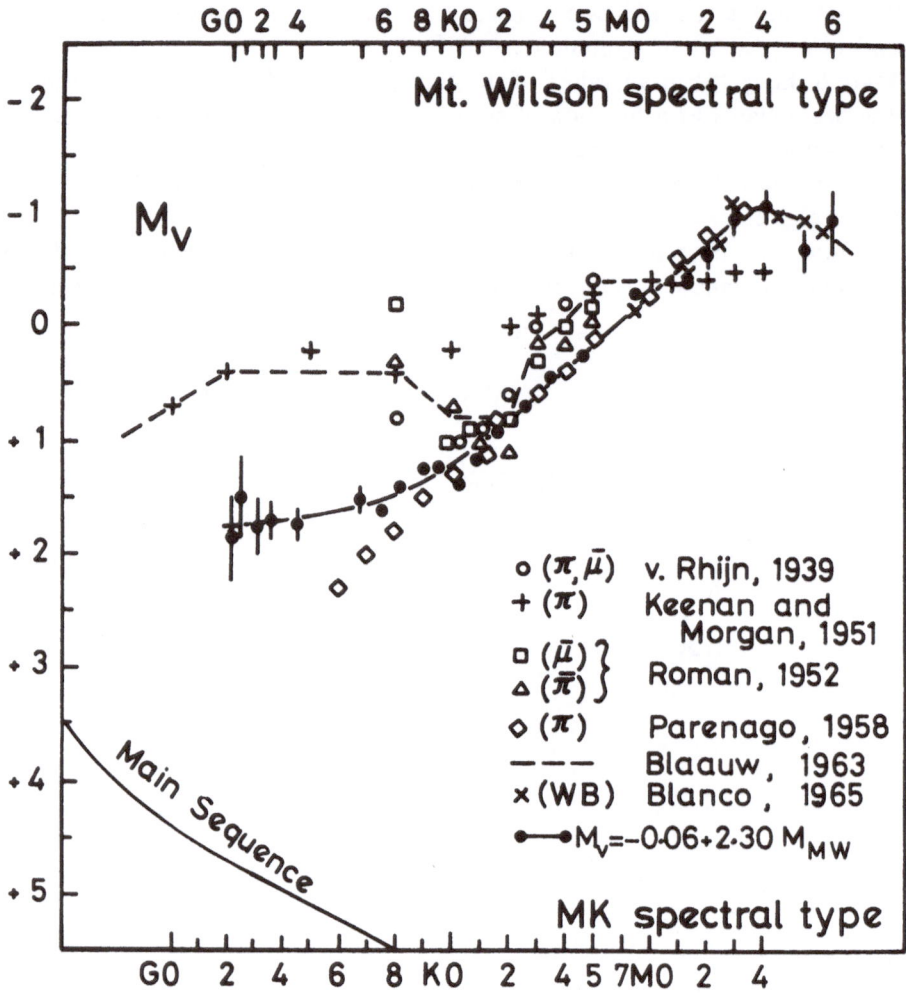

Fig. 6.   Mean absolute magnitudes of giant branch stars.

like, full line; M_v=corrected Mt. Wilson magnitude). The standard deviation is $\pm 0^m\!.73$, which is only slightly larger than the observational error of $\pm\ 0^m\!.6$, which was obtained from the scatter in Figure 1. This would indicate that the cosmic scatter of the absolute magnitudes of field giant branch stars cannot be considerably larger than about $\pm 0^m\!.5$. The dashed and the dash-pointed lines are Gaussian profiles with $\sigma = \pm 0^m\!.73$ (observed distribution) and $\pm 0^m\!.44$ ('true' dispersion).

References

Adams, W. S., Joy, A. H., Humason, M. L., and Brayton, A. M.: 1935, *Astrophys. J.* **81**, 187.
Blaauw, A.: 1963, *Stars and Stellar Systems* **3**, 383.

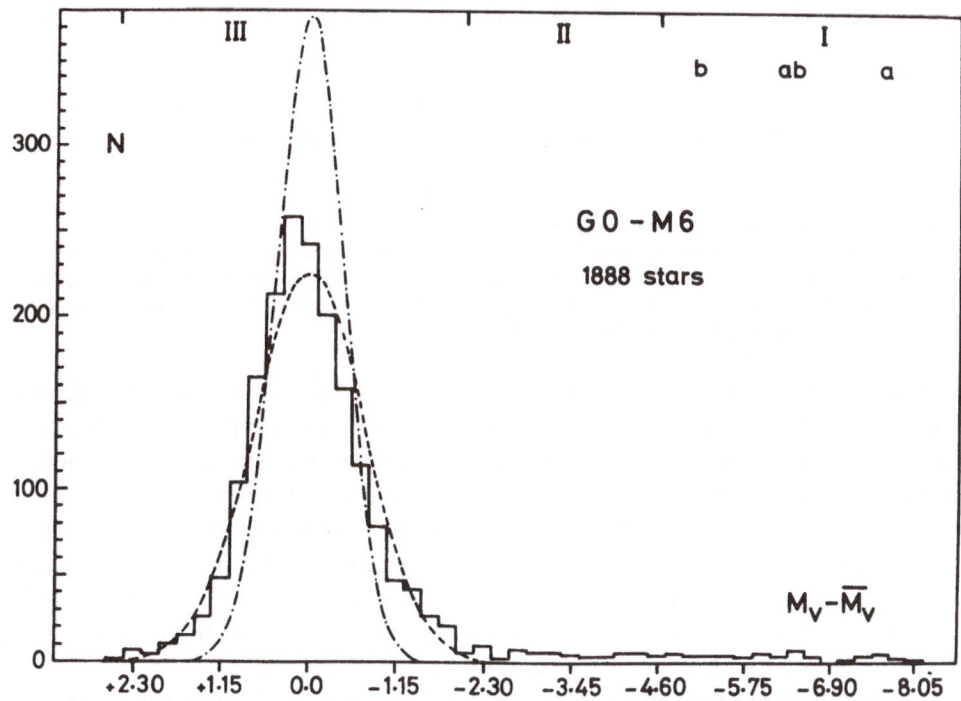

Fig. 7.   Distribution of giant and supergiant stars around mean position of the giant branch.

Blanco, V. M.: 1965, *Stars and Stellar Systems* **5**, 241.

Blanco, V. M., Demers, S., Douglas, G. G., and FitzGerald, M. P.: 1970, *Photoelectric Catalogue,* Publ. U.S. Naval Obs., 2nd series, Vol. 21.

FitzGerald, M. P.: 1970, *Astron. Astrophys.* **4**, 234.

Keenan, P. C.: 1963, *Stars and Stellar Systems* **3** 78.

Keenan, P. C. and Morgan, W. W.: 1951 in J. A. Hynek (ed.), *Astrophysics*, p. 17.

Parenago, p.p.: 1958, *Soviet Astron. A. J.* **2**, 151.

Rhijn, van P. G.: 1939, Publ. Capteyn Astron. Lab. Groningen, No. 49.

Roman, N. G.: 1952, *Astrophys. J.* **116**, 122.

Wilson, O. C.: 1959, *Astrophys. J.* **130**, 499.

Wilson, O. C. and Bappu, M. K. V.: 1957, *Astrophys. J.* **125**, 661.

DISCUSSION

Pecker: The difference between your III α, β classes could be due (and one of your slides is clearly pointing to this interpretation) to circumstellar reddening in the case of G–M giants where mass ejection could well be a general phenomenon.

Steinlin: Neckel's Figure 7 cannot show distinctly the supergiant classes II to Ia because it shows the difference M_v (III) to M_v (II) etc., and this varies slightly with spectral type. So plotting all spectral types in one diagram will necessarily smear out the higher luminosity classes.

Blaauw: Is it correct to say that the photometric differences between your classes IIIα and IIIβ indicate properties already brought out in the Mt. Wilson spectroscopic classification, independent of the re-scaling of the absolute magnitudes?

Neckel: Yes, it is correct. But these photometric differences confirm the distinctness of the Mt. Wilson magnitudes with regard to the giants.

ON THE K-LINE WIDTH-ABSOLUTE MAGNITUDE RELATION

M. K. V. BAPPU

Indian Institute of Astrophysics, Kodaikanal, India

Abstract. Bright fine mottles seen on the solar disk are identified as the agency that locates the Sun on the K-line width-absolute magnitude relation of Wilson and Bappu. The contribution of the supergranular network emission and the plage emission from centres of activity tends to upset the uniqueness of the relation. The line width-absolute magnitude relation is thus a characteristic of the convection zone underlying the stellar atmosphere.

Perhaps the most striking characteristic of the K-line emission from stellar chromospheres is the relationship between the width of the line and the visual absolute magnitude of the star that is valid over a range of about fifteen magnitudes. Indeed, in itself it is very suggestive of a common astrophysical characteristic operative in these atmospheres over the entire range that experiences an increase in magnitude of its impulsive scope as one goes from stars of fainter visual luminosity to the brighter ones. The Sun fits this relation well and since it is the only star which presents a disk for detailed scrutiny, much interest centres on an identification on the solar disc of that feature which gives rise to the K-line width-absolute magnitude relation.

A calcium spectroheliogram of the solar disc shows primarily three principal characteristics. If we have a centre of activity on the visible portion at the time, it will manifest itself with enhanced calcium emission over a sizeable region and this we term a calcium plage. Then, there is the network of calcium emission which has been shown by Leighton to coincide with the boundaries of the supergranulation. And within the configuration of a network cell, we come across localized bright points of emission which we recognize as bright fine mottles. The boundary of the supergranule has enhanced emission with a twofold contribution. The fine mottles are there too and some of them clump together to form a coarse mottle of roughly 7000 km diameter. There is also enhanced emission at the boundary by magneto-hydrodynamic heating accentuated by the piling up of the magnetic field swept up by the horizontal flow within the domain of the supergranule. The three principal sources of calcium emission are thus the plage, the network boundaries and the fine mottle.

In a recent paper (Bappu and Sivaraman, 1971) we have carried through a study that shows which of these three features is responsible for the K-line width-absolute magnitude relation of Wilson and Bappu (1957). We carried out this study from a single frame of excellent quality which we had obtained in a K-line time sequence of a specific region at the centre of the disc. A spectroheliogram taken soon after this exposure showed us the location of the slit in terms of the K_{232} network and other features seen in the light of ionized calcium. The dispersion of the spectrogram was 9.4 mm Å^{-1}, almost a hundredfold larger than what Wilson and I had used earlier in the stellar case! It is well to realize however, that the K emission feature in the case of the Sun is seldom seen at dispersions lower than about 4 Å mm^{-1}. At the large values of dispersion

B. Hauck and B. E. Westerlund (eds.), Problems of Calibration of Absolute Magnitudes and Temperature of Stars, 64–67.
All Rights Reserved. Copyright © 1973 by the IAU.

employed currently for the study of solar inhomogeneities, one sees a large amount of detail, specially when the spectrogram is obtained during the moments of good seeing. One sees numerous emission streaks of sizes in the range 1–2″, mostly with emission greater in the violet portion of K_2 than in the red or $I_{K2V} > I_{K2R}$. About one out of every five such emission streaks have $I_{K2V} < I_{K2R}$. About one in twenty have the two portions equal and it is very seldom indeed that both K_{2V} and K_{2R} are absent simultaneously. Our study shows a typical characteristic of K_2 emission from a quiet region of the Sun to be the case $I_{K2V} > I_{K2R}$. We interpret the changes from this normal aspect as due to the action of two types of absorbing components that are present on the same scale of inhomogeneity in sizes as the bright features. One component with minor Doppler displacements acting on the normal K_{232} profile produces the changes observed between I_{K2V} and I_{K2R} relatively enhancing one at the expense of the other. The other component arises from what we call 'dark condensations' with down flowing velocities of 5–8 km s^{-1} and sizes of about 5000 km. These dark condensations can be seen easily on any good K_3 spectroheliogram and in our opinion give rise to the situation $K_{2R} = 0$.

These details of intensity fluctuations are of interest to the stellar case where we necessarily measure the integrated characteristics of K_2 emission and all those features that upset the intensity ratio between I_{K2V} and I_{K2R}. In cases like α Bootis it is well to remember this aspect as one possibility that we could utilize to explain changes in the relative intensities of the K_2 emission components.

Sivaraman and I have also measured the widths along the dispersion of all the double peaked emission features that we have on this particular frame. We measured the separations of the emission peaks as well as the separation of the minima between the emission feature and the regular K absorption feature. We call the former a K_2 width and the latter a K_1 width. When we plot a histogram of the frequency of the K_2 widths we find a remarkable uniqueness of the value of the K_2 width about a particular value of 26 km s^{-1}. The K_1 widths do not show such a sharp clustering about a single value; however, they also have a mean characteristic spacing giving rise to a K_1 witdh of 41 km s^{-1}. Since the intensities in the solar case of I_{K2V}, I_{K2R} are close to I_{K1V} and I_{K1R} we get a width at half intensity of 33.5 km s^{-1}, by taking a straight mean of the values given above. Compare this with the value of Wilson and Bappu (1957) obtained from an 'averaged' disc spectrogram of 34 km s^{-1} and by a micrometer setting on the emission feature. Since the sun fits well on the Wilson-Bappu relation with a measured width of 34 km s^{-1} it is clear that bright points of K2 emission are the principal contributors to the line width-absolute magnitude relation. The intensity scans perpendicular to the dispersion show that they have typical sizes of the order of 1–2″. Hence, we conclude that these bright points that enable the sun to follow the line width-absolute magnitude relation are the fine mottles seen on a good quality spectroheliogram. We stabilize this identification from a comparative study of the auto correlation function obtained from intensity scans perpendicular to the dispersion as well as that obtained on a two dimension scale from a high quality K_{232} spectroheliogram. Both these give full widths at half maximum of 7000 km. We conclude that the average spacing between the bright streaks on the spectrogram are the same as that of the fine mottles on the two dimension spectroheliogram. The argument thus secures our identification.

The value of 7000 km is the spacing between the brightest features within the supergranular network and also that between the bright streaks on the spectrogram. The spacing between these streaks also shows the emission peaks but of lower intensity. We may ascribe these to an unresolved background of fine mottles that awaits detection with the improved resolution techniques of the future.

The life times of the emission streaks on the spectrogram and of the fine mottles on spectroheliograms provide an additional confirmation of identity. Both have values around 200 s and hence both are identical.

The behaviour of emission peak separations on plage regions have been known for a very long time. From the stand point of our terminology the K_2 widths decrease with enhanced plage intensity and proximity to the seat of the centre of activity. Elske Smith (1960) has expressed this quantitatively. She also finds a minor change in the widths of the emission feature as one goes to plages with enhanced magnetic fields. In terms of the discussion above, it is seen that the plage regions offer little hope of being the source of a unique value for the K_2 separations similar to the case of the fine mottles.

A similar situation prevails at the boundary of the supergranular network. Calcium emission here is greatly enhanced and has a noticeable contribution from the magnetic fields accumulated by virtue of the supergranular flow. The K_2 width here is in the neighbourhood of 20 km s^{-1} instead of a value of 26 km s^{-1} for the fine mottles. It is clear that supergranular boundaries do not have any appreciable contribution to the unique value of the solar case in the relation of Wilson and Bappu (1957).

We are thus left with the bright fine mottle as the principal contributor to the K-line width-absolute magnitude relation. The fine mottle is not known to display any appreciable longitudinal field greater than the limit of photographic detectability. The bright mottle is thus, by some cause as yet unknown, a manifestation in the chromosphere of the mechanical energy dissipation from the convective layers below. It has characteristics prescribed by a combination of the fundamental parameters of the star.

In the calibration of the K-line width-absolute magnitude relation, it is necessary to assign a value for the sun that is truly representative of it in integrated light. Wilson and Bappu had 'averaged' a sample of the disc that gave them the value of 34 km s^{-1} utilized in the calibration of the relation. Such a spectrum would yield a totality of bright mottle spectral features with the systemic velocities of the different mottles all added together to give an integrated profile. Sivaraman and I have obtained several K-line spectra at high dispersion in truly integrated light with the same spectrograph used for the high resolution study. The integrated spectrum thus obtained has an emission line width in excess of the corresponding value obtained by averaging the mottle spectra. Clearly we see here the contribution by solar rotation to the integrated characteristics. The emission line width in the solar case is closer to 36 km s^{-1} than the previously utilized value by Wilson and Bappu (1957).

Calcium emission in the stars can be produced in several different ways. In binaries, we have often the case of emission from the gas flow within the system. Magnetic fields in star spots could produce enhanced calcium emission in star plages considerably more virulent than in the solar case. Also, from an extrapolation of the solar case we have the field of the bright mottle that is a feature of the convection below the stellar photosphere and which gives rise, according to our identification, to the Wilson-Bappu relation. The emission by magnetic field stimulated plages can be quite different from the emission of the quiet background, and cause an apparent deviation from the line width-M_v relation. There can also be gaseous outflow which gives a deeper K_3 and which indicates an abnormality in the emission origin. The case of Canopus seems to fall in line with such a possibility. K-line emission in the borderline cases of the earlier spectral types should be interpreted with caution, especially when they occur in the spectral domain where convection does not have a good role to play.

References

Bappu, M. K. V. and Sivaraman, K. R.: 1971, *Solar Phys.* **17**, 316.
Smith, E. V. P.: 1960, *Astrophys. J.* **132**, 202.
Wilson, O. C. and Bappu, M. K. V.: 1957, *Astrophys. J.* **125**, 661.

DISCUSSION

Blaauw: To what extent might variations in the phenomena *causing the Ca+ reversals* affect the widths and hence the derived luminosities?

Bappu: The presence of centres of activity and hence Ca+ plages would undoubtedly enhance the overall intensity of the emission. Such enhancements independently or coupled with a dominant flow of absorbing gas outward or inward are likely to produce changes with time of the relative intensities of the emission components K_{2V} and K_{2R}. The contribution to the widths as measured presently for luminosity determinations are likely to be small; they fall within the present limits of uncertainty. The case of α Bootis supports such a conjecture since its K-line width values do not seem to be affected by the changes seen in the K_2 emission.

LUMINOSITY CALIBRATION OF GIANTS AND SUPERGIANTS, G0–M5

PH. C. KEENAN*

Perkins Observatory, The Ohio State and Ohio Wesleyan Universities, Ohio, U.S.A.

Abstract. Calibration curves giving M_v for stars of luminosity classes III, II, Ib, Iab and 0 are derived and shown graphically in the HR diagram. There are serious gaps in which the calibration needs to be improved.

The luminosity calibrations that I have to discuss here do not involve any new methods of estimating absolute magnitudes. They do, however, meet the condition that virtually all the stars included had their spectral classification checked or revised by inter-comparison of a nearly uniform set of spectrograms. The two main reasons for imposing this requirement were:

(1) The need to come as close as we can to eliminating stars with spectral peculiarities, which can systematically affect both temperature types and luminosity classes. The resulting calibration is limited to stars of roughly solar composition.

(2) The desirability of reducing statistical corrections by making the variances within each group as small as possible.

The table of revised standard types for the stars is given in a review article on classification by Margon and Keenan (1973).

1. Class III – Giants

For the main giant branch the calibration of luminosity classes in terms of visual absolute magnitudes was based on mean trigonometric parallaxes, using the method of reduced parallaxes (Russell and Moore, 1938). As far as possible the class III stars were separated into the three subclasses IIIa, IIIb, IIIab, and the IIIa and IIIb stars were not used in the calibration. The stars that could be rejected as definitely brighter than the central giant branch, and, hence, classified as IIIa or IIIb were a minority of the giants, for all doubtfull cases were retained and classified as merely class III. This at least served to eliminate distortion of the means by such stars as β Gem, with its very bright apparent magnitude, which would dominate any weighted mean. From several good spectrograms β Gem could be classified as K0 IIIb, and the large trigonometric parallax of $+0''.093$ gives the good value of $+1.0$ for M_v. By comparison with Figure 1 it is evident that with good spectrograms one can just detect a luminosity difference of about a half magnitude in K-type giants.

By using only stars brighter than the fifth visual magnitude it was possible to derive

---

* Visiting Astronomer, Cerro Tololo Inter-American Observatory, which is operated by the Association of Universities for Research in Astronomy, Inc., under contract with the National Science Foundation.

B. Hauck and B. E. Westerlund (eds.), Problems of Calibration of Absolute Magnitudes and Temperature of Stars, 68–77.
All Rights Reserved. Copyright © 1973 by the IAU.

satisfactory values of $\langle M_v \rangle$ in spite of the small samples – averaging 12 or 13 stars for each group. Not all the eligible stars brighter than $V = 5.0$ have been re-classified, particularly in the southern hemisphere, and the resulting imcompleteness factor so nearly balanced the Malmquist correction that the latter could be omitted except for the last group (M1–M2). Most of these M-stars were fainter than $V = 4.5$, and a correction of $+0.2$ mag. was estimated and applied to this group. The general correction $\langle M \rangle - M (M_\pi) = 5(\langle \log \pi \rangle - \log \langle \pi \rangle)$ ranged from -0.2 to -0.7 mag.

Good arguments can be advanced for preferring either weighted or unweighted means. In the plots of group means in Figure 1 both solutions are given, and in the

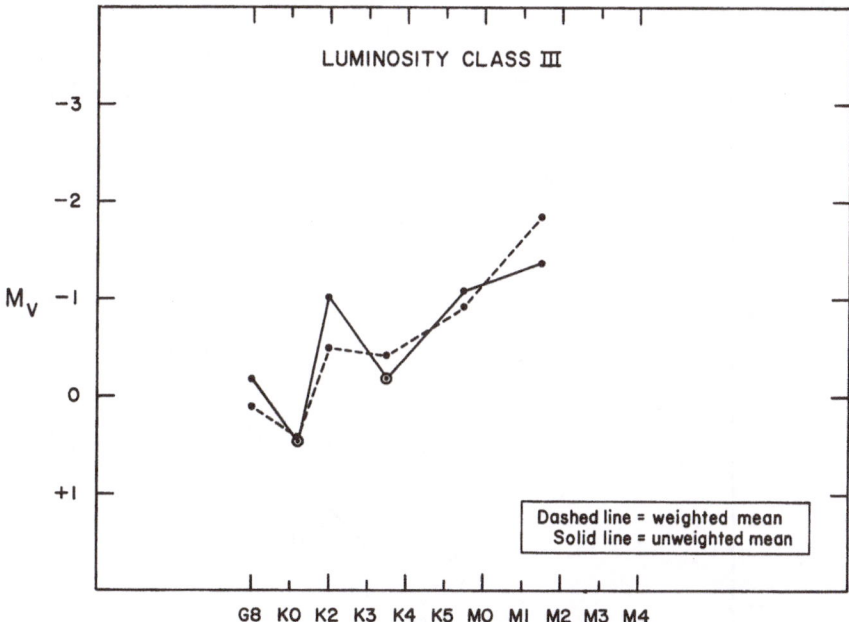

Fig. 1.   Calibration of giants by mean trigonometric parallaxes. Stars with $V < 5.0$.

final calibration the two sets were averaged. The two encircled points represent direct unweighted means of the absolute magnitudes for the groups in which no negative parallaxes occurred. Although their almost exact agreement with the mean parallax solutions must be fortuitous, it seems evident that no serious systematic error was involved in the corrections applied to the latter.

The zigzag pattern in Figure 1 is due, of course, to the small size of the samples, and there seems to be no reason not to draw a smooth curve through the band of points. This was done to give the final calibration of class III in Figure 6. From the deviations of the group means from this curve the mean error of the smoothed relation is 0.27 mag. from the unweighted means and 0.17 mag. from the weighted ones.

The choice of trigonometric parallaxes to calibrate class III was not made to disparage the value of statistical parallaxes, but I felt it desirable to have this independent solution to permit comparison with the luminosities derived from radial velocities and proper motions. In Figure 2 the new luminosities are plotted against the values found by Jung (1970) for the stars in the *Yale Catalogue of Bright Stars*. The dashed line has a 45° slope. No marked systematic difference is present but Jung's final adopted values were not smoothed as much as mine, and his groups means show greater oscillations, particularly at K1, where he obtained a luminosity lower by 0.6 mag. Of course, neither the sample population nor the breadth of the band of giants averaged was the same in the two solutions. Nevertheless, the overall agreement appeared close enough to justify the use of Jung's values to extend our solution in Figure 6 by the dashed line as far as type G5.

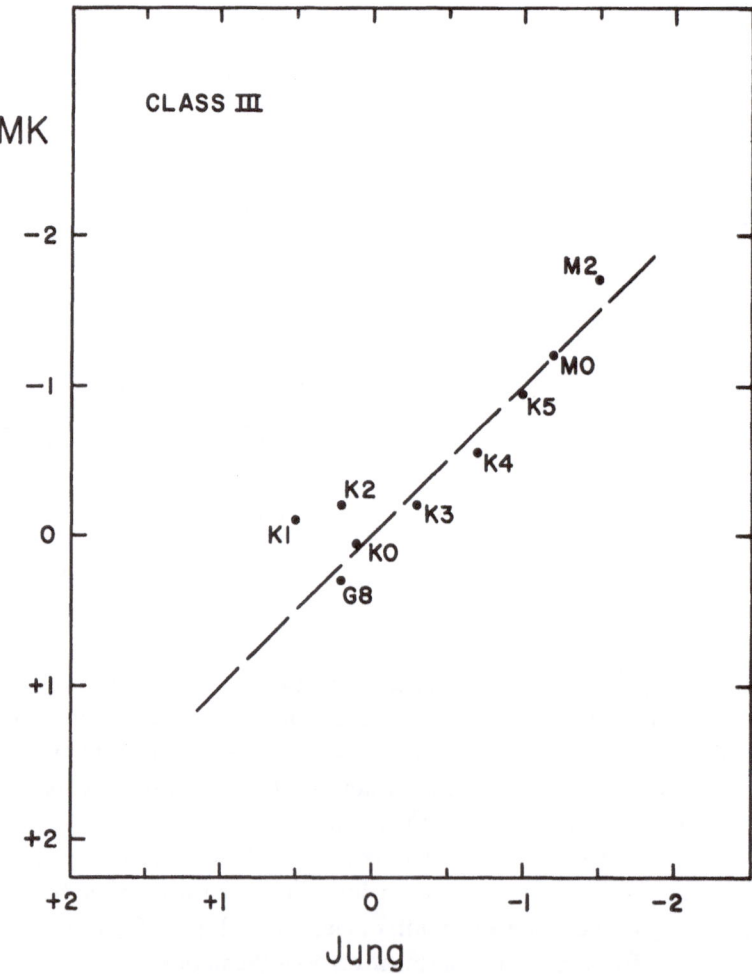

Fig. 2.   Comparison of smoothed mean trigonometric values of M_v with Jung's values derived from statistical parallaxes. The dashed line has a slope of 45°.

2. Class II – Bright Giants

When we go above the giant branch even the mean trigonometric parallaxes become too small to be meaningful. At the same time the number of well-classified stars is so small that one must lump a number of spectral classes together to have a large enough sample to give decent statistical parallaxes. There is an alternative means of calibration, however. The linear relation between K-line emission widths and absolute magnitudes that was found by Olin Wilson allows the values of M_v for class II stars to be derived by interpolation. Since his calibration (Wilson, 1970), was tied to the Sun, the Hyades main sequence and giants, and the Perseus cluster supergiants, it is independent of the class II stars, and can therefore be applied to them without any circularity in the argument. Dr Wilson is currently preparing a new catalogue of K-line luminosities and kindly provided me with the revised data for a number of stars having luminosity classes near II. His coudé spectrograms were limited for the most part of stars with $m_v \leqslant 7.0$, and since neither his sample nor mine was complete for this range, the elimination of stars classified IIa or IIb leaves only the few points plotted in Figure 3 for the center of the bright-giant group at the present time. The apparent dispersion in the diagram is too great to be accounted for by the uncertainty in classification alone.

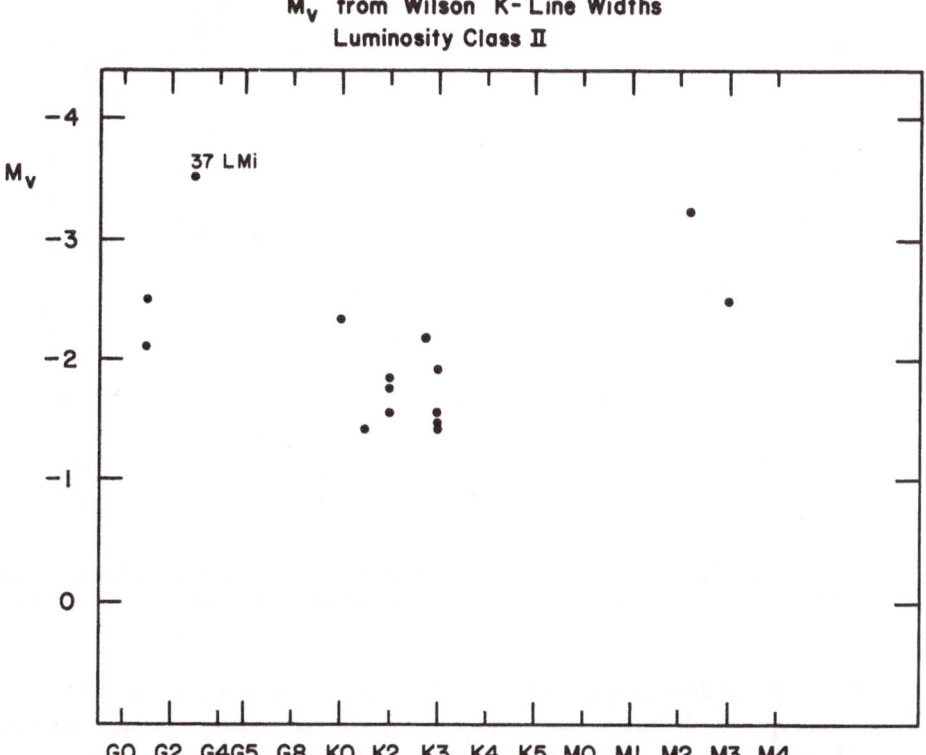

Fig. 3.   Luminosities of class II stars given by K-line calibration of Wilson.

If we regard the K-line emission width as another luminosity criterion more or less independent of the usual criteria used on small-scale spectrograms, the scatter in Figure 2 suggests that the several criteria do not correlate perfectly – which is not surprising when we consider that the different spectroscopic features will not necessarily respond in exactly the same way to individual variations in such physical characteristics as chromospheric activity. One striking discrepancy is shown by the well-known bright giant ζ Cyg, which nearly all recent observers have assigned to luminosity class II. The K-line luminosity is +1.1, which would put it at the lower edge of the giant branch and entirely off the diagram. ζ Cyg does have a slight enhancement of Ba II, but is not a real barium star and this degree of peculiarity does not seem sufficient to explain the large discrepancy.

In order to take account of all the stars near luminosity class II, and the evidence from membership in open clusters and binary systems, the stars between types G8 and K4 have been grouped together in Figure 4. For the two binaries near Ib, ε Peg and

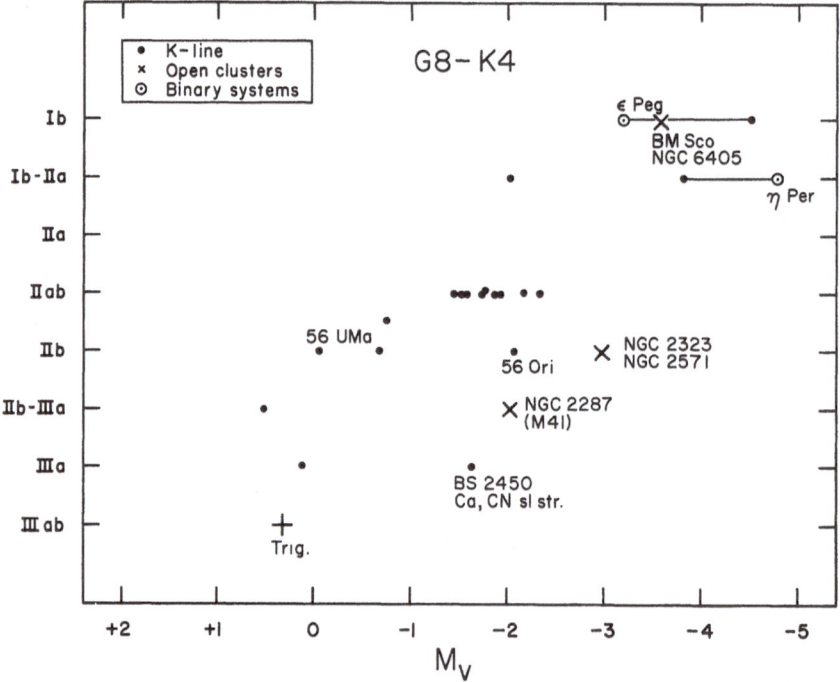

Fig. 4.   Comparison of K-line absolute magnitudes with those derived from open clusters and binary systems. The + shows the mean trigonometric absolute magnitude for class III, types G8–K4.

η Per, the K-line luminosity and that derived from the spectral classification of the early-class companion have been jointed by a bar. For these two stars the differences between the two methods is in the opposite sense. The class IIb, IIab, and IIa stars in clusters, however, appear to have higher luminosities than those given by the K-line widths. The adopted mean curve in Figure 6 represents a compromise between the two sets of data. As soon as a sufficient body of accurate types is available for class II stars – perhaps down to $V=8$ – it will be especially desirable to derive statistical parallaxes for this luminosity range.

3. Classes Ib, Ia and O – Supergiants

Membership in groups of stars at known distances is recognized as the best means of calibrating the luminosities of late-type supergiants. Both the membership of individual red stars in clusters and associations, and the distances to these aggregates, have been

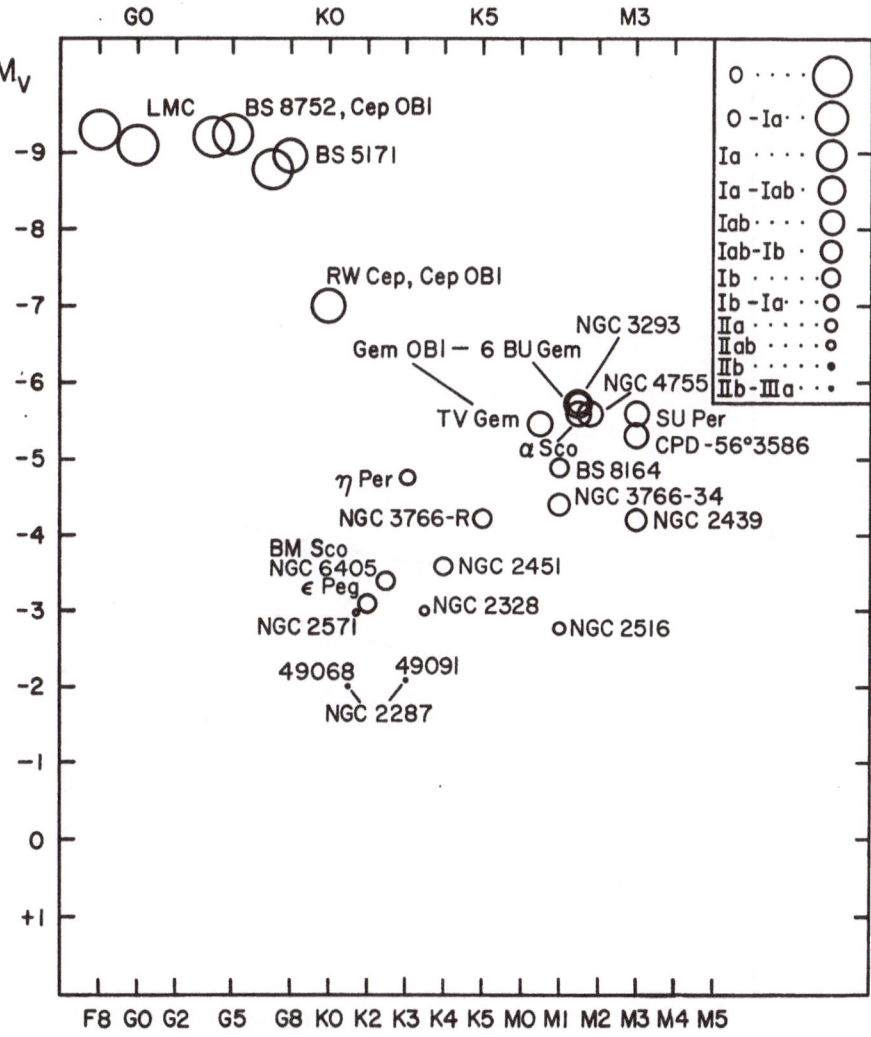

Fig. 5. Luminosities of supergiants given by their membership in binary systems, open clusters or associations.

PH. C. KEENAN

reviewed carefully by Schmidt-Kaler (1961), Hagen (1970), Humphreys (1970), Humphreys *et al.* (1971, 1972), Schild (1970), Stothers (1969, 1972), Stothers and Leung (1971), and others. These data, combined with our revised types, are summarized in Table I.

The only systematic changes that I have introduced followed naturally the recognition of the brightest supergiants of the LMC as defining luminosity class 0. For the distance modulus of 18.6 for the LMC (Wesselink, 1971), the visual absolute magnitudes of the four reddest 'super-supergiants' defined by Feast and Thackeray (1956) remain in the range −8.8 to −9.3 mag., very close to their original estimates. With the brighter M-type supergiants in h and χ Persei retained as defining class Iab, the other classes of supergiants can be assigned by interpolation.

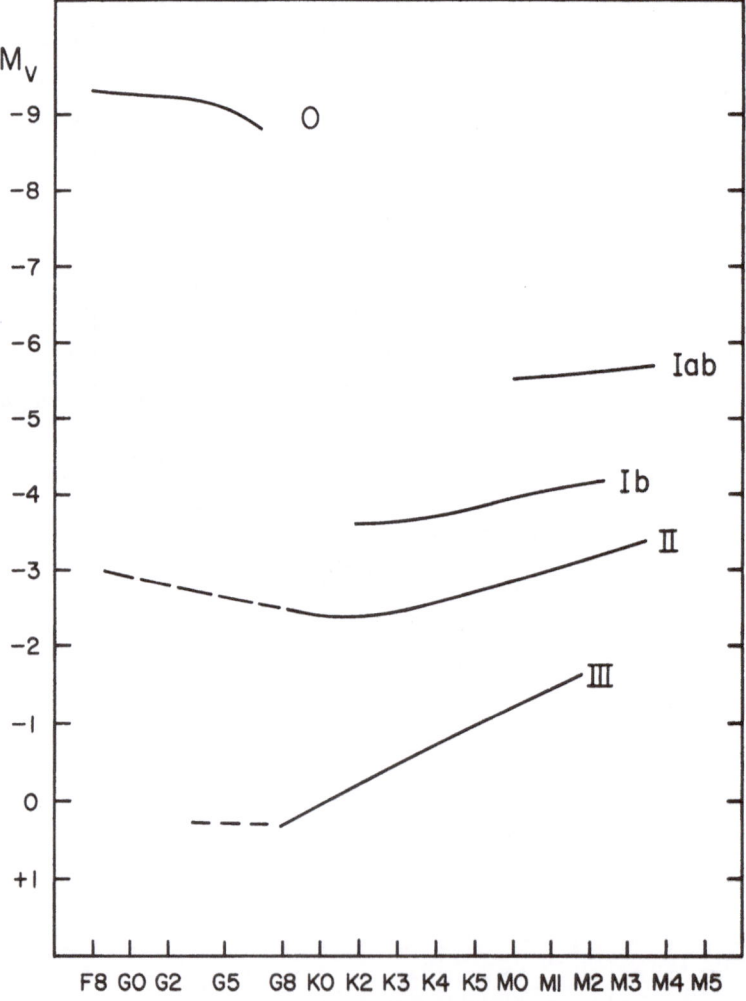

Fig. 6.   Adopted luminosity mean absolute magnitudes for several luminosity classes. The dashed lines indicate extensions of lower weight.

TABLE I

Adopted luminosities of late-type stars in clusters or binary systems

Star	System	Type	$(m-M)_0$	M_v	Sources	Remarks
SU Per	h & χ Per	M3–4 Iab	11.5	−5.6	Crawford et al (1969)	
η Per	Binary	K3–Ib–II	8.5	−4.7		Companion classified B8 V by Dr Slettebak [1]
CPD −31°1790	NGC 2439	M3 Iab–Ib	11.7	−4.2	Becker (1963)	
HD 52938	NGC 2323	K3.5 IIb		−3.0		
BS 3153	NGC 2516	M1 IIa	8.3	−2.8	Morgan and Abt (1969)	Astron. J. 74, 813
CoD −29°5941	NGC 2571	K1.5 IIb	11.4	−3.0	Lindoff (1967) Hagen (1971)	Arkiv. Astron. 4, 587.
CPD −57°3502	NGC 3293	M1.5 Iab–Ib	11.5	−5.7	Feast (1958)	Monthly Notices Roy. Astron. Soc. 118, 618
CPD −60°3621	NGC 3766	M1–Iab–Ib	11.1	−4.4	Kraus (1967)	Astron. Nachr. 289, 285 [2]
CPD −60°3636	NGC 3766	M0 Ib		−4.2:		
CPD −59°4459	NGC 4755	M2–Iab	11.8	−5.7	Feast (1961)	IAU Symp. 20, 22 [3]
α Sco	Binary	M1–M2 Iab	8.0	−5.6	Stone and Struve (1954)	
BM Sco	NGC 6405	K2.5 Ib CN+1	8.5	−3.4	Talbert (1965)	[4]
ε Peg	Binary	K2 Ib	5.5	−3.1	Keenan	

[1] NGC 2439. In this field, and close to the M3 supergiant, is found R Pup, for which the revised type is G2 O from its close match with HDE 269953 in the LMC. Accordingly, a luminosity of $M_v \approx -9$ should be expected, which would place R Pup at about twice the distance of the cluster. Hence, although the radial velocities of the two supergiants appear to be comparable on the basis of visual inspection of the low-dispersion spectrograms, I regard the membership of R Pup in NGC 2439 as very doubtful.

[2] NGC 3766. The distances found by Sher (1962), Ahmed (1962) and Kraus (1967) have been averaged.

[3] α Sco. From membership in the association OB2 Stothers and Leung (1971) derive the slightly lower luminosity, $M_v = -5.2$. The star varies appreciably in light and temperature type.

[4] NGC 6405. The moduli of Rohlfs et al. (1959), Eggen (1961) and Talbert (1965) have been averaged. BM Sco appears to have unusually strong CN bands. The apparent brightness has varied considerably at a slow rate.

Most of the best calibrated objects from Table I are plotted as an HR diagram in Figure 5. Two VV Cep-class close pairs are CPD −56°3586 and BS 8164, and their absolute magnitudes are found from their B-type secondaries (Keenan, 1970). The point for α Sco (M1–M2, Iab) also is based upon the spectral types of its companion (Stone and Struve, 1954; Garrison, 1967). The star BS 5171, shown in the midst of the LMC stars, is the G-type supergiant to which attention was directed by Humphreys *et al.* (1971), who derived an absolute magnitude of −8.9 from their classification of B0 Ibp for its visual companion, using Blaauw's luminosity calibration for that type. My type for the primary component, G8 0–Ia, agrees essentially with theirs and is based on a spectrogram kindly loaned by Dr Humphreys. Since a value of $M_v = -9.0$ is obtained if one uses the calibration of Weaver and Ebert (1964) for B-stars, it is possible that my luminosity class is too low.

For nearly all the late-type supergiants both the temperature type and the luminosity vary appreciably, and the points plotted in Figure 5 are averages. The most accordant results are given by the five points for Iab stars, which cluster close to $M_v = -5.6$ over the observed range from M0 to M3 in type. This agrees with the result of Stothers (1972). Additional points for several luminosity classes could be added by including stars in other associations, besides OB1 and Gem 1, but the evidence for both membership and distance for objects assigned to associations is so much less definite than for stars in clusters that the latter should be given much higher weight in calibrations.

The calibrations adopted on the basis of the discussions in this paper are shown graphically by the curves in Figure 6. The lines extend only as far as we have reasonably good absolute magnitudes at the present time. The gaps are serious, but there is hope of filling them – at least in part.

References

Ahmed, F.: 1962, *Publ. Roy. Obs. Edinburgh* 3, 57.
Blaauw, A.: 1963, *Stars and Stellar Systems* 3, 383.
Crawford, D. L., Glasdey, J. W., and Perry, C. L.: 1970, *Astron. J.* 75, 822.
Eggen, O. J.: 1961, *Quart. J. Roy. Astron. Soc.* 2, 163.
Feast, M. W. and Thackeray, A. D.: 1956, *Monthly Notices Roy. Astron. Soc.* 116, 587.
Garrison, R. F.: 1967, *Astrophys. J.* 147, 1003.
Hagen, G. L.: 1970, *Publ. David Dunlap Obs., Toronto* 4.
Humphreys, R. M.: 1970, *Astrophys. J.* 160, 1149.
Humphreys, R. M., Strecker, W., and Ney, E. P.: 1971, *Astrophys. J.* 167, L35.
Humphreys, R. M., Strecker, W., and Ney, E. P.: 1972 (in press).
Jung, J.: 1970, *Astron. Astrophys.* 4, 53.
Keenan, P. C.: 1970, *Astrophys. J.* 162, 199.
Kraus, B.: 1967, *Astron. Nachr.* 289, 285.
Morgan, W. W. and Keenan, P. C.: 1973, *Ann. Rev. Astron.*, to be published.
Rohlfs, K., Schrick, K. W., and Stock, J.: 1959, *Mitt. Sternw. Hamburg-Bergedorf* 23, No. 269.
Russell, H. N. and Moore, C. E.: 1938, *Astrophys. J.* 87, 389.
Schild, R. E.: 1970, *Astrophys. J.* 161, 855.
Schmidt-Kaler, T.: 1961, *Z. Astrophys.* 53, 1.
Sher, D.: 1962, *Observatory* 82, 63.
Sher, D.: 1965, *Monthly Notices Roy. Astron. Soc.* 129, 237.
Stone, S. N. and Struwe, D.: 1954, *Publ. Astron. Soc. Pacific* 66, 191.
Stothers, R.: 1969, *Astrophys. J.* 155, 935.

Stothers, R.: 1972, *Publ. Astron. Soc. Pacific* **84**, 373.
Stothers, R. and Leung, K. C.: 1971, *Astron. Astrophys.* **10**, 290.
Talbert, F. D.: 1965, *Publ. Astron. Soc. Pacific* **77**, 19.
Weaver, H. and Ebert, A.: 1964, *Publ. Astron. Soc. Pacific* **76**, 6.
Wesselink, A. J.: 1971, *Monthly Notices Roy. Astron. Soc.* **152**, 159.
Wilson, D. C.: 1970, *Publ. Astron. Soc. Pacific* **82**, 865.

DISCUSSION

Van den Bergh: Stars of luminosity class III represent a mixture of objects of differing age and composition. The absolute magnitude calibration for giants is therefore function of the data selected for the calibration. A good example is provided by the M giants in the nuclear bulge of the Galaxy for which $\langle M_v \rangle \simeq 0$ compared to $\langle M_v \rangle \simeq -1.5$ for the M giants near the Sun.

Schmidt-Kaler: I would say that perhaps the greatest merit of the MK-system is that it represents a reference system for about 95% of the stars. So it may serve as a guide to all subsequent work.

CALIBRATION EN MAGNITUDES ABSOLUES DE LA CLASSIFICATION BCD. APPLICATION À LA DÉTERMINATION DU MODULE DE DISTANCE DU GRAND NUAGE DE MAGELLAN

L. DIVAN

Institut d'Astrophysique, Paris, Observatoire de Haute-Provence,
European Southern Observatory, La Silla

Abstract. The calibration of the BCD stellar classification in absolute magnitudes (classification in three parameters λ_1, D, Φ_b) was used to determine the distance modulus of the Large Magellanic Cloud. The described method makes use of the spectrophotometric parameters of B and A supergiants of the Large Cloud and gives a value of the distance modulus which is independent from other determinations, in particular from those which are based on the RR Lyrae and the Cepheid variables. The value found for the distance is slightly smaller than those generally admitted. The results are still based only on a limited number of measurements and new observations are in process; however, it seems doubtful that one will obtain much larger values. In other respects, the observations have shown that the parameter λ_1 was still sensitive to luminosity for the B and A stars brighter than $M = -8$ and that the calibration of the $\lambda_1 D$ diagram in absolute magnitudes can be extended up to $M = -9$.

Résumé. La calibration en magnitudes absolues de la classification stellaire BCD (classification à trois paramètres λ_1, D, Φ_b) a été utilisée pour déterminer le module de distance du Grand Nuage de Magellan. La méthode qui est décrite fait intervenir les paramètres spectrophotométriques de supergéantes B et A du Grand Nuage et donne une valeur du module de distance indépendante des autres déterminations, en particulier de celles qui sont basées sur les variables RR Lyrae et les Céphéïdes. La valeur trouvée pour la distance est un peu plus faible que celles généralement admises. Elle ne repose encore que sur un petit nombre de mesures et de nouvelles observations sont en cours; il semble cependant douteux que l'on arrive à des valeurs beaucoup plus grandes. Par ailleurs, les observations ont montré que le paramètre λ_1 était encore sensible à la luminosité pour des étoiles B et A plus brillantes que $M = -8$ et la calibration du diagramme $\lambda_1 D$ en magnitudes absolues peut être prolongée jusqu'à $M = -9$.

1. Introduction, calibration de la classification BCD en types spectraux, gradients et magnitudes absolues

Les méthodes spectrophotométriques décrites dans quatre articles (Baillet *et al.*, 1952; Laffineur, 1952; Chalonge et Servigne, 1952; Chalonge et Divan, 1952) ont permis d'établir une classification stellaire à trois paramètres mesurables: λ_1, D et Φ_b. Plusieurs applications de cette classification ont été décrites, en particulier par Chalonge (1958, 1966). Les méthodes et instruments ont été progressivement perfectionnés et l'étude homogène de plusieurs centaines d'étoiles a permis d'obtenir un certain nombre de résultats. Ces résultats, déjà présentés à l'Assemblée Générale de Brighton en Septembre 1970, sont actuellement en cours de publication (Chalonge et Divan, 1973). Dans ce qui suit, nous résumons quelques points importants qui serviront à interpréter les mesures faites sur les étoiles supergéantes du Grand Nuage de Magellan.

B. Hauck and B. E. Westerlund (eds.), Problems of Calibration of Absolute Magnitudes and Temperature of Stars, 78–85.

A. CALIBRATION DU DIAGRAMME $\lambda_1 D$ EN TYPES SPECTRAUX

Dans le diagramme $\lambda_1 D$, les courbes séparant les différents types spectraux et classes de luminosité ont été améliorées, principalement dans la région des étoiles supergéantes; la Figure 1 reproduit les nouvelles courbes.

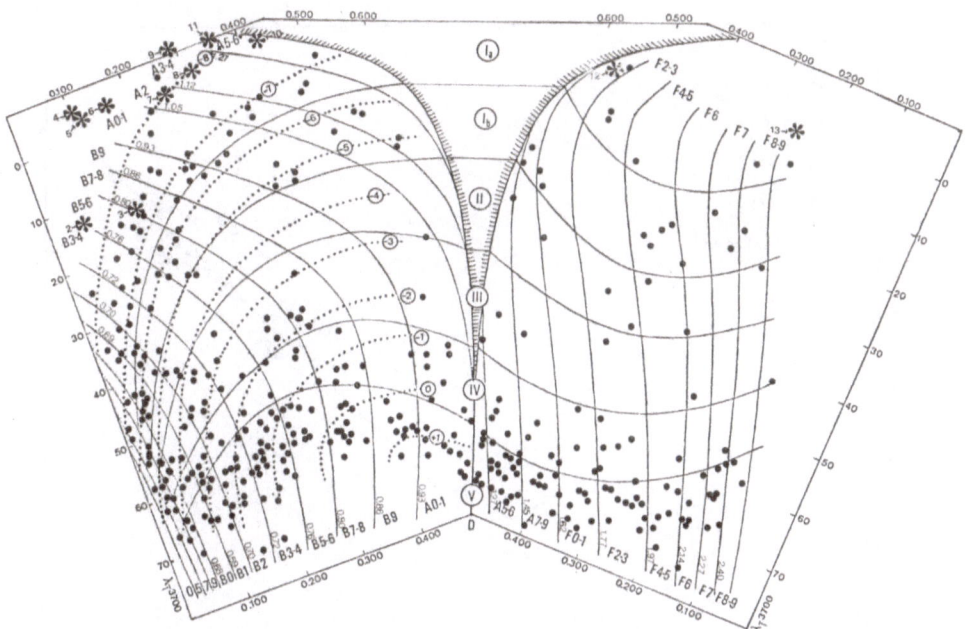

Fig. 1. Diagramme $\lambda_1 D$. ● étoiles galactiques; ✳ étoiles du Grand Nuage de Magellan; — lignes d'égal type spectral et lignes d'égale classe de luminosité (classification MKK); .... lignes d'égale magnitude absolue.

B. CALIBRATION DU DIAGRAMME $\lambda_1 D$ EN GRADIENTS

Avec une marge possible d'incertitude de ± 0.10 les courbes d'égal type spectral sont également des courbes d'égal gradient bleu (domaine spectral 4600–4000 Å); la valeur de ce gradient est inscrite sur chacune des courbes dans la Figure 1.

C. CORRECTIONS D'ABSORPTION INTERSTELLAIRE

Grâce à la calibration précédente, la connaissance de $\lambda_1 D$ pour une étoile permet d'obtenir la valeur de son gradient intrinsèque (celui que l'on observerait en l'absence de matière interstellaire) Φ_{ob}. La valeur observée Φ_b donne alors l'excès de gradient $e_b = \Phi_b - \Phi_{ob}$, excès dont on peut déduire l'absorption A_v pour la magnitude V si l'on connaît la forme de la loi d'absorption interstellaire. Avec une loi d'absorption normale, $A_v = 1.9 e_b$.

D. CALIBRATION DU DIAGRAMME $\lambda_1 D$ EN MAGNITUDES ABSOLUES

La mesure de $\lambda_1 D \Phi_b$ pour des étoiles naines et supergéantes situées dans quelques amas et associations (Ori I, Pléïades. Per II, Per III, Per I) dont les distances photo-métriques sont considérées comme bien connues, a permis de placer ces étoiles dans le diagramme $\lambda_1 D$, puis de calculer pour chacune d'elles la valeur de A_v par la formule $A_v = 1.9e_b$; on en déduit alors V_o puis, à partir du module de distance photométrique

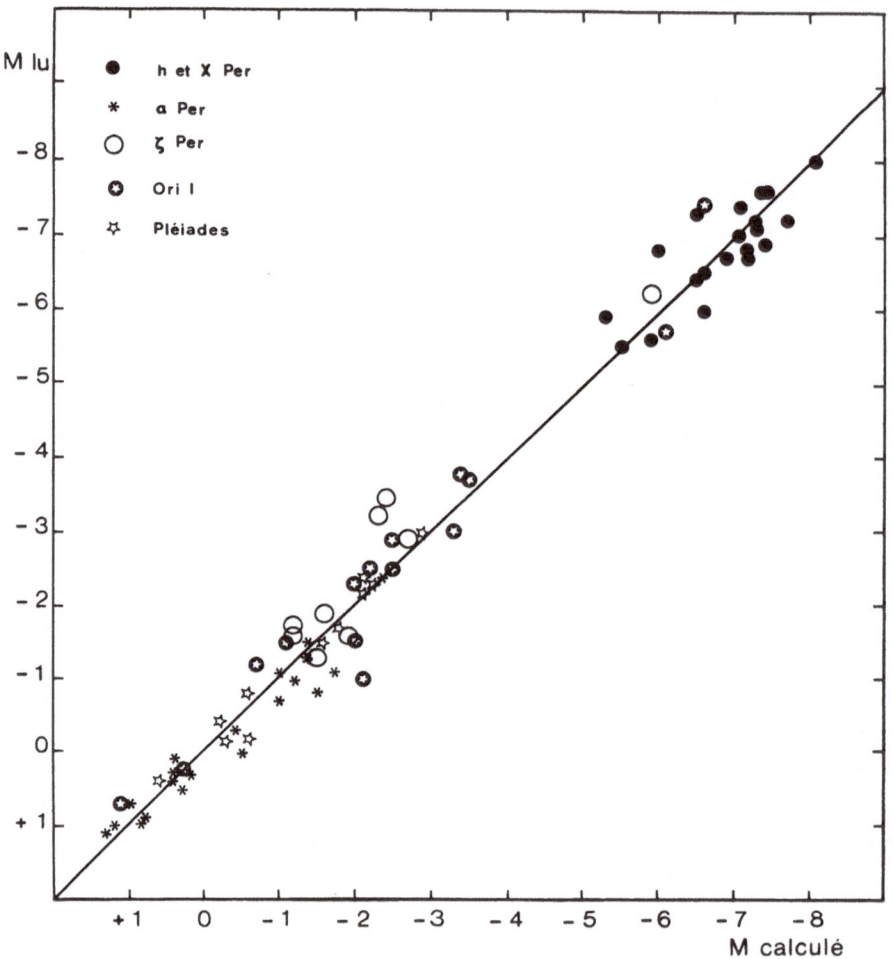

Fig. 2. Comparaison entre les valeurs calculées pour les magnitudes absolues et celles lues sur le réseau des courbes de la Figure 1, pour les étoiles de cinq amas ou associations.

de l'amas, la magnitude absolue M. Les magnitudes absolues ainsi obtenues ont été inscrites aux points correspondants du diagramme $\lambda_1 D$ et il a été possible de tracer un réseau régulier de courbes d'égale magnitude absolue, de $M = +1$ à $M = -8$ (Figure 1), avec un écart moyen entre les valeurs précédemment calculées de M et celles lues sur le réseau inférieur à 0.5 mag. (voir Figure 2).

2. Etude de quelques supergéantes du Grand Nuage de Magellan (GNM)

A. GÉNÉRALITÉS

L'étude dans le système BCD des supergéantes du GNM est intéressante à plusieurs points de vue:

(i) le paramètre λ_1 très sensible à la luminosité pour les étoiles galactiques les plus brillantes ($M \sim -8$) permet encore de classer les supergéantes du GNM pour lesquelles les magnitudes absolues atteignent -9 et même -10.

(ii) la détermination des paramètres $\lambda_1 D \Phi_b$ pour des étoiles du GNM moins lumineuses et se situant en $\lambda_1 D$ dans les limites de la calibration en magnitudes absolues décrite plus haut, permet d'obtenir une valeur de module de distance du GNM indépendante de celles obtenues par l'intermédiaire des variables RR Lyr ou des Céphéides.

B. LES OBSERVATIONS

Elles ont été faites en Janvier 1971 au télescope de 152 cm de l'ESO à La Silla, avec l'équipement spectrophotométrique de l'institut d'Astrophysique de Paris (spectro-graphe Chalonge à chassis oscillant ... etc) suivant les techniques exposées par Chalonge et Divan (1973). En particulier l'absorption atmosphérique et la quantité d'ozone étaient déterminées pour chaque nuit d'observation.

Les quantités $\lambda_1 D \Phi_b$ ont été obtenues pour 13 supergéantes de types B, A et F choisies
 (i) parmi les plus lumineuses
(ii) dans un groupe d'étoiles plus faibles que les précédentes de 1 à 1.5 magnitude.
Les résultats sont donnés dans le Tableau I. Les valeurs de V, $B-V$, $U-B$ sont de Ardeberg et al. (1972).

Φ_{rb}, Φ_b et Φ_{uv} sont les gradients moyens obtenus pour les domaines spectraux 6200–4000 Å, 4600–4000 Å et 3700–3130 Å.

Les points représentatifs des étoiles du Tableau I sont portés sur la Figure 1, sauf pour R 81 dont le continu de Balmer est en émission; cette émission modifie les valeurs de D et λ_1 et enlève toute signification à la position du point correspondant dans le diagramme $\lambda_1 D$.

C. LES RÉSULTATS

(i) *Sensibilité du paramètre λ_1 aux grandes luminosités*

L'examen du Tableau I et de la Figure 1 montre une bonne corrélation entre la position des étoiles du GNM par rapport à la courbe $M = -8$ et leur magnitude apparente: les étoiles les plus brillantes sont au-dessus de cette courbe, les moins brillantes au-dessous. Ceci sera précisé plus loin lorsque les corrections d'absorption interstellaire seront apportées aux magnitudes apparentes considérées ici, mais on peut déjà dire que le paramètre λ_1 est encore sensible à la luminosité pour des étoiles nettement plus lumineuses que les supergéantes galactiques. La classification $\lambda_1 D$ des étoiles du GNM a donc une signification même pour les plus brillantes.

TABLEAU I

Paramètres spectrophotométriques des supergéantes étudiées dans le GNM

No.	HD			Type Ardeberg et al. (1972)	V	B − V	U − B	Φ_{rb}	Φ_b	Φ_{uv}	D	λ_1-3700	Poids	Type $\lambda_1 D$
1	269128	R 81		pec B3 Ia	10.52	0.00	− 0.75	1.36		1.12	0.022	− 12	1	
2	269546	R103		B3 puis B5 Ia	10.4			1.23		1.18	0.052	13	2	B4–B5 Ia
3	269644	R107	G335	B7 Ia	11.12	0.00	− 0.62	1.22		1.30	0.150	14	1	B7 Ia
4	32034	R 62		B9 Iae:	9.69	+ 0.09	− 0.61	1.39		1.59	0.103	− 05	2	A0 Ia
5	269781	R118		A0 Iae	9.8	+ 0.10	− 0.48		1.39	1.62	0.114	− 03	2	A0 Ia
6	268946	R 75		A0 Ia	10.24	+ 0.12	− 0.37		1.40	1.68	0.163	− 04	3	A1 Ia
7	269787	R119	G390	A0 Ia0:	10.81	+ 0.19	− 0.18		1.37	1.67	0.259	− 02	2	A2 Ia
8	269651		G339	A2 Ia0:	10.73	+ 0.19	− 0.24		1.37	2.07	0.315	− 04	1	A3 Ia
9	33579	R 76		A3 Ia0	9.12	+ 0.25	0.00		1.55	1.92	0.292	− 09	4	A4 Ia
10	269331		G253	A3 Ia0 puis A5 Ia	10.30	+ 0.21	− 0.16		1.63	2.29	0.425	− 05	2	A5 Ia
11	269541		G297	A8:Ia0	10.40	+ 0.41	+ 0.30		1.51	1.89	0.355	− 08	1	A5 Ia
12	269697		G352	F6 Ia	10.28				1.99	2.38	0.561	04	1	contradiction entre type $\lambda_1 D$ et type raies.
13	269542		G296	F6 Ia	9.90	+ 0.42	− 0.10		2.15	1.78	0.250	02	3	

(ii) *Classification spectrale*

En extrapolant les courbes d'égal type spectral définies à l'aide des étoiles galactiques ayant des types MKK, on obtient les classification inscrites dans la derniére colonne du Tableau I; elles sont en bon accord avec celles que Prévot et Maurice ont déterminées (Tableau I, colonne 5) et publiées dans Ardeberg *et al.* (1972).

(iii) *Corrections d'absorption interstellaire*

Pour les étoiles du GNM, l'absorption interstellaire est faible et la forme de la loi absorption adoptée pour la corriger importe peu; nous avons adopté une absorption normale pour laquelle $A_v = 1.9\, e_b$ (ou $1.7\, e_{rb}$).

Le calcul des excès de gradient a été fait comme dans le cas des étoiles de h et χ Per qui ont servi à calibrer en magnitudes absolues la partie du diagramme $\lambda_1 D$ la plus proche de celle des étoiles du GNM étudiées ici: on a adopté la valeur de Φ_{ob} (égale à Φ_{orb} pour les étoiles B) lue par interpolation sur les courbes d'égal type spectral (ou sur leur extrapolation) dans le diagramme $\lambda_1 D$. Ces valeurs sont données par les colonnes 7 et 10 du Tableau II; les colonnes 8 et 11 donnent les valeurs de A_v en magnitudes et la colonne 12 la valeur V_o de la magnitude V après correction de l'absorption interstellaire.

On peut remarquer que la valeur moyenne des A_v que nous obtenons (0.6 mag. environ) est un peu supérieure à la valeur moyenne généralement admise (0.3 mag.)

(iv) *Module de distance du GNM*

Parmi les étoiles du Tableau II, certaines se situent dans la partie du diagramme $\lambda_1 D$ déjà calibrée en magnitudes absolues et on peut lire les valeurs de M correspondantes sur la Figure 1 (Tableau II, colonne 13) puis faire le calcul du module de distance, $V_o - M_{lu}$, du GNM (Tableau II, colonne 14); les 4 valeurs obtenues se situent entre 17,8 et 18,4 avec une moyenne de 18,1. Cette valeur plus faible que celles généralement admises, est tout à fait provisoire car le nombre d'observations est encore insuffisant. L'intérêt de la méthode est que le résultat est indépendant des étoiles RR Lyr et des Céphéïdes.

On peut remarquer également que les hypothèses faites sur les couleurs intrinsèques des supergéantes du GNM entre $M = -7$ et $M = -8$, hypothèses qui entraînent une valeur un peu supérieure à celle généralement admise pour l'absorption interstellaire, sont sans influence sur la valeur obtenue ici pour le module de distance; en effet les mêmes hypothèses ont été faites pour le calcul des magnitudes absolues des étoiles galactiques situées dans la même région du diagramme $\lambda_1 D$ et les changer reviendrait à changer de la même quantité les valeurs de M pour les étoiles de la galaxie et pour le GNM; les courbes d'égale magnitude absolue se déplaceraient un peu dans le plan $\lambda_1 D$ mais la valeur du module de distance du GNM resterait le même.

Par contre toute modification du module de distance que nous avons admis pour h et χ Per (11.8) entraînerait une modification correspondante de la distance du GNM et on peut dire que la détermination faite ici est celle de la différence entre les modules de distance de h et χ Per et du GNM.

TABLEAU II

Calcul des corrections d'absorption interstellaire et module de distance du GNM

No.	HD		V	Φ_{rb}	Φ_{0rb}	$1.7\,e_{rb}$ $(=A_v)$	Φ_b	Φ_{ob}	$1.9\,e_b$ $(=A_v)$	V_o	M_{lu}	$V_o - M_{lu}$	M (en admettant $V_o - M = 18.1$)
1	269128	R 81	10.52	1.36						9.6			-8.5
2	269546	R103	10.4	1.23	0.76	0.80				10.44	-7.4	17.8	-7.7
3	269644 G335	R107	11.12	1.22	0.82	0.68				9.03			-9.1
4	32034	R 62	9.69	1.39	1.00	0.66				9.0			-9.1
5	269781	R118	9.8				1.39	0.98	0.78	9.56			-8.5
6	268946	R 75	10.24				1.40	1.04	0.68	10.28	-8.0	18.3	-7.8
7	269787 G390	R119	10.81				1.37	1.09	0.53	10.39	-8.0	18.4	-7.7
8	269651 G339		10.73				1.37	1.19	0.34	8.57			-9.5
9	33579	R 76	9.12				1.55	1.26	0.55				
10	269331 G253		10.30				1.63	1.40	0.44	9.86	-7.9	17.8	-8.2
11	269541 G297		10.40				1.51	1.32	0.36	10.04			-8.1
12	269697 G352		10.28				1.99			(9.8)		moyenne $\rightarrow$ $=18.1$	-8.3
13	269542 G296		9.90				2.15			9.4			-8.7

(v) *Extension de la calibration du plan $\lambda_1 D$ en magnitudes absolues*

Dans la dernière colonne du Tableau II on a inscrit pour chaque étoile la valeur de la magnitude absolue en admettant 18.1 comme module de distance du GNM. On voit facilement, en reportant ces magnitudes absolues sur la Figure 1, que les étoiles plus brillantes que $M = -8$ (à la seule exception de G 253) sont bien situées au delà de la courbe $M = -8$ et que l'on pourrait tracer une nouvelle courbe d'égale magnitude $M = -9$ nettement séparée de la courbe $M = -8$.

(vi) *Cas des étoiles F6 Ia*

Nous avions inclus dans le programme d'observation G 352 et G 296 classées toutes deux F6 Ia par Prévot et Maurice, car leurs positions dans le diagramme *UBV* différaient notablement, sans que l'on puisse dire laquelle était anormale. Les spectres obtenus par nous donnent bien, en ce qui concerne les raies, un type F6 Ia et les deux étoiles ont des spectres très analogues dans le domaine 6200–4000 Å. Par contre leurs discontinuités de Balmer diffèrent considérablement: 0.561 et 0.250; il en résulte que l'une d'elles se classe F0 Ia en $\lambda_1 D$ et l'autre G0 Ia; elles sont donc toutes les deux anormales. Ce phénomène qui n'a jamais été observé pour les étoiles F6 Ia de la galaxie est probablement fréquent dans le GNM car plusieurs étoiles de ces types se placent en *UBV* entre les position extrêmes occupées par G 352 et G 296. Aucune explication satisfaisante n'a être trouvée.

Bibliographie

Ardeberg, A., Brunet, J. P., Maurice, E., et Prévot, L.: 1972, *Europ. Southern Obs.*, preprint No. 1.
Baillet, A., Chalonge, D., et Cojan, J.: 1952, *Ann. Astrophys.* **15**, 144.
Chalonge, D.: 1958, *Rich. Astron.* **5**, 345.
Chalonge, D.: 1966, in K. Lodén, L. O. Lodén and U. Sinnerstad (eds.), 'Spectral Classification and Multicolour Photometry', *IAU Symp.* **24**, 77.
Chalonge, D. et Divan, L.: 1952, *Ann. Astrophys.* **15**, 201.
Chalonge, D. et Divan, L.: 1973, *Astron. Astrophys.* **23**, 69.
Chalonge, D. et Servigne, M.: 1952, *Ann. Astrophys.* **15**, 151.
Laffineur, M.: 1952, *Ann. Astrophys.* **15**, 154.

DISCUSSION

Van den Bergh: Could your indices be affected by small differences in composition between the supergiants in the Galaxy and the Large Cloud?

Divan: Our indices depend only on the continuum of the stars and are very little affected by variations in chemical composition for the region of the $\lambda_1 D$ diagram in which those B and early A supergiants are located. Moreover, the few chemical analysis of LMC stars done until now have revealed no really significant differences between LMC stars and the Sun.

Jaschek: In the last years it has become evident that even among the early B-type supergiants there are peculiarities like the N anomaly discussed by Walborn and the Jaschek's and the O anomaly found by the Jaschek's. It is therefore important to watch for possible difference in the M_v's, before lumping together the M coming from different groups.

THE $(\log g, \log T_{\text{eff}})$-DIAGRAM, A FUNDAMENTAL HR-DIAGRAM

E. B. NEWELL

Yale University Observatory, U.S.A.

Abstract. Spectroscopic and photometric parallaxes are based on the assumption that to each point in the relevant region of $(g, T_{\text{eff}}, \boldsymbol{a})$-space there corresponds a unique mass. The implications of a breakdown in this assumption are discussed briefly and an alternative approach to problems that traditionally have required spectroscopic or photometric parallaxes is presented. This alternative approach involves the use of the $(\log g, \log T_{\text{eff}})$-diagram as an HR-diagram.

I would like to draw your attention to a source of errors in spectroscopic and photometric parallaxes that we have not yet discussed at this meeting. That is, the possible failure of the basic assumption that gravity-sensitive spectral features are uniquely related to the absolute visual magnitude M_v.

For the purposes of this discussion let us define *spectroscopic observations*, to be all those in which the quantities measured, S, are selected attributes of a star's emergent flux distribution F_v. Note that this definition of the term 'spectroscopic' is broader than usual; it covers the full range of available spectral resolution, from high-dispersion spectrophotometry to broad-band photometry, and includes visual spectral classification.

It is well known that, to the extent that the plane-parallel assumption is valid, the structure of a star's atmosphere, and hence its emergent flux distribution F_v, can be parameterized in terms of g, T_{eff}, and surface chemical composition $\boldsymbol{a}$ (for the purpose of this discussion we can ignore the possible influence on F_v of non-thermal velocity fields, magnetic fields, etc., although these effects can in principle be included [see, for example Peat 1969]). Thus a given spectral feature S can be written as a function of the above three variables:

$$S = fn(F_v) = f(g, T_{\text{eff}}, \boldsymbol{a}). \tag{1}$$

It follows that g, T_{eff} and $\boldsymbol{a}$ are the principal arguments of *all* such observations.

From the preceeding argument it is obvious that the basic physical parameters of a star $(\mathfrak{M}, L, R)$ cannot be determined directly from spectroscopic data. At most, one can derive a mass-luminosity ratio $\mathfrak{M}/L$ (see, for example, Keenan, 1963). This implies, in particular, that the determination of spectroscopic parallaxes is a process of limited validity; i.e. the observational parameters employed in the derivation of a spectroscopic parallax are not explicit functions of the quantity being determined (the ordinate of the conventional HR-diagram, M_v). Typically, a *surface gravity sensitive* parameter, such as a line-strength ratio or the equivalent-width of Hγ, is calibrated against M_v. This spectroscopic parameter is then called, incorrectly, a 'luminosity criterion'.

If a heterogeneous stellar sample is being investigated, spectroscopic parallaxes

B. Hauck and B. E. Westerlund (eds.), Problems of Calibration of Absolute Magnitudes and Temperature of Stars, 86–89.
All Rights Reserved. Copyright © 1973 by the IAU.

are liable to be misleading. This can be seen from Equation (3) (below); two stars can have identical (g, T_{eff})-values, and therefore display identical 'luminosity criteria', but have different masses and, hence, different luminosities. Clearly, the basic assumption involved in the method of spectroscopic parallaxes is that there is a unique, single-valued relation between g and M_v. This is equivalent to the assumption that to each $(g, T_{eff}, a, B.C.)$-value there corresponds a unique value of the total mass $\mathfrak{M}$ (here $B.C.$ denotes the bolometric correction). This must not be regarded as a second order effect; may I remind you that the variation in M_v corresponding to a change $\Delta \log \mathfrak{M}$ in the mass is

$$\Delta M_v = 2.50 \cdot \Delta \log \mathfrak{M} \, (\text{mag.}). \tag{2}$$

Of course, the above assumptions are justified in many cases. Nevertheless, we must always exercise care; even in the best case (i.e., the study of open clusters) there are evolutionary phases during which the tracks of stars of different masses can cross. It is imperative, as we push our studies into the field, and, in particular, when we study faint samples that may contain objects that belong to the old disk or halo populations or when we study stars in external galaxies, that we keep the central assumption of the method of spectroscopic parallaxes firmly in mind. Let us consider two cases where this assumption breaks down: (i) in the study of the high-latitude faint blue stars one can easily encounter variations of mass by a factor of two at a given (g, T_{eff}, a)-value. Thus, ΔM_v can be of the order of $0^m.75$, and (ii) in the study of the distribution of A-stars perpendicular to the galactic plane one will begin to include appreciable numbers of blue horizontal-branch stars in the sample for $V \gtrsim 10^m.0$ – such stars will have their M_v–values overestimated by $\sim 1^m.5$ if a Population I calibration is adopted.

There is, nevertheless, an alternative approach that can, in some important contexts, eliminate the need for spectroscopic parallaxes. That is, to use g directly as an ordinate of the HR-diagram.

Let us define a *physical HR-diagram* to be one that (i) preserves the morphology of the original $(M_v$, spectral type)-diagram, and (ii) has arguments that are explicit functions of the fundamental *physical* characteristics of a star $(\mathfrak{M}, L, R)$. The most widely used physical HR-diagram is the $[\log L, \log T_{eff}]$-diagram. Nevertheless, surface gravity is an equally valid ordinate; from the definitions of g and T_{eff} we can derive the following relation between g and L:

$$g = [\text{constant} \cdot T_{eff}^4 \cdot \mathfrak{M}] \, L^{-1}. \tag{3}$$

From this equation it is obvious that, for a given value of $\mathfrak{M}$, g is uniquely related to L at each T_{eff}. Thus, under the condition of constant mass, a curve in the $(\log L, \log T_{eff})$-plane maps one-to-one into the $(\log g, \log T_{eff})$-plane. In most cases of practical interest, variations in $\mathfrak{M}$ do not destroy the correspondence.

We can readily appreciate, from a historical point-of-view, why the $(\log L, \log T_{eff})$-diagram is at present the most widely used physical HR-diagram. Nevertheless, the $(\log g, \log T_{eff})$-diagram is the fundamental diagram when one is dealing with spectro-

E. B. NEWELL

scopic observations. This fundamental nature of the $(\log g, \log T_{\text{eff}})$-diagram was first emphasized by Morgan (1937). Morgan noted that the features observed in stellar spectra are explicit functions of the two principal *physical* parameters of a stellar atmosphere, viz., the surface gravity g and the effective temperature T_{eff}. Thus, the spectral changes that form the basis of a visual classification scheme correspond, primarily, to changes in g and T_{eff}. As Morgan pointed out, a diagram that displays the variation of these two parameters along the principal stellar sequence is of central importance in attempts to understand stellar classification problems.

But the $(\log g, \log T_{\text{eff}})$-diagram can play a much wider role than that outlined above. For example, one of the most important applications of the $(\log g, \log T_{\text{eff}})$-diagram lies in the study of stellar evolution. Theoretical stellar evolution tracks can be presented just as readily in the $(\log g, \log T_{\text{eff}})$-diagram as in the conventional $(\log L, \log T_{\text{eff}})$-diagram. Provided that care is taken to ensure that the derived g and T_{eff} values are correctly related to the global quantities $\mathfrak{M}/R^2$ and L/R^2 (Newell *et al.*, 1969) then the predictions of stellar structure computations can be compared directly with observations, without reference to luminosities; i.e. *the comparison is independent of distances and bolometric corrections*. This technique is obviously one of considerable power in observational tests of stellar interior theory and in attempts to understand observed stellar sequences. In particular, it allows both cluster and *field* stars to be compared with each other and with theory in a manner that is not possible in the conventional HR-diagram. Such comparisons can lead to luminosity-independent determinations of stellar ages. Examples of comparisons of this type are to be found in the work of Hyland (1967), Norris (1971), Osborn (1971), Gross (1972), and Newell (1973).

In summary, those involved in the calibration of spectral features against absolute magnitude are urged to keep the central assumption of their approach firmly in mind and to ensure, for each group of stars studied, that they are justified in assuming a unique, single-valued relationship between g and M_v. The $(\log g, \log T_{\text{eff}})$-diagram provides an approach that is independent of spectroscopic parallaxes and, in my opinion, is valuable enough to justify considerable effort being spent on the problem of calibrating our spectroscopic data against g.

Acknowledgements

I am indebted to the faculty and graduate students of the Yale Astronomy Department for many useful discussions and to Dr W. W. Morgan for his enthusiastic support of this study.

References

Gross, P. G.: 1972, unpublished Thesis, Yale Univ. Obs.
Hyland, A. R.: 1967, in R. C. Cameron (ed.), *The Magnetic and Related Stars*, Mono Book Corp., Baltimore, p. 311.
Keenan, P. C.: 1963, *Stars and Stellar Systems* 3, 78.
Morgan, W. W.: 1937, *Astrophys. J.* **85**, 380.

Newell, E. B.: 1973, *Astrophys. J. Suppl.* (in press).
Newell, E. B., Rodgers, A. W., and Searle, L.: 1969, *Astrophys. J.* **156**, 597.
Norris, J.: 1971, *Astrophys. J. Suppl.* **23**, 213.
Osborn, W. H.: 1971, unpublished Thesis, Yale Univ. Obs.
Peat, D. W.: 1961, in O. Gingerich (ed.), *Third Harvard-Smithsonian Conference on Stellar Atmospheres*, MIT Press, p. 55.

DISCUSSION

Pecker: The problem with the s-shaped tracks at the hydrogen-exhaustion phase may not be as serious as you have indicated. A recent study by Lesh and Aizenmann places the ultra-short period variables in this part of the diagram and thus we may be able to distinguish between stars undergoing this rapid evolution and stars that are not.

Newell: Such a discriminant can alert us to the fact that we are working in a region where the mass of a star may be uncertain, but it cannot remove the problem with spectroscopic parallaxes.

Demarque: On the question of the identification of β Cephei variables with the hydrogen exhaustion phase, it has, some years ago, been looked into by Percy. On the basis of the frequency of β Cephei stars, Percy concluded that they are too numerous to be associated with such a rapid phase of evolution, and must rather be near the end of their core burning phase.

PART IV

ABSOLUTE MAGNITUDE DETERMINATIONS FROM HYDROGEN-LINE PHOTOMETRY

ABSOLUTE MAGNITUDE DETERMINATIONS FROM HYDROGEN-LINE PHOTOMETRY

D. L. CRAWFORD

Kitt Peak National Observatory*, U.S.A.

Abstract. The history and status of Balmer hydrogen-line photometry are briefly reviewed, and problems associated with calibrations in general are commented on.

The calibration work at Kitt Peak is described in some detail for B, A, and F type stars. The data used, the determination of intrinsic colours and colour excesses, photometric classification, and the relation of our indices to other people's are reviewed. With this background, the procedure used in establishing the absolute magnitude — Hβ calibration is given, and the preliminary calibration is presented. Finally, the work remaining to be done before the calibration is final is noted, and a comparison is given to Blaauw's zero-age main sequence calibration.

1. History

The history of the use of Balmer-line strengths for the determination of stellar luminosities has been a rich one over the past 50 yr, and I cannot hope to review it all, or even mention it all, here today. I will content myself, and I hope you, by briefly noting some of the past work that I feel has been particularly influential, by summarizing the current status of Balmer hydrogen-line work, and by describing in some depth the work I am most acquainted with: my own calibration efforts with the β and *uvby* systems.

Pre-1950, the major work was by Lindblad (1922, 1925, 1926), Anger (1931, 1932), Öhman (1935), and Williams (1936), who measured Balmer lines on spectral plates and discussed the relation of the derived hydrogen-line strengths to stellar luminosities. Data for stars in clusters were especially useful in such discussions, as the relative luminosities of the stars were well known.

Later, Petrie and his collaborators at Victoria (Petrie, 1950, 1953, 1956, 1965; Petrie and Maunsell, 1950; Petrie and Moyls, 1956) developed and applied the photographic technique to determine hydrogen-line equivalent widths from spectral plates for B-type stars. The technique and calibrations are reviewed and discussed in depth by Petrie (1965) in a paper giving the revision to his earlier calibration. In a paper later today, Crampton will summarize the Victoria system as it now stands.

Others who lately have been in photographic hydrogen-line efforts are Hack (1953), Kopylov (1958), Sinnerstad (1954, 1961a, b), Beer (1961), and Furenlid (1971).

Soon after Petrie's extensive effort, photoelectric techniques to measure a parameter related to hydrogen-line strengths were developed and used by Strömgren (1951, 1952, 1956a, b, 1958), Crawford (1958), Hoag (1965), Bappu *et al.* (1962), Johnson and Iriarte (1958), Beer (1964), Andrews (1968), Graham (1967), and others.

Absolute magnitude calibrations of the β index, that we use, have been given

* Operated by the Association of Universities for Research in Astronomy, Inc., under contract with the National Science Foundation.

by Hardie and Crawford (1961), Fernie (1965), and Crawford (1970), but were based on considerably less material than the calibration that I'll present later in this paper.

Strömgren (1963) gave an excellent summary of the Balmer line work to that time, and he (Strömgren, 1966) again reviewed the status of hydrogen-line spectral-classification work. Both these review papers also discussed other narrow-band photometric classifications and calibration work.

2. Some Problems

Calibration of anything observed vs anything more-or-less physical for B-type stars is not an easy task, as the authors above have clearly shown.

What we'd really like to have are observed parameters, easy to measure for bright and faint stars, internally accurate, free of systematic effects, that are closely (even linearly) related to a physical parameter of the stars, very sensitive to the physical parameter (that is, giving a large range of the observed parameter), and completely free of the influence of any other physical parameter. Nature has not allowed any such situation to exist, and the best we can do is try to obtain parameters that will do as good a job as possible, so that we can use them with some confidence in our astronomical research. Sometimes we can measure an extra parameter, which, while useful in itself, will also allow us to correct one of the main parameters for an undesirable side effect.

An obvious example: We measure a colour index such as $(B - V)$, which is well related to effective temperature. Side effects are interstellar reddening, abundance effects, and even rotation velocity, magnetic field, and binary star effects. Any extra parameter to help correct for these side effects, such as $(U - B)$, has side effects of its own. And so it goes. Quite difficult.

In many cases, theory can help us observers quite a lot. We would be lost, or at the least inefficient, without theory to guide us. However, we must be extremely careful (as Morgan has often pointed out) not to force-fit or to let pre-conceived ideas mess us up. We are measuring observed parameters, and these we relate, or calibrate, to physical parameters. In my opinion, this philosophy is one of the keys that has led to the success of the MK system. It is a network (or two observed 'parameters') that 'classifies' or describes a star's spectrum. It is, in principle, even independent of the spectral dispersion used, for it is defined by standard stars. The observed parameters are then calibrated in terms of temperature and absolute magnitude. In fact, they have been re-calibrated many times. Such methods and philosophy are certainly valid for the best and most useful photometric systems too, I believe.

Enough of philosophy, let me now list some of the specific problems we have to watch out for, or allow for, in hydrogen-line photometry:

(a) accurate and precise standard systems, equipment, and techniques are needed to insure an accurate parameter free of systematic error,

(b) interstellar reddening corrections,

(c) rotational velocity effects,

(d) peculiar stars (however peculiar is defined),

(e) binaries,

(f) cluster membership,

(g) emission line stars, and

(h) accurate and precise absolute magnitudes to calibrate the observed parameter against.

3. The Kitt Peak Work

With this brief background, let me now go on to a detailed description of the work at Kitt Peak. The final calibrations are not quite finished, but little yet remains to be done, and I hope to finish it by the end of this year. The preliminary results presented today should differ little from the final ones.

A. THE DATA

Most of the discussion to follow will be devoted to parameters of (a) the *uvby* system, defined by Crawford and Barnes (1970c), that was originated by Strömgren and developed by him and by us at Kitt Peak, and (b) the β system defined by Crawford and Mander (1966).

The parameters used are:

V – an apparent magnitude, on the same system as the V of the UBV system;

$(b-y)$ – a colour index, freer of line blanketing effects than $(B-V)$ of the UBV system;

$(u-b)$ – a colour index;

m_1 – a colour difference, related to blanketing in the 4100 Å region;

c_1 – a colour difference, related to the Balmer discontinuity;

β – the hydrogen line-strength parameter.

These parameters are discussed in the standard-system papers mentioned above, and by Strömgren (1963, 1966).

We also use UBV and MK data taken from the literature, especially for comparison purposes. Not all the *uvby* or β data are our own: some have been taken from the literature and some have been kindly given to me in advance of publication. I wish to thank very much those who have supplied their unpublished data to me.

Data for the following stars have been used:

(a) northern hemisphere O–B5 stars brighter than $V=6.5$, Crawford *et al.* (1971b);

(b) southern hemisphere O to G0 stars brighter than $V=5.0$, Crawford *et al.* (1970);

(c) southern hemisphere O–B5 stars $V=5.0$ to 6.5, Crawford *et al.* (1971a);

(d) northern hemisphere A2–G0 stars brighter than $V=6.5$, Strömgren and Perry (1965) and Crawford *et al.* (1966);

(e) northern hemisphere A0-type stars brighter than $V=6.5$, Crawford *et al.* (1972);

(f) northern hemisphere B8- and B9-type stars, Crawford *et al.* (unpublished);

(g) northern hemisphere O-type stars fainter than $V=6.5$, Crawford and Golson (unpublished);

and for the following clusters and associations:
- (a) Hyades, Crawford and Perry (1966);
- (b) Praesepe, Crawford and Barnes (1969b);
- (c) Coma, Crawford and Barnes (1969a);
- (d) NGC 752, Crawford and Barnes (1970a);
- (e) IC 4665, Crawford and Barnes (1972);
- (f) Pleiades, Crawford (unpublished);
- (g) α Per, Crawford and Barnes (unpublished);
- (h) h and χ Per, Crawford et al. (1970);
- (i) NGC 6231, Crawford et al. (1971);
- (j) IC 2602, Hill and Perry (1969);
- (k) IC 2391, Perry and Hill (1969);
- (l) Sco-Cen, Hardie and Crawford (1961), Glaspey (1971);
- (m) Orion, Crawford (1958), Crawford and Barnes (1966);
- (n) III Cep, Crawford and Barnes (1970b);
- (o) NGC 2362, Perry (unpublished);
- (p) II Per, Crawford (1958);
- (q) NGC 6871, Cohen (1969);
- (r) NGC 6910, Crawford and Barnes (unpublished);
- (s) NGC 6913, Crawford and Barnes (unpublished);
- (t) NGC 6611, Crawford and Barnes (unpublished);
- (u) NGC 5460, Clariá (1971);
- (v) NGC 2264, Strom et al. (1971); and
- (w) NGC 2244, Heiser (unpublished).

B. INTRINSIC COLOURS AND COLOUR EXCESSES

For the A- and F-type (those stars cooler than about A2, the location of maximum hydrogen absorption), we determine the intrinsic colour from the following equation:

$$(b - y)_0 = a - b\,\beta - c\,\delta c_1 - d\,\delta m_1.$$

For A-type stars ($\beta = 2.890$ to 2.720), the constants used are $a = 2.943$, $b = 1.000$, $c = 0.100$, $d = 0.100$. The resulting mean error, for one star, as determined from stars within 100 pc, is $\pm 0\overset{\text{m}}{.}011$. This scatter includes effects of duplicity, rotation velocity, and so forth, as essentially no data were eliminated from the least squares solution. It can be seen that β is an effective parameter in predicting intrinsic colour, for c and d are small; i.e., little luminosity or abundance effects exist.

For F-type stars ($\beta = 2.720$ to 2.600), the coefficient b increases as we go to later types (smaller β), as β begins to lose sensitivity to temperature near G0 while $(b - y)$ does not; the coefficient c is 0.1 or a bit smaller; and the coefficient d increases toward later spectral type, as blanketing effects become larger.

Details and limitations of the calibrations for A- and F-type stars will be discussed fully in a forthcoming paper; a summary has been given by Crawford (1970).

For the B-type stars, we have determined colour excesses, $E(U - B)$, and intrinsic colours $(U-B)_0$, for those stars with available UBV data, by a procedure described by Crawford (1958). A linear reddening slope has been assumed: $E(U-B)/E(B-V)$ $=0.72$. This use of a linear slope may not be justified in all cases, but should be adequate for the discussions to follow. I prefer the use of $(U-B)_0$ to $(B-V)_0$ for B-type stars, as the former has a considerably larger range (about 4 times larger than $(B-V)_0$). I also prefer it over the parameter Q (Johnson and Morgan, 1953) as it is a more natural parameter, i.e., the intrinsic colour, and is no more difficult to determine than is Q.

For an overall discussion of effects of interstellar reddening on the UBV parameters, I believe that the investigation by Fitzgerald (1970) is the most complete.

For investigation of reddening effects on the $uvby$ system, I have used the data noted above. Figures 1, 2, and 3 show the relation between the observed $(u-b)$, m_1, and c_1 indices with respect to $(b-y)$ for O-type stars. Separate symbols are used in

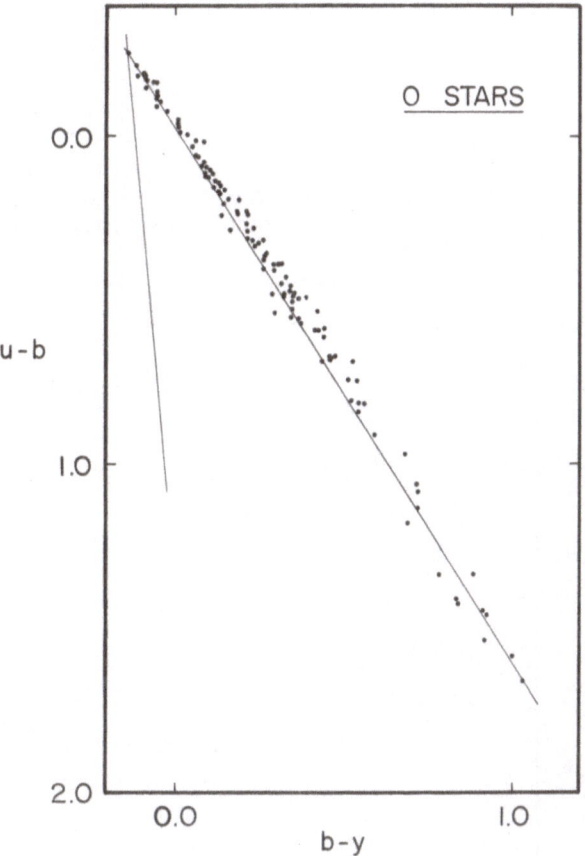

Fig. 1. The $(u-b)$ vs $(b-y)$ relation for a number of O-type stars observed from Kitt Peak. A linear line of slope 1.61 is shown through the data points. The line to the left traces the intrinsic colour line for B-type stars.

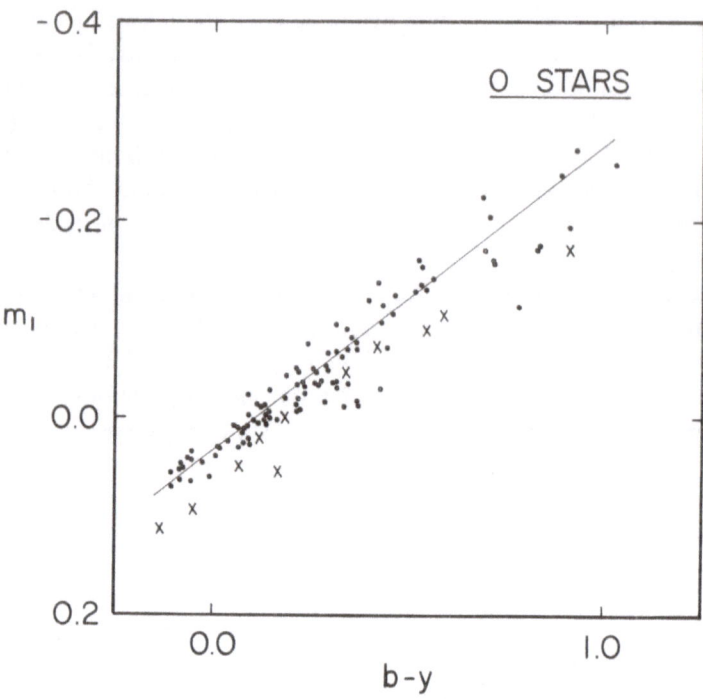

Fig. 2. The m_1 vs $(b-y)$ relation for a number of O-type stars. A reddening line, with slope near -0.3, is shown. Crosses denote Of-type stars.

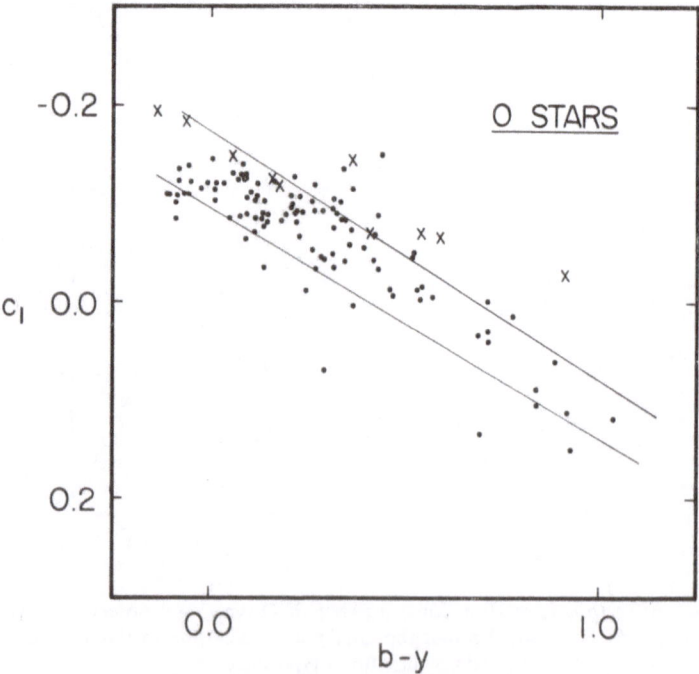

Fig. 3. The c_1 vs $(b-y)$ relation for a number of O-type stars. Two reddening lines, with slopes near 0.2, are shown. Crosses denote Of-type stars.

Figures 2 and 3 for the known Of type stars. Allowing for some luminosity and spectral type effects, I have derived the following reddening relations:

$$E(u - b) = 1.6 E(b - y),$$
$$E(m_1) = -0.3 E(b - y), \quad \text{and}$$
$$E(c_1) = 0.2 E(b - y).$$

One can also derive $E(b-y) = 0.73 E(B-V)$ from a plot of $(b-y)$ vs $(B-V)$ for the same stars. I have used these four reddening relations in the work described below.

Figure 4 shows the relation between the observed c_1 and $(b-y)$ indices for the brighter O to B5-type field stars. A well defined, left-hand envelope is evident, and it

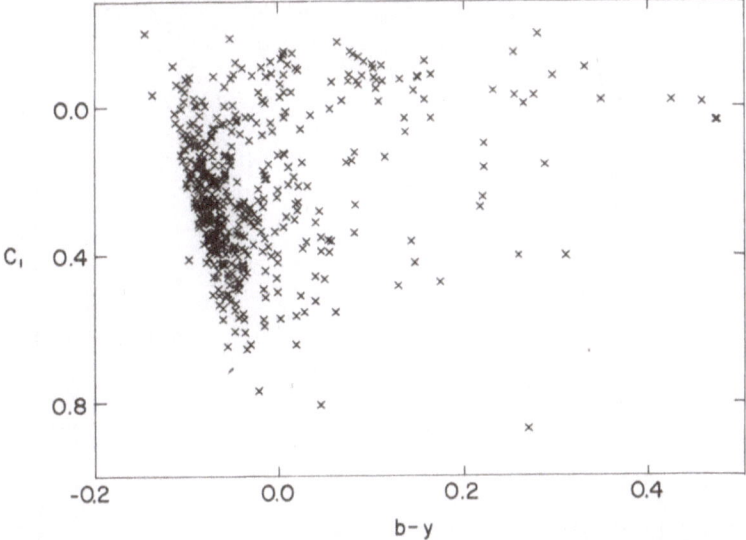

Fig. 4. The c_1 vs $(b-y)$ diagram for O-to-B5 type stars brighter than $V = 6.5$. Reddening lines are nearly horizontal in the diagram, with the intrinsic colour line as the left-hand envelope of the points.

may be defined as the intrinsic colour relation (at least to a first approximation). For the preliminary calibration, I have assumed $(b-y)_0 = -0.116 + 0.097c_0$, where c_0 is the unreddened c_1 index. The final calibration will allow for non-linearities and for small luminosity effects evident both in the observed data for bright stars and in theoretical model atmospheres. Checks have been made by calculating colour excesses for clusters in which the reddening is either zero or essentially uniform. (See, for example, Hill and Perry (1969) and Crawford *et al.* (1970).)

A comparison of the reddening values so obtained with values obtained from *UBV* data (the procedure described above) and from MK types and $(B-V)$ values shows that the agreement (where data are good) is excellent. For example, a detailed comparison of the three methods is given by Crawford *et al.* (1970) for the stars in the h and χ Per clusters.

In all the work to follow, we have assumed the ratio of total-to-selective absorption to be $A_V/E(B-V)=3.2$; i.e., $A_V/E(b-y)=4.3$.

C. HYDROGEN-LINE DATA, AND PHOTOMETRIC TWO-DIMENSIONAL CLASSIFICATION FOR B-TYPE STARS

In Figure 5, I show the relation between our observed β values and Petrie's Hγ values for B-type stars with data in common. The Hγ values have been taken from a recent compilation by Crampton, who has kindly sent me his data. Only a sampling of data

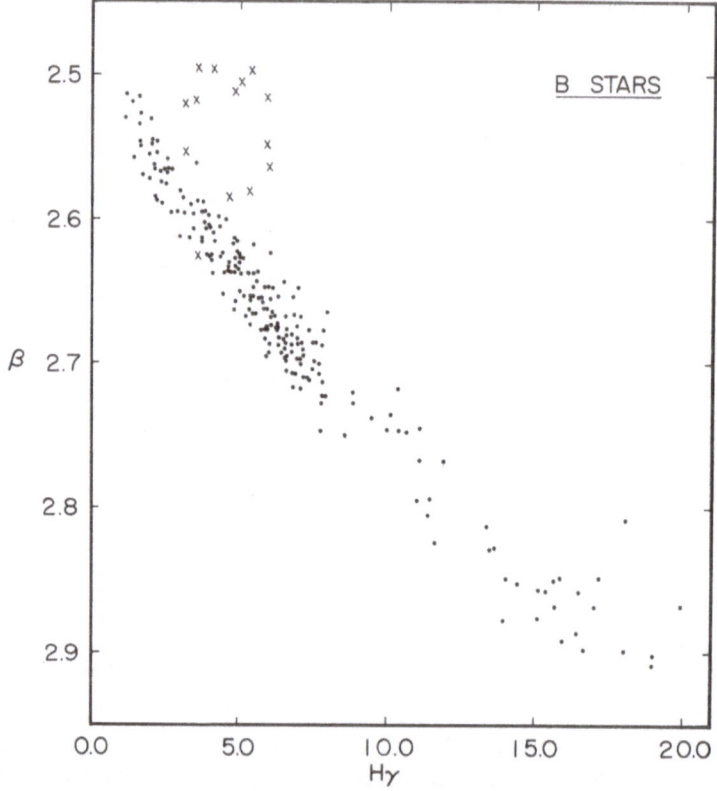

Fig. 5.   The relation between the β parameter and Petrie's Hγ values, for B-type stars. Crosses denote known emission line objects; points for some such objects would lie off of the top of the figure. Only a sampling of points for late B-type stars are shown.

for the later B-type stars has been plotted, to show the trend. Emission line objects are generally well separated from the main relation. No scatter due to rotation velocity effects is apparent in the data, and the overall scatter in the relation is close to that expected from observational error alone.

In Figure 6, we show the relation between two well-observed photoelectric parameters. The data plotted are for the β system standard stars, the β's taken from

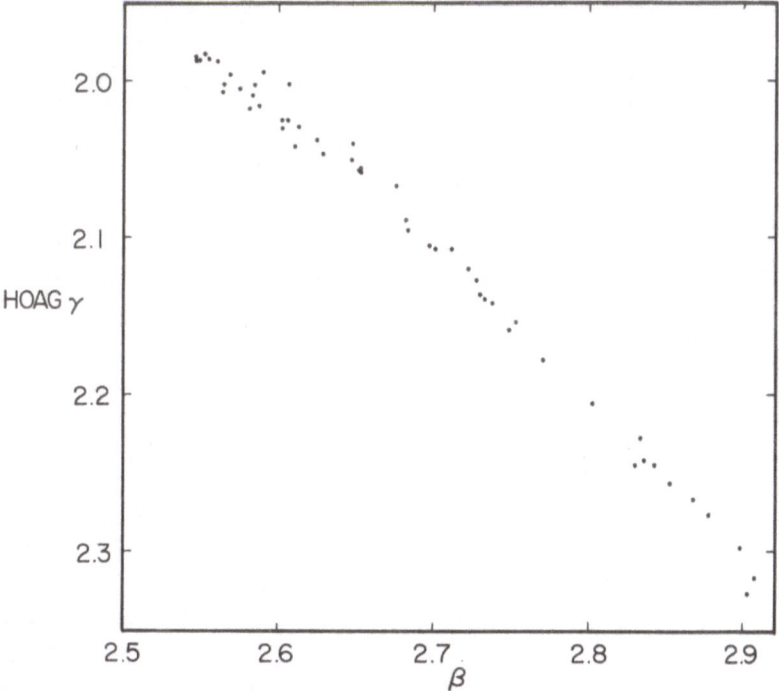

Fig. 6.   The relation between the β parameter and Hoag's photoelectric Hγ parameter, for standard stars of the β system.

Crawford and Mander (1966) and the Hγ values from Hoag's unpublished γ values for the same stars. We wish to thank him for permission to show his data. As can be seen, the scatter is quite small.

In Figure 7, we show our β data plotted against Andrews' (1968) Hα data, for B-type stars. The points above the average relation are due to emission line stars. Again, as in the Hγ vs β plots, points for emission line stars are well separated from the main relation, and no rotational velocity effects are evident in the relation. Many supergiant stars have hydrogen emission, but the data for them lie along the main relation (upper left). The separation of such stars from main sequence emission line objects is rather easy. The turn-up for the brighter B-type stars is due to non-LTE effects, leading to emission (see Mihalas, 1972, and the references given there). The transition to emission is smooth, as theory predicts. An Hα parameter should thus be better than an Hβ parameter for absolute magnitude determinations for the brightest B-type stars. For later B-type stars, Hβ is probably better, and the two together can separate out 'emission-line stars'. Neither will work well if the emission is variable.

Photometric classification is possible with the parameters we have measured or derived above. The parameters measuring the Balmer discontinuity $[(U-B)_0, c_0]$ relate to effective temperature, the hydrogen-line parameters $[\beta, \gamma, \text{or } \alpha]$ relate to luminosity, or absolute magnitude. A (β, c_0) diagram, therefore, is rather like an HR

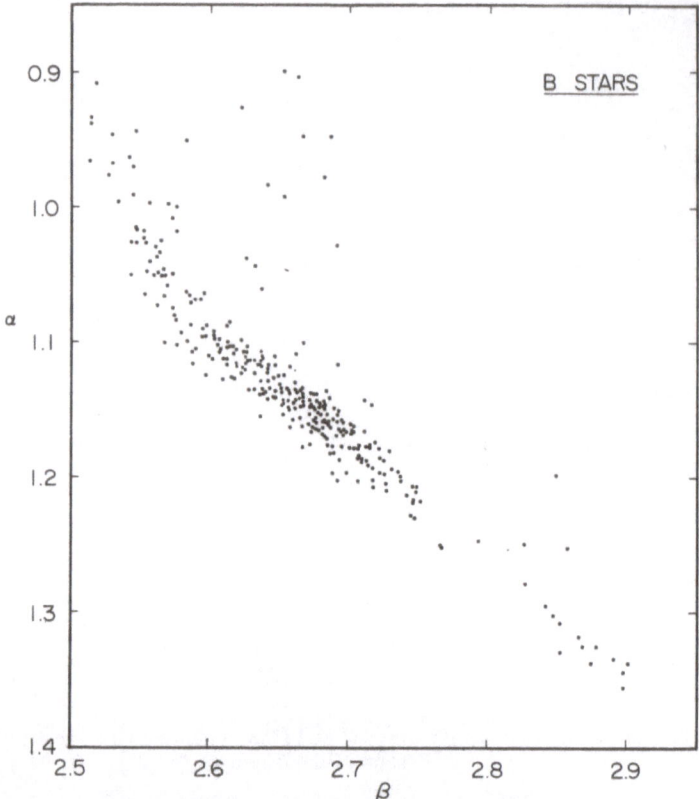

Fig. 7. The relation between the β parameter and Andrews' photoelectric Hα parameter. Emission line objects lie above the average relation. Points for B-type supergiants lie in the upper left of the diagram. Only a sampling of points for late B-type stars are shown.

diagram, or a colour-magnitude diagram. Furthermore, 'boxes' can be drawn in such a diagram relating to MK types. In fact, the relation of the photometry to the spectral types is very good; see, for example, Crawford (1958). Exceptions are usually 'peculiar' stars; for example, see Garrison (1967) and Cowley and Crawford (1971).

In Figure 8, we show the (β, c_0) data for the O–B5 stars brighter than $V=6.5$. A lower envelope is apparent. This lower boundary defines our 'zero-age main sequence' (ZAMS), and we interpret the scatter above it (except for that due to observational error!) to be due to the stars having evolved above the ZAMS and hence having greater luminosities for a given temperature. Such an interpretation is good only to a first approximation, of course, as the parameters are not ideal ones, as described in Section 2 above. In the final calibration, we will use a parameter $\delta\beta$ defined as

$$\delta\beta = \beta\,(\text{ZAMS}) - \beta\,(\text{observed}),$$

and relate this $\delta\beta$ to δM_V, where

$$\delta M_V = M_V\,(\text{ZAMS}) - M_V\,(\text{'observed'}).$$

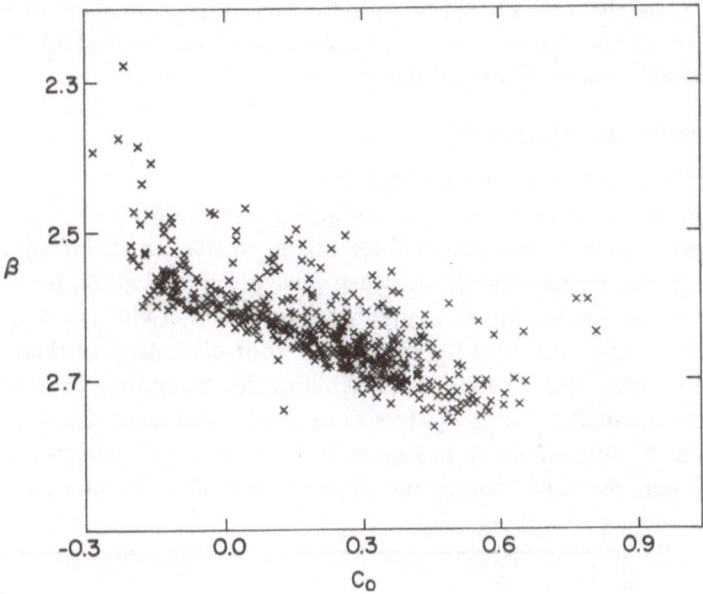

Fig. 8.   The β vs c_0 diagram for O- to B5-type stars brighter than $V = 6.5$. We define the lower envelope as the 'zero-age main sequence' (ZAMS).

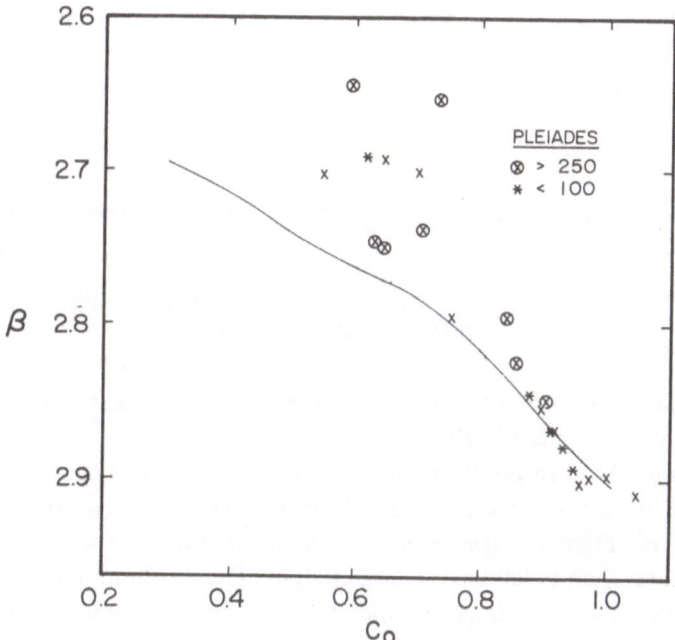

Fig. 9.   The β vs c_0 relation for B-type stars that are members of the Pleiades cluster. Different symbols denote stars with large or small $V \sin i$ values (in km s^{-1}). The line drawn in the diagram is the ZAMS.

In Figure 9, we show the (β, c_0) relation for the B-type stars in the Pleiades cluster. The separation of the brighter stars from the ZAMS line is evident. Little, if any, effect is noticeable due to $V \sin i$ differences.

D. THE ABSOLUTE MAGNITUDE CALIBRATION

We do the calibration by several distinct steps:

(1) Determine the shape of the ZAMS relation between M_V and β for the A and F stars, by observations in clusters. In three of the clusters used, Pleiades, α Per, and IC 4665, the A and F stars should be nearly unevolved; therefore, no correction for any δM_V above the ZAMS has been applied. For the other clusters, a correction has been applied to those stars with significant δc_1 (For discussion of these corrections, see Stromgren, 1966, and Crawford, 1970.). For A-type stars $8\ \delta c_1 = \delta V_0$ has been added to the individual V_0's; for F stars $11\ \delta c_1 = \delta V_0$ was used. The V_0 vs β relation for the stars in IC 4665 is shown in Figure 10, as an example. Diagrams for the separate clusters were overlayed, sliding the diagrams vertically along lines of equal β, and

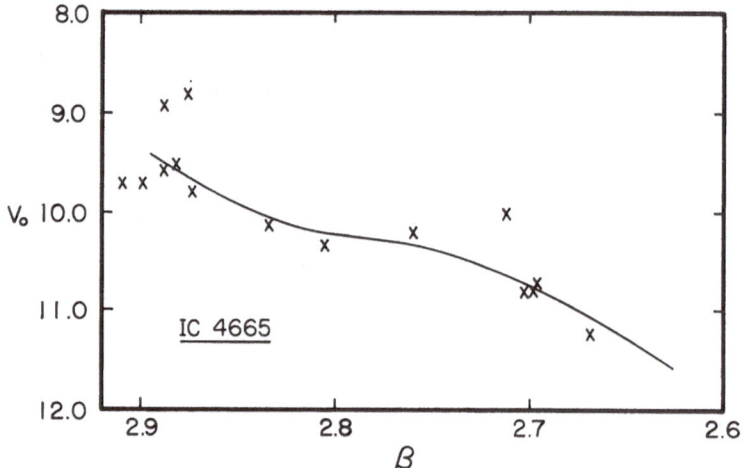

Fig. 10. The V_0 vs β relation for A, F-type members of the cluster IC 4665. A mean relation is drawn through the points.

a smooth mean relation best fitting the individual relations was drawn. The Hyades and Praesepe were *not* used in this determination.

(2) Determine the zero-point of the relation from a fit to trigonometric parallax stars. Absolute magnitudes were calculated for those stars with parallaxes greater than $0''.100$, in the *Yale Trigonometric Parallax Catalog* (Jenkins, 1963), and for those with parallaxes greater than $0''.060$ having large weight. The resultant absolute magnitudes were plotted in a (M_V, β) diagram, and the corrections for δM_V applied whenever δc_1 was not equal to zero. The mean slope determined above, in Step 1, was then fitted to the corrected points. The result is shown in Figure 11. It can be seen that the mean cluster slope fits the data for the parallax stars quite well.

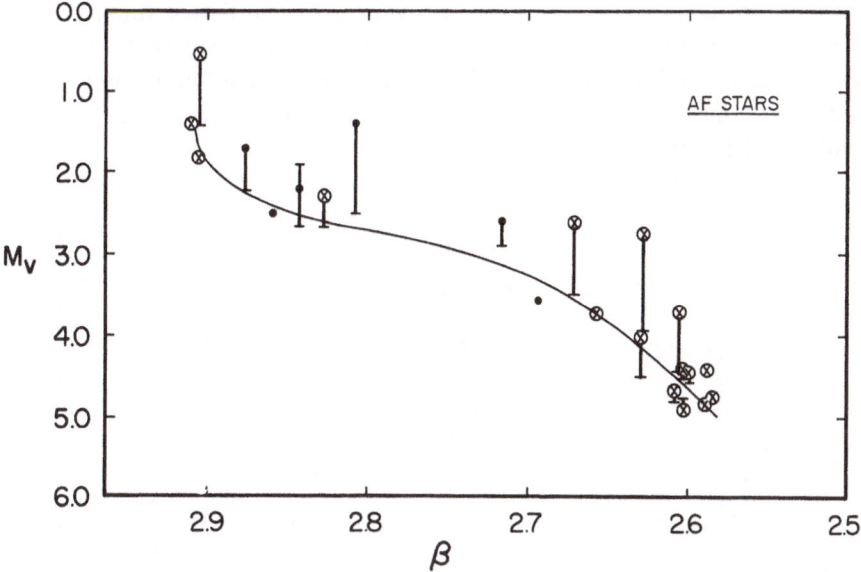

Fig. 11.   The M_V vs β relation for A, F-type stars with large trigonometric parallaxes. The vertical line with the cross-bar shows the correction for evolutionary effects (see text). The mean relation from clusters is drawn in the figure, as best fitting the points for the parallax stars. The line, therefore, is the calibration valid for the ZAMS for the A- and F-type stars.

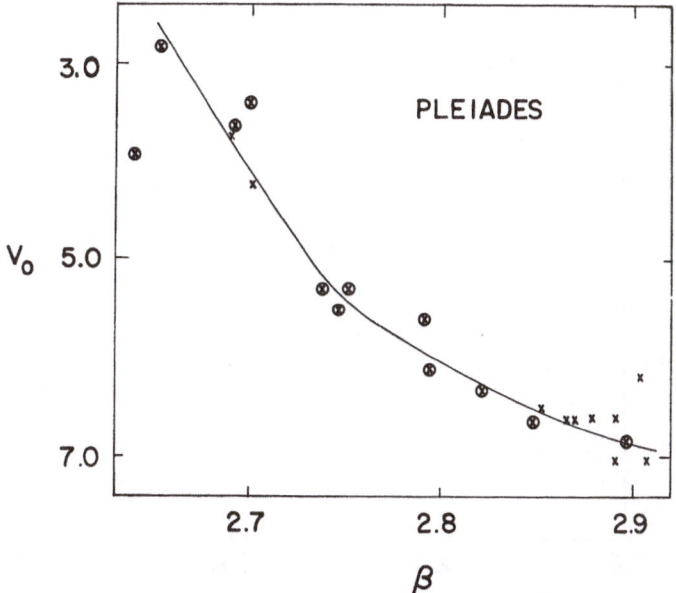

Fig. 12.   The V_0 vs β relation for B-type members of the Pleiades cluster. The crosses with circles about them denote stars with $V \sin i$ greater than 200 km s^{-1}. The line is an eye estimate of the best fit to the points.

We therefore define this best fit as the ZAMS for A- and F-type stars, as a function of the parameter β. Furthermore, if the observed data for a given star, whether cluster member or field star, has a non-zero δc_1, we correct the ZAMS absolute magnitude for this 'evolutionary' effect. That is, $M_v = M_V \text{(ZAMS)} - f \delta c_1$, where $f = 8$ for an A star and $f = 11$ for an F star.

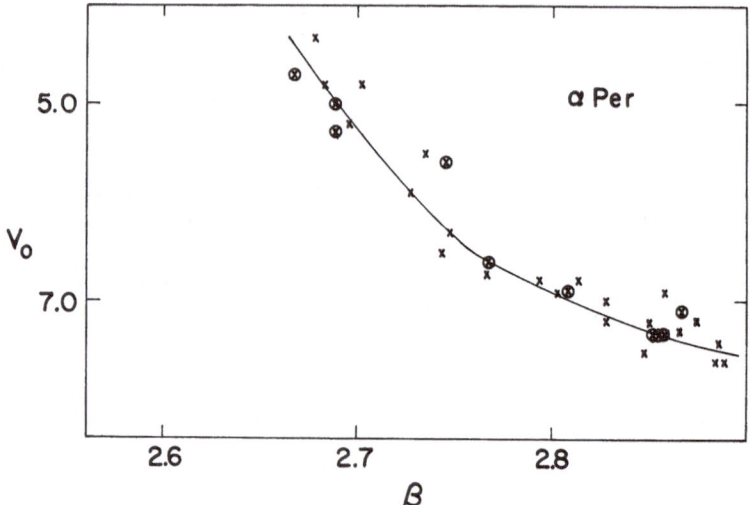

Fig. 13.   As in Figure 12, but for members of the α Per cluster.

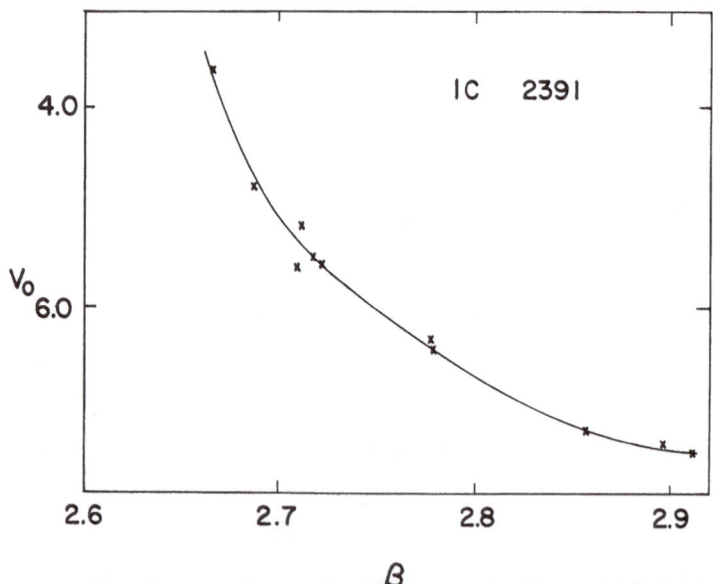

Fig. 14.   The V_0 vs β relation for B-type members of the cluster IC 2391. Data from Perry and Hill (1969). The line is an eye estimate of the best fit to the points.

A 'by-product' of this preliminary calibration is the distance modulus to each of the clusters used in the fitting process. In particular, we find $5^{m}\!.5$ for the Pleiades, $6^{m}\!.2$ for the α Per cluster, and $7^{m}\!.5$ for IC 4665. We will use these values to fix the zero-point for the absolute magnitude calibration for the B-type stars.

(3) Determine the V_0 vs β relation for clusters containing B-type stars. Data for each of the clusters referenced above were used. We show in Figures 12 to 16 the relations for several of the clusters: the Pleiades, α Per, IC 2391, Orion, and NGC 6231.

(4) Overlay these V_0 vs β diagrams, sliding along lines of equal β, so as to determine the best fitting mean relation for all clusters. This procedure is quite similar to that used by Petrie for his Hγ calibration. He had considerably less data to use, however. In doing this overlay, I was impressed that little evolutionary or $V \sin i$ effects appear to be present.

(5) Determine the zero-point for the resultant mean relation (that is, change the V_0 scale to M_V by forcing the distance moduli of the three clusters of Step 2 to agree with the calibration). The resulting, preliminary, calibration is shown in Figure 17, as a smooth line. The symbols show Fernie's (1965) calibration, based on earlier, less complete, data.

The calibration also fits well the points for Sirius and Vega (from trigonometric parallaxes) and for Spica (from the interferometric work of Herbison-Evans *et al.* (1971)).

(6) Check the preliminary calibration for systematic errors to due age differences,

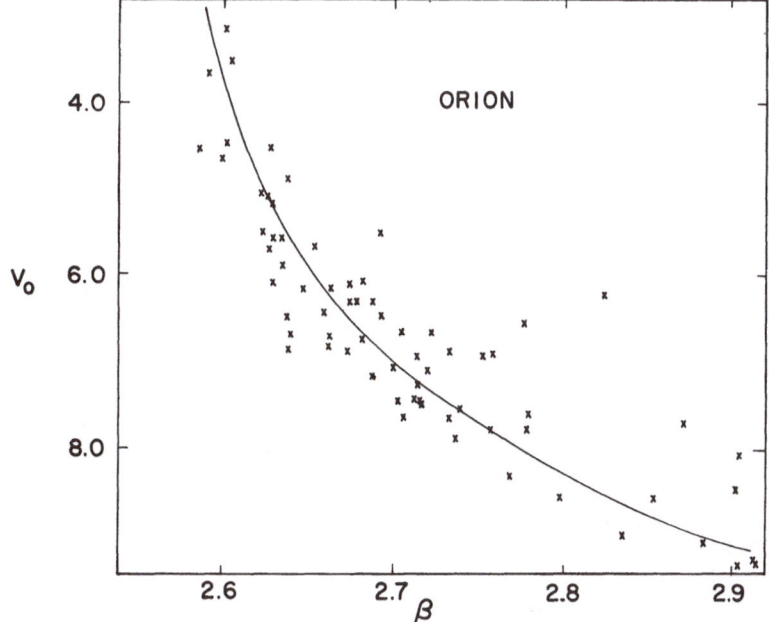

Fig. 15.   The V_0 vs β relation for B-type stars in the Orion association. The line is an eye estimate of the best fit to the points.

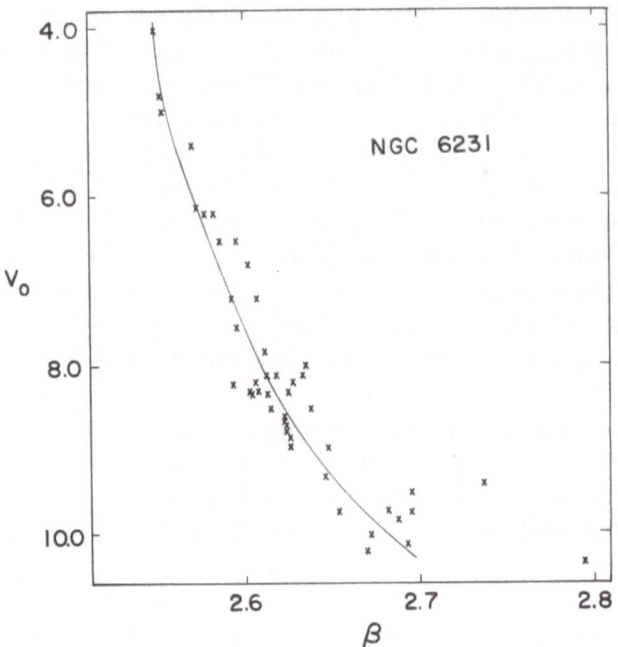

Fig. 16. As in Figure 15, but for members of the cluster NGC 6231.

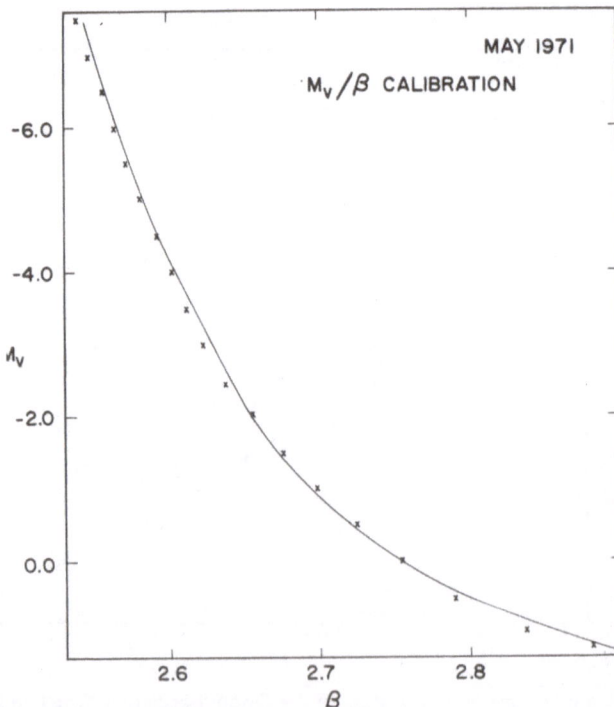

Fig. 17. The preliminary M_V vs β calibration. The crosses indicate the calibrations of Fernie (1965). The preliminary calibration is valid for stars on or near the ZAMS, but should be useful for evolved stars as well (see the text).

rotational velocity effects, frequency-of-binary differences from cluster to cluster, spectral type effects, emission line stars, etc.

In general, we find few significant effects. In particular, there appear to be no systematic effects due to differences in $V \sin i$ from star to star, as Petrie (1965) also concluded from his photographic Hγ work.

Small age effects, or spectral type effects, depending on your point of view, do exist, but they are less than in the cluster fitting techniques previously used, for example, by Johnson (1957) or Blaauw (1963).

The last two figures are propaganda for the hydrogen-line technique and calibration. Figure 18 shows the $(V_0 - M_V)$ vs V_0 relation for the Pleiades, where evolutionary effects are certainly present. In the top part of the diagram, M_V's were determined using Blaauw's M_V (ZAMS) vs $(U - B)_0$ calibration. In the bottom half, the M_V's were determined using the M_V vs β calibration of Figure 17. Figure 19 shows the equivalent diagram for the α Per cluster. Clearly the latter technique is to be preferred (at least, I think so!), especially for field stars.

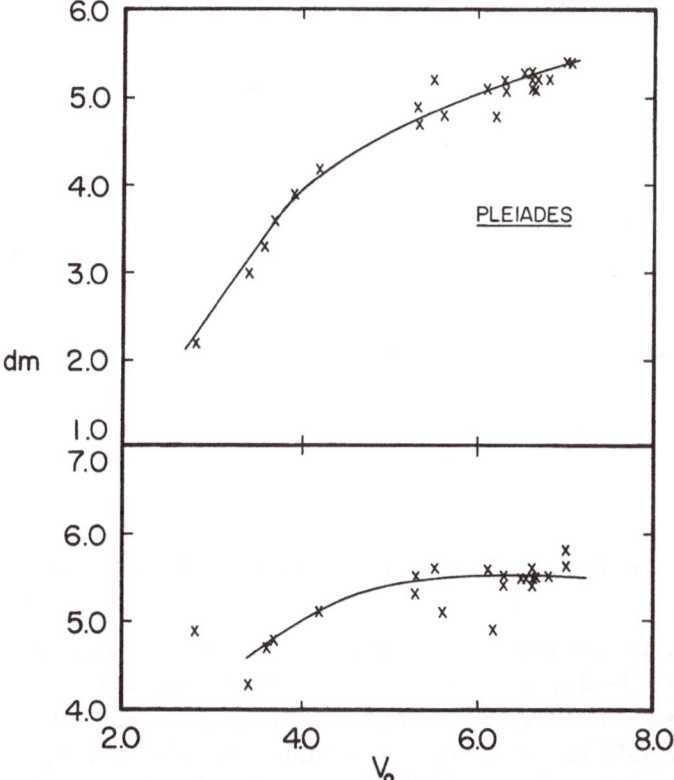

Fig. 18. The calculated distance modulus for B-type members of the Pleiades plotted versus their V_0 magnitude. Points on the top diagram were calculated using Blaauw's (1963) calibrations of M_V (ZAMS) vs $(U - B)_0$; points in the bottom diagram using the calibrations shown in Figure 17. Curvature indicates 'evolutionary effects'. (see text).

D.L.CRAWFORD

The final calibrations, nearly completed, will allow for these evolutionary effects
via a δM_V correction, in terms of a $\delta\beta$ above the ZAMS in the β vs c_0 diagram.

Things remaining to be done before I am willing to label the calibration as final are:

(a) remaining checks for systematic effects,

(b) averages for each MK spectral type,

(c) final determination of the $\delta\beta$ factor as a function of spectral type, and

(d) comparisons to other absolute magnitude calibrations.

I would like to conclude by showing one of the comparisons; Table I summarizes
the comparison. For stars of a given MK spectral type, I have determined the average

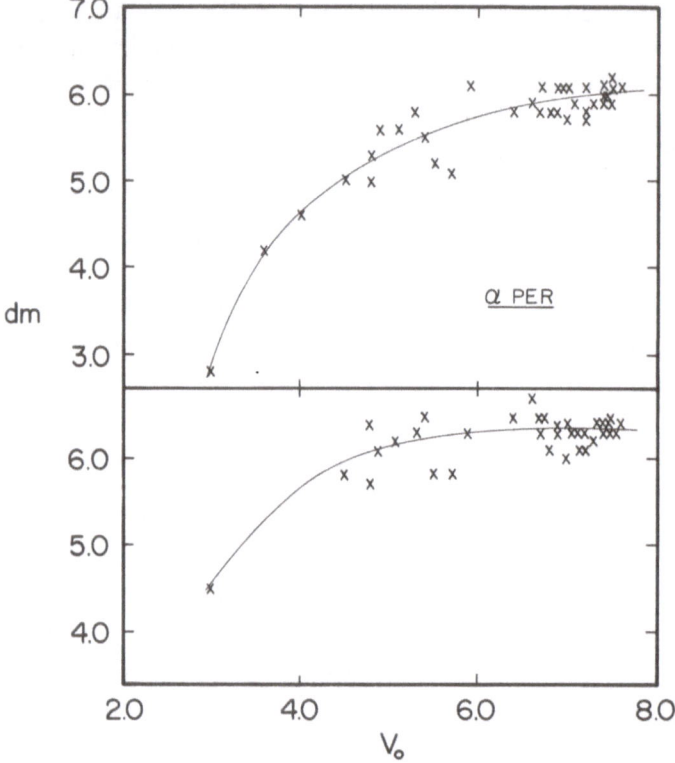

Fig. 19. As in Figure 18, but for members of the α Per cluster.

c_0 (there were about 20 stars in each sub-type). I then read off plots of c_0 vs $(U-B)_0$
the equivalent $(U-B)_0$ for each type. The resultant values agree closely with other
author's average values for each sub-type, for example, Schmidt Kaler's (1965). I also
read off the ZAMS line in the c_0 vs β diagrams the value of β equivalent to each
average c_0. This β then gives us a M_V value from the calibration, valid for the ZAMS.
From Blaauw's M_V (ZAMS) vs $(U-B)_0$ calibrations, I also obtained an M_V value
valid for the ZAMS. Each of these values is given in Table I, and the agreement is
excellent, better than I would have expected, perhaps. In any case, I think one can

TABLE I

Comparison of the β, M_V calibration with Blaauw's $(U-B)_0$,
M_V calibration, both valid for the ZAMS

MK type	c_0	$(U-B)_0$	β(ZAMS)	$M_V(\beta)$	$M_V(U-B)_0$
O9	-0^m12	-1^m10	2^m590	-4^m6	-4^m5
B0	-0.07	-1.05	2.608	-3.9	-3.9
B1	$+0.02$	-0.96	2.629	-2.9	-2.8
B2	$+0.15$	-0.84	2.658	-1.9	-1.8
B3	$+0.33$	-0.67	2.701	-1.0	-0.9
B4	$+0.37$	-0.63	2.709	-0.8	-0.8
B5	$+0.42$	-0.59	2.720	-0.6	-0.6
B6	$+0.48$	-0.55	2.735	-0.3	-0.3

confidently use the (β, M_V) calibration, especially for stars near the ZAMS, and, with care, even for evolved stars.

I hope to have the final calibration done and in press shortly (particularly if the 4-m telescope program goes smoothly this fall and winter!). I would be most happy to receive constructive criticisms both now and after the meeting, before I get the final calibration finished.

Acknowledgments

I wish to thank very much those who have allowed me to use their data in advance of publication. I also wish to thank Dr Bengt Strömgren for his encouragement and advice throughout all phases of the work, Mrs Jeannette Barnes, and Mr John Golson, without whom my research would have withered more than it has during my large telescope adventures, and those many astronomers with whom I have had many enjoyable and profitable discussions about photometry and calibrations.

References

Andrews, P. J.: 1968, *Mem. Roy. Astron. Soc.* **72**, 35.
Anger, C. J.: 1931, *Harvard Circ.* No. 362.
Anger, C. J.: 1932, *Haward Circ.* No. 372.
Bappu, M. K. V., Chandra, S., Sanwall, N. B., and Sinvhal, S. D.: 1962, *Monthly Notices Roy. Astron. Soc.* **123**, 521.
Beer, A.: 1961, *Monthly Notices Roy. Astron. Soc.* **123**, 191.
Beer, A.: 1964, *Monthly Notices Roy. Astron. Soc.* **128**, 261.
Blaauw, A.: 1963, *Stars and Stellar Systems* 3, 383.
Clariá, J. J.: 1971, *Astron. J.* **76**, 639.
Cohen, H. L.: 1969, *Astron. J.* **74**, 1168.
Cowley, A. P. and Crawford, D. L.: 1971, *Publ. Astron. Soc. Pacific* **83**, 296.
Crawford, D. L.: 1958, *Astrophys. J.* **128**, 185.
Crawford, D. L.: 1970, in A. Slettebak (ed.), *Stellar Rotation*, Reidel Publishing Co., Dordrecht, p. 204.
Crawford, D. L. and Barnes, J. V.: 1966, *Astron. J.* **71**, 610.
Crawford, D. L. and Barnes, J. V.: 1969a, *Astron. J.* **74**, 407.
Crawford, D. L. and Barnes, J. V.: 1969b, *Astron. J.* **74**, 818.
Crawford, D. L. and Barnes, J. V.: 1970a, *Astron. J.* **75**, 946.
Crawford, D. L. and Barnes, J. V.: 1970b, *Astron. J.* **75**, 952.

Crawford, D. L. and Barnes, J. V.: 1970c, *Astron. J.* **75**, 978.
Crawford, D. L. and Barnes, J. V.: 1972, *Astron. J.* **77**, 862.
Crawford, D. L., Barnes, J. V., Faure, B. Q., Golson, J. C., and Perry, C. L.: 1966, *Astron. J.* **71**, 709.
Crawford, D. L., Barnes, J. V., Gibson, J., Golson, J. C., Perry, C. L., and Crawford, M. L.: 1972, *Astron. Astrophys. Suppl.* **5**, 109.
Crawford, D. L., Barnes, J. V., and Golson, J. C.: 1970, *Astron. J.* **75**, 624.
Crawford, D. L., Barnes, J. V., and Golson, J. C.: 1971a, *Astron. J.* **76**, 621.
Crawford, D. L., Barnes, J. V., and Golson, J. C.: 1971b, *Astron. J.* **76**, 1058.
Crawford, D. L., Barnes, J. V., Hill, G., and Perry, C. L.: 1971, *Astron. J.* **76**, 1048.
Crawford, D. L., Glaspey, J. W., and Perry, C. L.: 1970, *Astron. J.* **75**, 822.
Crawford, D. L. and Mander, J.: 1966, *Astron. J.* **71**, 114.
Crawford, D. L. and Perry, C. L.: 1966, *Astron. J.* **71**, 206.
Fernie, J. D.: 1965, *Astron. J.* **70**, 575.
FitzGerald, M. P.: 1970, *Astron. Astrophys.* **4**, 234.
Furenlid, I.: 1971, *Astron. Astrophys.* **10**, 321.
Garrison, R. F.: 1967, *Astrophys. J.* **147**, 1003.
Glaspey, J. W.: 1971, *Astron. J.* **76**, 1041.
Graham, J. A.: 1967, *Monthly Notices Roy. Astron. Soc.* **135**, 377.
Hack, M.: 1953, *Ann. Astrophys.* **16**, 417.
Hardie, R. H. and Crawford, D. L.: 1961, *Astrophys. J.* **133**, 843.
Herbison-Evans, R., Hanbury Brown, R., Davis, J., and Allen, L. R.: 1971, *Monthly Notices Roy. Astron. Soc.* **151**, 161.
Hill, G. and Perry, C. L.: 1969, *Astron. J.* **74**, 1011.
Hoag, A. A. and Applequist, N. L.: 1965, *Astrophys. J. Suppl.* **12**, 215.
Jenkins, L. F.: 1963, *General Catalogue of Trigonometric Stellar Parallaxes*, Yale University Observatory, New Haven.
Johnson, H. L.: 1957, *Astrophys. J.* **126**, 121.
Johnson, H. L. and Iriarte, B.: 1958, *Lowell Obs. Bull.* **4**, 47.
Johnson, H. L. and Morgan, W. W.: 1953, *Astrophys. J.* **117**, 313.
Kopylov, I. M.: 1958, *Izv. Krymsk. Astrofiz. Obs.* **20**, 156.
Lindblad, B.: 1922, *Astrophys. J.* **55**, 85.
Lindblad, B.: 1925, *Nova Acta Rej. Soc. Sci. Upsala, Ser.* 4, **6**, No. 5.
Lindblad, B.: 1926, *Medd. Astr. Obs. Uppsala*, No. 11.
Mihalas, D.: 1972, *Astrophys. J.* **176**, 139.
Öhman, Y.: 1935, *Stockholm Obs. Ann.* **12**, No. 1.
Perry, C. L. and Hill, G.: 1969, *Astron. J.* **74**, 899.
Petrie, R. M.: 1950, *Publ. Dominion Astrophys. Obs.* **8**, 319.
Petrie, R. M.: 1953, *Publ. Dominion Astrophys. Obs.* **9**, 251.
Petrie, R. M.: 1956, *Vistas in Astronomy* **2**, 1346.
Petrie, R. M.: 1965, *Publ. Dominion Astrophys. Obs.* **12**, 317.
Petrie, R. M. and Maunsell, C. D.: 1950, *Publ. Dominion Astrophys. Obs.* **8**, 253.
Petrie, R. M. and Moyls, B. N.: 1956, *Publ. Dominion Astrophys. Obs.* **10**, 287.
Schmidt-Kaler, Th.: 1965, in K. H. Hellweg (ed.), *Landolt-Börnstein*, Springer-Verlag, Berlin and New York, New Ser., Group 6, Vol. 1, p. 284.
Sinnerstad, U.: 1954, *Stockholm Obs. Medd.*, No. 82.
Sinnerstad, U.: 1961a, *Stockholm Obs. Ann.* **21**, No. 6.
Sinnerstad, U.: 1961b, *Stockholm Obs. Ann.* **22**, No. 2.
Strom, K. M., Strom, S. E., and Yost, J.: 1971, *Astrophys. J.* **165**, 479.
Strömgren, B.: 1951, *Astron. J.* **56**, 142.
Strömgren, B.: 1952, *Astron. J.* **57**, 200.
Strömgren, B.: 1956a, in J. Neyman (ed.), *Proceedings of the Third Berkeley Symposium*, University of California Press, Berkeley and Los Angeles, Vol. 3, p. 49.
Strömgren, B.: 1956b, *Vistas in Astronomy* **2**, 1336.
Strömgren, B.: 1958, in D. J. K. O'Connell (ed.), *Stellar Populations*, Interscience Publishers, New York, p. 385.
Strömgren, B.: 1963, *Stars and Stellar Systems* **3**, 123.
Strömgren, B.: 1966, *Ann. Rev. Astron. Astrophys.* **4**, 433.

Strömgren, B. and Perry, C.: 1965, unpublished report, Institute for Advanced Study, Princeton, N. J.
Williams, E. G.: 1936, *Astrophys. J.* **83**, 279.

DISCUSSION

Crampton: In the case of known binary stars did you make correction for duplicity?

Crawford: We did not make any correction for duplicity. The binaries are generally included in the discussion.

Blaauw: How large would you estimate the probable error of the main sequence fit to the trigonometric parallax stars, i.e., the p.e. of the zero point of the newly derived M_V system?

Crawford: Something like 0^m1 or 0^m2.

Wesselink: Could you use your M_V vs β technique to the non emission (apparently fainter) B stars in the Magellanic Clouds with consequent result for the distance modulus?

Crawford: Yes, we are observing just such stars at the present time.

Schmidt-Kaler: You showed a diagram Hα vs Hβ with quite a few emission B stars, and a diagram Hβ vs Hγ with very few. Did you put the same stars in both diagrams? Or does this mean that you find a discontinuous Balmer jump?

Crawford: Most of the stars are the same. Many more stars show emission at Hα than at Hβ or Hγ.

Jaschek: Did you observe in Orion stars which show helium line anomalies? Did you exclude B-type peculiar stars or, in general, peculiar stars?

Crawford: We observed a few of the stars in Orion that you refer to. Some look odd in our photometry, some do not. In general, we include peculiar stars in our work. The Am stars fit the calibrations for the A stars quite well. Most Ap stars look like B stars to me.

Maeder: With regard to the position of the Of stars in some of your diagrams one should note than, according to Walborn, the f-characteristics may be identified with a luminosity effect. One may show that the different position of the Of stars in $(U - B)$ vs $(B - V)$ diagram is in complete agreement with the luminosity effect predicted by the recent non-LTE models of Auer and Mihalas. This may be considered as a supplementary support to the hypothesis that the Of stars are intrinsically brighter that the so-called normal O-type stars. The observed difference is not due to the contribution of the emission lines in the filters ($< 0^m003$) but to a change produced by the luminosity effect in the energy distribution.

Blaauw: Do stars in the Taurus stream, which is associated with the Hyades, behave similar to the Hyades proper in the c_1 vs β diagram?

Crawford: Yes, most of them look like stars in the Hyades cluster, according to Eggen.

Jones: I find that a back-warming correction is required in β when comparing stars of very different metal abundance. It amounts to roughly 0^m05 between stars of 0.01 the solar abundance and those of solar abundance.

Murray: Could the Hyades discrepancy be accounted for by slight differences in the proper motion system depending on apparent magnitude, leading to systematic differences in the absolute magnitudes, also depending on apparent magnitude?

Crawford: I don't think so; if anything, the photometry would indicate that there are no such difficulties.

Garrison: Isn't it disturbing that the two clusters which you find do not fit are the only old, rich clusters?

Crawford: Yes, however, the statistics of small numbers allow all sorts of puzzles.

Hauck: I just want to mention that all published measurements in the $uvby\ \beta$ system have been compiled here by Lindemann and myself. It is possible to obtain the tape at the Centre de Données Stellaires in Strasbourg. Now we have 6000 stars and we hope to publish as soon as possible a general catalogue with homogeneous data.

A PRELIMINARY TEST OF THE CALIBRATION OF THE Hβ SYSTEM IN TERMS OF ABSOLUTE MAGNITUDES AND INTRINSIC COLOURS FOR EARLY-TYPE STARS

A. F. J. MOFFAT, Th. SCHMIDT-KALER, and N. VOGT

Astronomisches Institut der Ruhr-Universität Bochum, F.R.G.

Abstract. The dependency of the Hβ-index on *two* physical parameters is briefly discussed. The previous calibration of $M_v(\beta)$ (Fernie, 1965 or Crawford, 1971) is confirmed for main sequence O and B stars; however, the β-values of the later B-supergiants tend to be too large for their absolute magnitudes. The problem of filtering out stars with marginal emission is emphasized.

The existing calibrations of the Hβ-index as a function of absolute magnitude M_v (Graham, 1967; Fernie, 1965; Crawford, 1971) neglect the fact that β depends on at least two parameters; e.g. temperature and pressure. In addition, certain areas of the Hertzsprung-Russell-diagram show discrepancies; for example the β-values of A-type supergiants are too large for their absolute magnitudes (see, e.g., Figure 1).

The fundamental objective of Hβ-measurements is the determination of distances especially of field stars with spectral types O, B, A. For this purpose photometry in at least three colours is also necessary (e.g. *UBV*). Since the relations

$$\beta\text{-index} \qquad \beta = f_1\left(M_v, T_{\text{eff}}\right),$$
$$\text{intrinsic colour } C_0 = f_2\left(M_v, T_{\text{eff}}\right)$$

hold, it is clear that the final calibration must be carried out *simultaneously* in terms of the photometric quantities *UBV* and β if the spectral type of the star is not known. Since the first equation has two solutions of M_v for a given β it is necessary to distinguish between O–B9 and A0–F stars. Also stars with shells show emission lines, and must be considered separately. A calibration system which satisfies the above conditions is in preparation.

Here we wish only to present a preliminary test of Crawford's (1971) calibration $M_v(\beta)$. Fernie's (1965) calibration is practically equivalent to Crawford's. For this we have taken observations of stars in the following young open clusters: the nuclei of h and χ Persei (Crawford, Glaspey and Perry, 1970), III Cep (Crawford and Barnes, 1970), I Sco (Crawford *et al.*, 1971) and our observations of 60 stars in 30 southern clusters. This large number of clusters has the advantage that systematic errors which may arise in individual cases are randomized. In all cases the distance moduli have been determined *independently* from Hβ using *UBV* photometry. For h and χ Persei, III Cep and I Sco, the values $V_0 - M_v = 11.7$, 9.3 and 11.5, respectively, have been used; for the 30 southern clusters the distance moduli have been derived by Moffat and Vogt (1973) and Vogt and Moffat (1972, 1973). All data have been reduced to $R = A_v/E_{B-V} = 3.2$ for A0-stars. The internal errors of our Hβ-values are $\pm 0^{\text{m}}.008$ for $V < 10^{\text{m}}$; external errors are less than $0^{\text{m}}.005$. The results are presented in Figure 1.

B. Hauck and B. E. Westerlund (eds.), Problems of Calibration of Absolute Magnitudes and Temperature of Stars, 114–116.

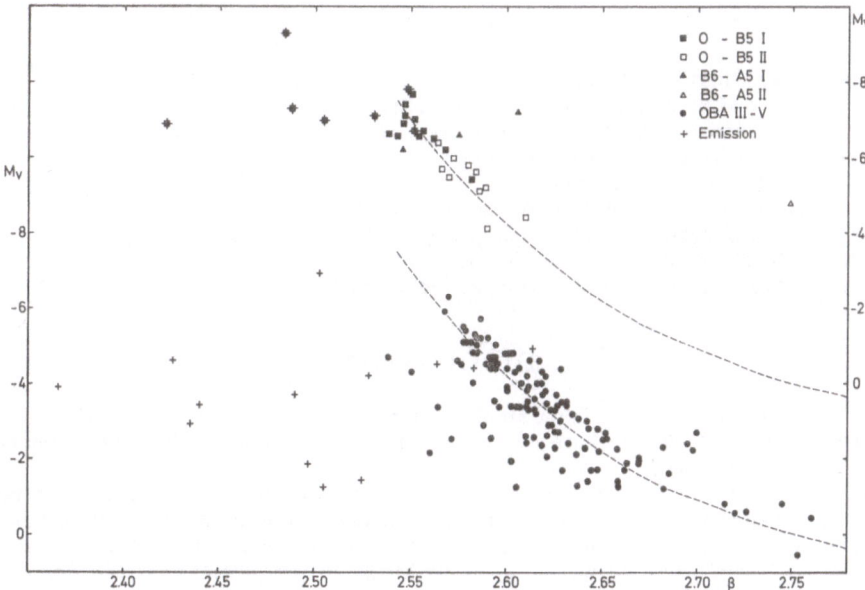

Fig. 1.   The calibration of the visual absolute magnitude M_v as a function of the β-index. The diagram is separated into two parts: supergiants and class III–V stars. The dotted curves represent the calibration of Crawford (1971).

The known emission B-stars (shown by crosses) occupy the region to the left of the calibration curve, as expected. The frequency distribution about the calibration curve suggests two components: a symmetrical component centered on the curve, and an asymmetrical component corresponding to the emission stars. Therefore, in computing the dispersion we have omitted nine stars in addition to the spectroscopically known emission stars. The resulting scatter is

Range	Spectraltype	σ_{M_v}	Number
$\beta < 2.61$	O–B5 I–II	$\pm 0^m53$	22
$\beta \leqslant 2.63$	O–B1 III–V	± 0.58	82
$2.63 < \beta < 2.68$	B2–5 III–V	± 0.54	27

Supergiant emission B-stars tend to be 0^m5 brighter than non-emission stars of the same spectral type. Taking into account the errors of the distance moduli and the interstellar absorption, the scatter is reduced to 0^m4. This value includes the effects of observational errors, rotation and spectroscopic binaries.

We confirm Crawford's and Fernie's $M_v(\beta)$ calibrations. Stars with $\beta < 2.54$ are certainly emission objects; emission stars between this line and the calibration curve clearly present a problem when they are field OB-stars in case no sufficient spectroscopic information is available.

References

Crawford, D. L., Glaspey, J. W., and Perry, C. L.: 1970, *Astron. J.* **75**, 822.
Crawford, D. L. and Barnes, J. V.: 1970, *Astron. J.* **75**, 946.
Crawford, D. L., Barnes, J. V., Hill, G., and Perry, C. L.: 1971, *Astron. J.* **76**, 1048.
Crawford, D. L.: 1971, private communication.
Fernie, J. D.: 1965, *Astron. J.* **70**, 575.
Graham, J. A.: 1967, *Monthly Notices Roy. Astron. Soc.* **135**, 377.
Moffat, A. F. J. and Vogt, N.: 1973, *Astron. Astrophys. Suppl.*, in press.
Vogt, N. and Moffat, A. F. J.: 1972, *Astron. Astrophys. Suppl.* **7**, 133.
Vogt, N. and Moffat, A. F. J.: 1973, *Astron. Astrophys. Suppl.* **9**, 97.

DISCUSSION

Schild: Your value for net scatter in the absolute magnitude calibration of ± 0.4 is strikingly small; from the statistics of double stars, have you been able to assess the contribution to this scatter caused by spectroscopic binaries?

Schmidt-Kaler: We did not correct for binaries. As to their percentage I might quote Petrie's result that around half of the early type B stars are binaries. A large proportion of the companions would, however, be too faint to affect appreciably the calibration. On the other hand we do not want to correct for binaries since we wish to apply the results to very distant and very faint clusters.

Crampton: It wasn't clear to me whether your results are based only on the three northern clusters or on more extensive data?

Schmidt-Kaler: We have used the MK-classified stars only of the clusters observed by Crawford *et al.* as a skeleton, the rest is from 30 southern open clusters which had thus far never been observed photometrically.

THREE-DIMENSIONAL REPRESENTATION OF A0–G5 STARS

B. HAUCK

Institut d'Astronomie de l'Université de Lausanne et Observatoire de Genève, Suisse

Abstract. It is possible to obtain for the stars of the spectral type included between A0 and G5 three parameters respectively correlated with the effective temperature, the luminosity and the blanketing. A method to determine the absolute magnitude is given.

Résumé. Il est possible d'obtenir pour les étoiles de type spectral compris entre A0 et G5 trois paramètres corrélés respectivement avec la température effective, la luminosité et le blanketing. Une méthode de détermination de la magnitude absolue est donnée.

1. Introduction

Dans un article précédent (Hauck, 1968) nous avions montré qu'il était possible d'obtenir une représentation tri-dimensionnelle des étoiles A0–G5. A la suite de l'étude de l'effet du rougissement interstellaire par Goy (1971), le coefficient rendant les paramètres indépendants du rougissement a été modifié. Dans le cas qui nous intéresse, ces paramètres sont:

$$B_2 - V_1 \qquad \text{température effective}$$
$$d = (U - B_1) - 1.430 (B_1 - B_2) \qquad \text{magnitude absolue}$$
$$m_2 = (B_1 - B_2) - 0.457 (B_2 - V_1) \qquad \text{blanketing}$$

TABLEAU I
Séquences de référence

Type	$B_2 - V_1$	d	M_v	p_0	m_2
A 2	−0.100	1.390	1.80	6.0	−0.503
A 3	−0.075	1.360	1.88	6.0	−0.482
A 4	−0.065	1.350	1.93	6.0	−0.473
A 4	−0.050	1.325	1.98	6.0	−0.467
A 5	−0.025	1.280	2.05	6.0	−0.461
A 6	0.000	1.230	2.15	6.0	−0.461
A 8	0.050	1.148	2.50	6.0	−0.466
A 9	0.075	1.098	2.75	6.0	−0.470
F 0	0.100	1.040	2.95	6.0	−0.476
F 2	0.150	0.930	3.30	6.0	−0.478
F 3	0.175	0.880	3.45	7.5	−0.476
F 4	0.200	0.825	3.60	10.0	−0.473
F 5	0.218	0.795	3.75	11.0	−0.468
F 6	0.250	0.740	4.00	12.5	−0.460
F 8	0.300	0.670	4.35	15.0	−0.428
G 0	0.350	0.600	4.80	17.0	−0.390
G 2	0.400	0.530	5.40	18.5	−0.352
G 5	0.450	0.475	5.75	20.0	−0.313
G 8	0.500	0.410	6.15	21.5	−0.275
K 4	0.700	–	–	–	−0.119

B. Hauck and B. E. Westerlund (eds.), Problems of Calibration of Absolute Magnitudes and Temperature of Stars, 117–119.
All Rights Reserved. Copyright © 1973 by the IAU.

Il va sans dire que ces modifications apportées à la définition des paramètres ont nécessité une nouvelle calibration des séquences de référence (Tableau II.5 de l'article de 1968) et des corrections à apporter aux différents paramètres pour tenir compte, suivant le cas, des effets résiduels de blanketing ou de luminosité.

Les nouvelles séquences de référence sont données dans le Tableau I.

2. Paramètre de température

L'indice de couleur $B_2 - V_1$ peut être utilisé dans cet intervalle de types spectraux comme paramètre de température. Pour les étoiles des Hyades, nous avons établi la corrélation entre les valeurs θ_{eff} données par Oke et Conti (1966) et nous avons obtenu

$$\theta_{\text{eff}} = 0.727 (B_2 - V_1) + 0.649$$
$$\pm 0.002 \qquad \pm 0.017 .$$

3. Calibration en magnitude absolue

Pour la calibration en magnitude absolue, nous avons utilisé les magnitudes absolues de qualité A, B ou C catalogue de Gliese (1969) et celles des Hyades (V de Johnson et Knuckles 1955, module de distance de Heckmann et Johnson 1956). Connaissant les séquences de références dans les diagrammes d vs $B_2 - V_1$ et M_v vs $B_2 - V_1$, il est alors possible de calculer la magnitude absolue selon une méthode semblable à celle proposée par Strömgren (1963) pour le système $uvby$, soit

(a) déterminer d_0 (séquence de référence) pour la valeur $B_2 - V_1$ (avec la correction éventuelle de blanketing, cf. 5) de l'étoile à partir du diagramme d vs $B_2 - V_1$;

(b) calcul de $\Delta d = d - d_0$ (d avec la correction éventuelle de blanketing, cf. 5);

(c) détermination de $p_0 = (\Delta M_v / \Delta d)_0$ pour la valeur $B_2 - V_1$ envisagée;

(d) calcul de $\Delta M_v = \Delta d [p_0 + 20 \Delta d]$;

(e) détermination de $(M_v)_0$ d'après le diagramme M_v vs $B_2 - V_1$;

(f) calcul de $M_v = (M_v)_0 - \Delta M_v$.

Pour les étoiles de classe de luminosité V ou IV la précision est de l'ordre de ± 0.15 mag. tandis qu'elle est un peu plus faible pour celles de classe III.

4. Paramètre de blanketing

Nous avions défini la grandeur $\Delta m_2 = m_2$ (étoile) $- m_2$ (Hyades) comme significative de l'effet de blanketing.

La relation entre [Fe H] et Δm_2 est établie à partir des étoiles de Wallerstein (1961). Elle est valable pour des étoiles de type spectral compris entre F8 et G2. Avec la nouvelle définition de m_2 nous avons obtenu:

$$[\text{Fe H}] = \quad 6.830 \Delta m_2 + 0.203$$
$$\pm 0.16 \quad \pm 0.767 \qquad \pm 0.097 .$$

5. Effets résiduels de blanketing

Pour les étoiles ayant un type spectral plus tardif que F5, ou une valeur de $B_2 - V_1$ plus grande que 0.230, il faut tenir compte d'un effet résiduel de blanketing sur $B_2 - V_1$ et sur d.

L'effet résiduel sur $B_2 - V_1$ ne se produit que lorsque $\Delta m_2 \leqslant -0.060$ et il a alors pour valeur

$$\Delta (B_2 - V_1) = 1.20 (\Delta m_2^* + 0.060)$$

Il faut tenir compte de cet effet avant de calculer la magnitude absolue et Δm_2. La valeur Δm_2^* est donc celle obtenue avant la correction.

L'effet résiduel sur d a pour valeur

$$\Delta d = -0.4 \Delta m_2 \quad \text{pour} \quad \Delta m_2 \geqslant -0.060$$

et

$$\Delta d = -1.1 \Delta (B_2 - V_1) - 0.024 \quad \text{pour} \quad \Delta m_2 < -0.060.$$

6. Effet résiduel de luminosité

Pour les étoiles ayant un type spectral plus précoce que F5, ou une valeur de $B_2 - V_1$ plus petite que 0.230, il faut tenir compte d'un effet de luminosité sur le paramètre de blanketing ayant pour valeur

$$\Delta m_2 = -0.20 \Delta d$$

Références

Gliese, W.: 1969, *Veröffentl. Astron. Rechen-Inst. Heidelberg*, No. 22.
Goy, G.: 1971, *Publ. Obs. Genève*, No. 78.
Hauck, B.: 1968, *Publ. Obs. Genève*, No. 75.
Heckmann, O. et Johnson, H. L.: 1956, *Astrophys. J.* **124**, 477.
Johnson, H. L. et Knuckles, C. F.: 1955, *Astrophys. J.* **122**, 209.
Oke, J. B.: 1957, *Astrophys. J.* **126**, 509.
Oke, J. B.: 1959, *Astrophys. J.* **130**, 487.
Oke, J. B. et Conti, P. S.: 1966, *Astrophys. J.* **143**, 134.
Roman, N.: 1952, *Astrophys. J.* **116**, 122.
Rufener, F.: 1971, *Astron. Astrophys. Suppl. Ser.* **4**, 43.
Strömgren, B.: 1963, *Stars and Stellar Systems* **3**, 123.

THE Hγ-ABSOLUTE MAGNITUDE CALIBRATION

D. CRAMPTON

Dominion Astrophysical Observatory Victoria, B.C. Canada

Abstract. A new calibration of Hγ equivalent widths is presented. Unlike Petrie's method, the new calibration does not require an estimate of the luminosity class or spectral type of the star. A brief comparison with other calibrations is made.

This paper briefly describes an investigation of the Hγ calibration recently completed in collaboration with Mr L. Balona of Witwatersrand University.

Although Petrie's (1965) calibration of Hγ equivalent widths appeared to give cluster distance moduli in good agreement with the photometrically-determined moduli, Petrie himself noted that there was a difference of $+1.3$ mag, between his calibration and the MK (Blaauw, 1963) calibration for O6–8 stars and $+0.6$ mag. for O9–B2 main sequence stars. These differences led us to re-examine Petrie's calibration and to reconstruct a new calibration based on more extensive data.

We used Petrie's Hγ equivalent widths (W) of stars in the Pleiades, α Per, Ori OB1 and Per OB1, supplemented by our measurements of W in NGC 3293, NGC 4755, IC 2944, Sco-Cen and h Per. The Sco-Cen data provides a better overlap between the α Per and Ori OB1 sequences; the extension of the observations in h Per to fainter stars and the addition of NGC 3293 and NGC 4755 strengthened the early B and supergiant calibrations and the substitution of IC 2944 for Mon OB1 (used by Petrie) improved the O star calibration. The calibration was carried out independently by each of us using slightly different data; L. Balona (L. B.) preferred to correct each star individually for absorption while D. Crampton (D. C.) used mean cluster absorption wherever possible ($R=3$ was assumed). The Sco-Cen data was not used by D. C either.

Petrie demonstrated that the relationship between the absolute magnitude, M, and W is little affected by stellar rotation or moderate age effects so that M is primarily a function of W with a slight dependence on the spectral type, S. Instead of adopting, as Petrie did, 'spectral-type corrections' which differ for class V–II stars and for class I stars we adopted a method which allows a smooth variation with luminosity class, L, and substituted W itself as a quantitative estimate of L. In each cluster curves of constant S were drawn in the $V_0 - W$ diagram and V_0 was changed relative to the Pleiades to ensure the best continuation of the curves. This gives the distance moduli with respect to the Pleiades for each cluster and we decided to adopt 5.55 mag. as the distance modulus of the Pleiades. The resulting distance moduli are given in the second column of Table I.

For $W < 16$ it was found empirically that the interpolation formula

$$M = a + b \log_{10} W$$

fitted the data for a given S very well. We experienced considerable difficulty, however,

B. Hauck and B. E. Westerlund (eds.), Problems of Calibration of Absolute Magnitudes and Temperature of Stars, 120–123.

TABLE I

Distance moduli determined for clusters
used in the calibration

Cluster	Distance modulus from fitting procedure	Distance modulus from calibration	No. of stars
Pleiades	5.55	5.53 ± 0.08	17
α Per	6.20	6.21 ± 0.07	27
Sco-Cen	6.45	6.49 ± 0.07	65
Ori OB1	8.15	8.12 ± 0.07	43
h & χ Per	11.45	11.46 ± 0.07	56
NGC 4755	11.85	11.75 ± 0.08	22
NGC 3293	12.35	12.32 ± 0.09	23
IC 2944	12.00	12.04 ± 0.11	21

in attempting to fit a smooth continuous curve through the variation of the coefficients a and b with S. Two solutions were adopted; (1) two curves were derived for
a

$$\text{Viz; } a = -21.82 + 1.949S - 0.0696S^2 \quad S > 14$$
$$a = -9.28 + 0.257S - 0.0133S^2 \quad S \leqslant 14$$
$$b = 10.40 - 0.537S + 0.026S^2,$$

where $S = 6$ at O6 and increase by unity for each later spectral subclass reaching
$S = 23$ at A3 and (2) $(B-V)_0$ was substituted for S and an independent relation
between M and W was found:

$$M_v = -9.72 - 7.18(B-V)_0 + 9.88 \log_{10} W + 9.66(B-V)_0 \log_{10} W$$
$$\begin{pmatrix} W < 14A \\ (B-V)_0 < 0 \end{pmatrix}.$$

The two calibrations are, of course, not identical but the differences are small,
particularly for main sequence stars. The mean difference between calibrations 1 and
2 is -0.04 mag. with a dispersion of 0.2 mag. It should be noted that for most stars
$(B-V)_0$ can be derived from the UBV colours alone and then no knowledge of the
spectral type is required to use calibration 2.

The mean distance moduli, derived from the two calibrations, of the clusters used
in the calibration are listed the third column of Table I. The distance moduli are in
good agreement with those from other recent determinations with the exception of
Sco-Cen for which a value of 6.0 is more usual. There is considerable scatter among
the distance moduli of the individual members of this group but the reason for the
difference between our determination and the others is not understood. The average
dispersion in absolute magnitude about the calibration ranges from 0.3 mag. for late
B main sequence stars to 0.6 mag. for O, early B and supergiant stars; the mean dispersion for all stars is 0.5 mag.

The difference between the absolute magnitude derived from Petrie's calibration
and that from calibration 1 is shown plotted as a function of spectral type in Figure 1.

D. CRAMPTON

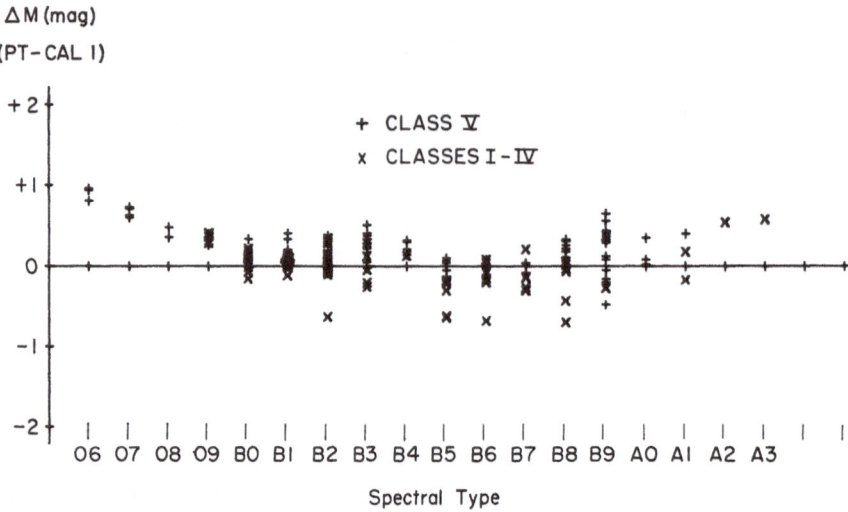

Fig. 1.   Difference between the absolute magnitude derived from Petrie's calibration and calibration
1 vs spectral type.

It may be readily seen that the absolute magnitudes derived from the new calibration are up to 1 mag. brighter for the O stars but the agreement is surprisingly good for all other spectral types.

A preliminary comparison with the MK calibration tabulated by Blaauw (1963) based on a large number of stars with measured W indicates that our calibration is (1) ~0.4 mag. fainter than the MK calibration for class III–V stars earlier than B2, (2) about −0.3 mag. brighter for B3–6 class V stars, and (3) brighter for the super-giants earlier than B3 (our calibration gives $M_v \sim -7.0$ for all B0–A1 class Ia stars). We are in excellent agreement with the recent calibration given by Walborn (1972) for stars of spectral type earlier than B3.

In summary, we have improved Petrie's calibration, particularly for the O stars, and have devised a new method of calibration which does not require an estimate of the luminosity class of the star or its spectral type in order to determine its absolute magnitude.

References

Blaauw, A.: 1963, *Stars and Stellar Systems* 3, 383.
Petrie, R. M.: 1965, *Publ. Dominion Astrophys. Obs.* 12, 317.
Walborn, N. R.: 1972, *Astron. J.* 77, 312.

DISCUSSION

Van den Bergh: Could the grey absorbing shells that Strom and others have recently found around *faint* young stars have affected the main sequence fitting used for your calibration of bright stars in very young clusters? In particular might this effect account for the observed discrepancy between the new distance modulus of Scorpio-Centaurus and previous values obtained from proper motions?

Crampton: In most of the clusters we used this effect is probably negligible. However, the systematic error in Petrie's calibration for the O stars may be partly due to such an effect in NGC 2244. I would not expect the effect to be large in the stars we observed in Scorpio-Centaurus.

Newell: Are all your spectra exposed for radial velocities and, if so, are they not too dense for the determination of spectrophotometric quantities?

Crampton: Hγ measurements are only made on spectra which are, in fact, suitable for spectrophotometry but, in general, the plates are not too dense.

Jones: Could you comment on the fact that while you place Scorpio-Centaurus 0ᵐ65 further away than I do, you find corrections to main-sequence B stars in exact agreement?

Crampton: The reason for this is not immediately obvious, however, members of the Scorpio-Centaurus association have very little weight in the determination of the mean absolute magnitude.

CALIBRATION OF MK TYPES BY FITTING
THE HR DIAGRAMS OF THREE MOVING CLUSTERS

R. F. GARRISON

David Dunlap Observatory, University of Toronto, Canada

Abstract. A composite HR diagram constructed from the main sequences of three moving clusters (Hyades, α Persei, and the Inner Region of Upper Scorpius) using MK classifications, is presented and discussed.

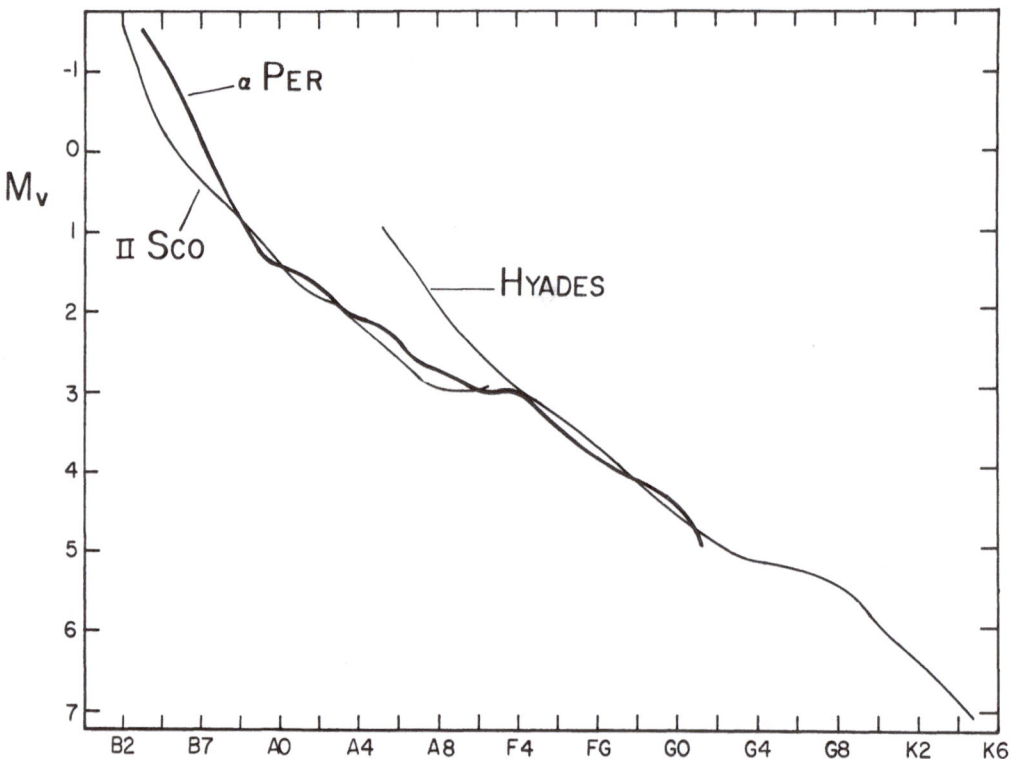

Fig. 1. Composite HR diagram for three moving clusters. The sequence labelled II Sco is for the Inner Region of Upper Scorpius.

Several years ago, Hiltner, Schild and I began a photometric (UBV) and spectroscopic (MK) study of galactic structure in the southern hemisphere. We have now completed the observations for all the OB stars south of $-30°$ and brighter than tenth magnitude which are listed in the Heidelberg objective prism catalogue (Klare and Szeidl, 1966).

Before these data can be used to study spiral structure, it is necessary to choose an absolute magnitude calibration. As a preliminary step, it is interesting to use the

B. Hauck and B. E. Westerlund (eds.), Problems of Calibration of Absolute Magnitudes and Temperature of Stars, 124–125.

cluster-fitting procedure for clusters whose distance moduli can be independently determined and for which reliable MK classifications are available. Three clusters which fit these criteria are the Hyades, the α Persei cluster and the Inner Region of Upper Scorpius. The reason for choosing the Inner Region of Upper Scorpius instead of the entire Scorpio-Centaurus Association is that the former has a well-determined, narrow main sequence extending to F0. It is fortunate that the α Persei cluster bridges the age gap between the Hyades and Scorpius and provides a good overlap with both main sequences.

The MK types for the Hyades are from Morgan and Hiltner (1965); for the α Persei cluster, they are from Morgan *et al.* (1971); and for the Inner Region of Upper Scorpius, they are from Garrison (1967). Peculiar stars and extremely rapid rotators have been excluded and the *V* magnitudes of double stars have been corrected. For the remaining stars in each cluster, a mean absolute magnitude was determined at each spectral type. The internal errors in the resultant composite diagram are quite small, as can be seen from the degree of agreement of the lines; when the main sequences are compared over the entire range where they overlap, it is obvious that a shift of a few tenths of a magnitude would result in a poorer fit.

In the area of overlap, there are no significant differences in rotation or in metal abundance. The suggestion was made yesterday that differences in metal abundance would cause a shift in the HR diagram. This is true for colours, but not for carefully determined spectal types. In the two dimensional MK system, differences in metal abundance would be apparent as peculiarities because the types obtained from the hydrogen lines, from the calcium lines and from the metallic lines would then not be consistent. Thus it is unlikely that a star would be misclassified earlier or later because of an abundance difference.

If it is assumed that the distance modulus of the Hyades is $3\overset{m}{.}0$, then the distance modulus of the α Persei cluster is 6.1 and that for the Inner Region of Upper Scorpius is 6.2. In the discussion yesterday (see page 40), these values were compared with the values obtained by the convergent point method.

One other way in which this composite main sequence can be tested is by means of the absolute magnitudes for stars with large parallaxes. All main sequence stars with large parallaxes fall near this composite main sequence with no systematic differences, indicating a high degree of consistency over a wide range of spectral types.

Support through the National Research Council of Canada is gratefully acknowledged.

References

Garrison, R. F.: 1967, *Astrophys. J.* **147**, 1003.
Klare, G. and Szeidl, B.: 1966, *Veröffentl. Landessternwarte Heidelberg-Koenigstuhl* **18**, 9.
Morgan, W. W. and Hiltner, W. A.: 1965 *Astrophys. J.* **141**, 177.
Morgan, W. W., Hiltner, W. A., and Garrison, R. F.: 1971, *Astron. J.* **76**, 242.

ABSOLUTE MAGNITUDES OF O STARS FROM A
STUDY OF VISUAL MULTIPLE SYSTEMS

M. L. BURNICHON

Institut d'Astrophysique de Paris, France

Abstract. From a careful study of physical multiple systems, it has been possible to derive differentially absolute magnitudes and intrinsic colours of O stars from the corresponding quantities, relatively well-known, of companions of these stars.

A study of wide physical multiple systems – whose brightest member is an O star or a blue supergiant, surrounded by less massive main sequence fainter companions of type B or sometimes A – was made in the Paris BCD spectrophotometric system (λ_1, D, φ_b, φ_{uv} parameters) and in the UBV photometric system (Burnichon, 1972).

One aim of this study was to determine the absolute magnitudes of bright components from those, which are much better known, of faint components. We assume that all stars of a same system are similarly reddened by interstellar absorption. This is justified because firstly, the reddening is not large (mean $E(B-V)=0.40$) as we chose the less reddened systems. Also most of the absorption is foreground absorption and it is hardly likely that it could be variable because of the small size of the systems studied. Moreover, no abnormal behaviour of reddening law was found spectrophotometrically. Finally, if the reddenings were different, the differences would be very small (some hundredths in $B-V$) and the error in magnitude found would only be of the order of 0.1.

The absolute magnitudes of the faint components were derived from their (λ_1, D) spectral classification and from the surface (Σ) luminosity calibration (Chalonge and Divan, 1973). The absolute magnitudes of the bright components were derived differentially.

For B supergiants, the results are in very good agreement with the surface (Σ) calibration.

For O stars, we find a very large dispersion (the absolute magnitudes approximately lie between -3.5 and -7) difficult to explain by uncertainties of the method used, and which is probably real.

Nevertheless, it must be emphasized that we observed very few systems (14 of which only 7 have a central star of type O), and we need more observations to confirm these results and to search for correlations with the precise O-type classification and with the 'f' characteristic.

Note. Another result I would only like to mention briefly concerns the intrinsic colours of O stars. The $(B-V)_0$ obtained for these stars lie between -0.29 and -0.25, that-is-to-say are redder than the classical values generally admitted (the mean color corresponds to the color of B1 main sequence stars).

Perhaps this result could be explained by circumstellar reddening...?

B. Hauck and B. E. Westerlund (eds.), Problems of Calibration of Absolute Magnitudes and Temperature of Stars, 126–127.
All Rights Reserved. Copyright © 1973 by the IAU.

References

Burnichon, M. L.: 1972, 'Contribution à l'étude des couleurs intrinsèques et des magnitudes absolues des étoiles bleues de grande luminosité à partir de l'observation de systèmes multiples', Thèse.
Chalonge, D. and Divan, L.: 1973, *Astron. Astrophys.* **23**, 69.

DISCUSSION

Jaschek: What was the average angular separation between the components of your pairs? Are you sure they are physical and not optical?

Burnichon: 15″ to 60″. They are not optical.

Schmidt-Kaler: If you make a plot (again I did so 11 yr ago) A_v vs distance for the late B stars you obtain a certain slope within, say, 500 pc. the line goes through zero with my intrinsic colour. If you do the same for the early B's you find the same result. If you plot however, the O-stars in the same way, you find the same slope, but the line does not go through zero. There is a 0^m3 residual. absorption, probably of circumstellar origin. This effect is also seen in a paper by K. H. Schmidt

PART V

GROUND BASED AND EXTRATERRESTRIAL OBSERVATIONS OF STELLAR FLUX

GROUND BASED AND
EXTRATERRESTRIAL OBSERVATIONS OF STELLAR FLUX

A. D. CODE

Washburn Observatory, University of Wisconsin, Madison, Wis., U.S.A.

Abstract. The significance of recent improvements made in the absolute monochromatic flux measurements of the Sun and Vega are discussed with a special emphasis on the absolute calibration of the V-magnitude of stars. The difficulties relating to the determination of vacuum ultra-violet fluxes are outlined and the improvements that can be achieved using synchroton radiation as the fundamental laboratory source are presented. Colour indices derived from OAO-2 observation are given for about 80 early type stars and the $(1700 - V)$ colour is discussed in terms of interstellar reddening and sensitivity to effective temperature. Basic problems requiring further investigation are finally outlined.

1. Introduction

About fifteen years ago I prepared a review of the then current status of observations of stellar flux. I would like to describe here some of the significant advances that have occurred since that time and conclude with several of the problems which still require attention.

The advances made in laboratory spectroradiometry have been important in improving our knowledge of stellar radiation. These include the more precise determination of the thermodynamic temperature scale and the establishment of spectral irradiance standards at the National Bureau of Standards in the United States and at Heidelberg, for example. Significant advances in instrumentation and development of reliable portable radiation sources such as the Copper black body furnace have resulted in improvement in accuracy. The careful measurements of Hayes and of Oke and Schild have taken advantage of these techniques for measurements in the visual region for stars. Our knowledge of the solar flux has been considerably improved by the ability to perform measurements from altitudes where variable components, such as water vapor and aerosols, are reduced to an insignificant level.

The extension of stellar measurements to the infrared and vacuum ultraviolet has opened up the opportunity to determine empirical bolometric corrections while substantially increasing our understanding of the stellar structure and of the interstellar and circumstellar medium. The intensity interferometer measurements of stellar angular diameters by Hanbury-Brown and associates has made possible the determination of the true emergent flux from these objects.

Until recently the measurement of absolute fluxes in the vacuum ultraviolet has been particularly difficult because of the lack of suitable radiation sources. The application of synchrotron radiation from a synchrotron storage ring has represented a major advance in vacuum ultraviolet energy calibration.

These advances in observational astronomy have been accompanied by equally impressive achievements on the theoretical side. The utilization of large digital computers

B. Hauck and B. E. Westerlund (eds.), Problems of Calibration of Absolute Magnitudes and Temperature of Stars, 131–145.
All Rights Reserved. Copyright © 1973 by the IAU.

has made possible the calculation of relatively sophisticated model atmospheres. Grids of models exist which include line blanketing from many individual lines as well as statistical line blanketing. In addition the effects of NLTE and of extended atmospheres have been investigated.

These advancements have clarified many features but also presented substantial new problems and some significant disagreements.

In what follows I shall describe some of the above investigations in more detail.

2. Solar Spectrum

Although the Sun has been studied extensively, observations of the Sun as a star have been limited. Until recently the solar flux was probably not as well determined as that of Vega and the best data on the sun were found using the monochromatic intensities at the center of the disk and the measured limb darkening. Furthermore, the determination of U, B, V magnitudes and MK spectral type are complicated by the inability to use the same instrumentation over this dynamic range of some 27 mag.

The most frequently employed determinations of the solar flux have been the compilation of Johnson (1954) and that of Nicolet (1951) derived using limb darkening data. More recently Labs and Neckel (1968) obtained measurements of the central intensity from 3300–12 500 Å from the Jungfraujoch Scientific Station, Switzerland (altitude 3.6 km) which were combined with center-limb variations and line blanketing coefficients to obtain a solar flux curve. All these studies include inaccuracies in the evaluation of the atmospheric attenuation and the inaccuracies in determining center-limb variations and the line blanketing coefficients.

Most compilations of the solar spectrum below 3330 Å have been based on NRL rocket measurements given by Tousey (1963) (cf. Furukawa et al., 1967). For the purpose of determining the total integrated solar flux or effective temperature, this data is completely satisfactory since the total flux shortward of 3330 Å is less than 3% of the total radiation. In studies of the ultraviolet albedo of planets, however, it is clear that these complications of the solar UV spectrum contain systematic errors (cf. Wallace et al., 1972) and that the extreme UV is variable. Recently Broadfoot (1972) has measured the solar flux between 2000–3000 Å with an aerobee rocket spectrometer and found good agreement with the earlier data except in the region of 2200 Å where the previous spectra have been depressed.

For the infrared between 8.6 and 13 μ the determinations by Saidy (1960) appear satisfactory. This region of the spectrum contains only 0.1% of the total energy and hence is primarily of importance for spectral investigation in this region.

Recently the solar flux from 3000 Å to 25000 Å has been carefully determined by direct high resolution measurements from a NASA Convair 990 aircraft (Arvesen et al., 1969). Eleven long duration flights between 11.5 and 12.5 km have been carried out using a particularly carefully designed instrumental technique. The results are independent of any assumptions about line blanketing or limb-darkening and free of atmospheric transmission variations due to aerosols or water vapor. The total flux value is considered good to 3%, while the monochromatic fluxes become uncertain to about 6% at 3200 Å, 7% at 3100 Å, and 25% at 3000 Å, the main source of uncertainty being due to the calibration of the standard lamp. The agreement with Labs and Neckel is good.

On the basis of existing measurements it would appear that the most reliable solar spectrum from 2000–3000 Å is the Broadfoot (1972) spectrum and from 3100–25000 Å the Arvesen (1969) spectrum, while longward of 25000 Å a 5800K grey body approximation joined to the Saidy (1960) spectrum should give an adequate representation of the solar continuum.

The total integrated flux of the Sun resulting from this spectrum is 1360 W m^{-2} (Duncan, 1969) which corresponds to an effective temperature for the Sun of 5770 K. It is more difficult to relate these results to measurements of stellar flux. The spectral type of G2 V for the Sun that has been determined by Morgan (Stebbins and Kron, 1957) is as well determined as that for other stars. However, the determination of the magnitude and color of the Sun has presented difficulties.

The determination of the visual magnitude and color of the sun by Stebbins and Kron (1957) was one of the most extensive direct comparisons of the Sun and stars. They obtained a visual magnitude $V = -26.73 \pm 0.03$ and a $B-V = +0.63$. Martinov (1959) rediscussed all data on the visual magnitude of the Sun and obtained a value of $V = -26.80 \pm 0.3$, while Johnson (1965) reviewed all recent data obtained by various indirect means along with the Stebbins and Kron measurements and arrived at a value of $V = -26.74$. Determinations of the $B-V$ color since Stebbins and Kron by indirect means have ranged from $+0.65$ to $+0.68$ and the most recent determination by Fernie et al. (1971) yields $(B-V) = +0.628$.

The Sun as a star is probably closely represented by a spectral type G2 V, a $B-V$ color of $+0.63$, and visual magnitude of $V = -26.74$. We shall return to this comparison, keeping in mind the variations described above after reviewing the status of stellar flux determinations.

3. Stellar Spectra

The energy distribution of Vega adopted by the author (Code, 1960) was a composite one including observations made at Jungfraujoch, by Kienle and his co-workers, Williams and Hall, as well as observations by Whitford and Code in the red. Subsequent observations by Bahner (1963) and others indicated that the Balmer jump was too small by about $0^m.13$. Furthermore the Paschen continuum could not be satisfactorily represented by model atmosphere calculations. Oke (1964) derived an energy curve for Vega based upon the best model representation known in 1964. Investigations by Gluschneve (1964), Kharitonov (1963), and Willstrop (1965) provided improved calibrations although discrepancies as great as 10% persisted.

A new and careful spectrophotometric calibration by Hayes (1967) resulted in a continuous energy distribution that was in good agreement with the predictions of model atmospheres and incidentally with the early work of Kienle. Oke and Schild (1970) carried out a program designed to measure the absolute monochromatic flux of Vega from 3300 Å to 10000 Å utilizing a telescope, scanner and photometric system specifically constructed for these observations. Three radiation sources – a tungsten ribbon lamp, a copper-point black body and a platinum-point black body – were directly compared with Vega. In the spectral interval from 4000–6000 Å they quote errors of ±2%. Beyond this range in the infrared and ultraviolet, scatter as high as 6% was found, but no systematic differences between different sets of data. In the region from 4000–6000 Å the agreement with Hayes is excellent. Figure 1 compares the

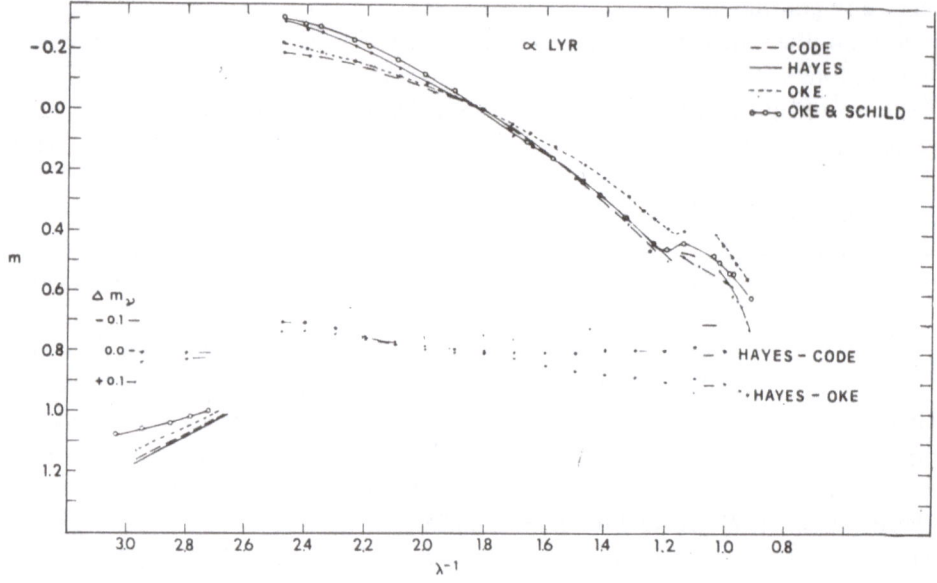

Fig. 1.   Comparison of different calibrations for Vega.

results of this investigation with several of the earlier energy curves of Vega. There is still a disquieting discrepancy shortward of the Balmer discontinuity and longward of the Paschen jump. I believe that these differences represent a realistic estimate of the overall uncertainties in the energy distribution of Vega and can suggest nothing better at the present time than to adopt the mean of Hayes (1970) and Oke and Schild (1970) as the best representation of the monochromatic flux of Vega in the spectral interval 3300 to 10000 Å.

On the basis of their investigation Oke and Schild find the absolute monochromatic flux at 5556 Å for α Lyr to be 3.36×10^{-9} ergs s^{-1} cm^{-2} Å^{-1}. Adopting an effective wavelength of 5480 Å for the V-filter of the UBV system they find a value of $3.64 \times \times 10^{-9}$ ergs s^{-1} cm^{-2} Å^{-1} for a star of visual magnitude $V = 0.00$. This number is an important quantity for providing a means of calibrating the V-magnitude of a star. The precise meaning of an energy curve derived from a wide band filter photometer is of course dependent upon the detailed energy distribution over the filter band pass and hence ambiguous. I shall, however, adopt the point of view that the V-magnitude of a star represents the integral effective intensity (Code, 1960) determined by a photometer with a band pass characteristic as tabulated for zero air mass by Matthews and Sandage (1963). This sensitivity function was derived by the author based on Johnson's (1955) tabulated filter and 1P21 response and the reflectivity of two aluminum mirrors. If one integrates over this band pass the Oke-Schild energy curve of Vega, the integral effective intensity corresponds to 3.58×10^{-9} ergs s^{-1} cm^{-2} Å^{-1}. It is of interest to compare this result with that found by integrating the Arvesen solar spectrum over the V band pass. If we adopt a visual magnitude for the Sun of $V_{\odot} = -26.74$ we find an integral effective intensity of 3.65×10^{-9} ergs s^{-1} cm^{-2} Å^{-1}, while the range of values for the visual magnitude of the Sun yield values from 3.68 to 3.45. A mean value for the integral effective intensity of a star of visual magnitude $V = 0.00$ as determined from the solar spectrum and from Vega is $3.61 \pm 0.10 \times 10^{-9}$ ergs s^{-1} cm^{-2} Å^{-1}. This result is not particularly sensitive to the detailed shape of the V filter sensitivity curve and within the uncertainty quoted is independent of any color term for solar type stars or earlier.

That there are no large systematic differences between the energy determinations of stars and of the Sun is shown in Figure 2, where the difference in magnitudes between

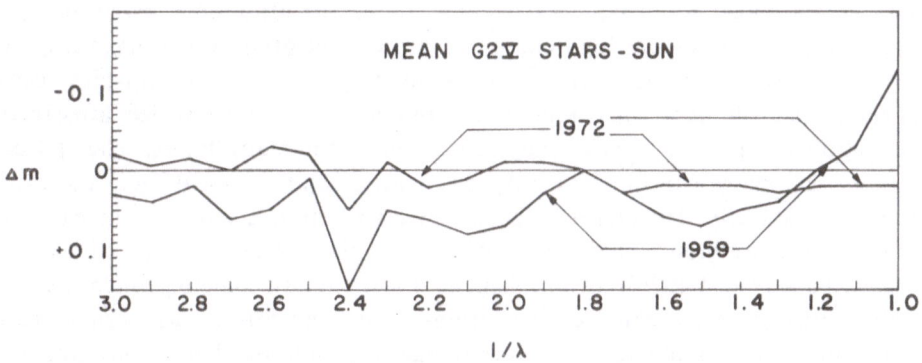

Fig. 2.   Differences between the energy determinations of stars and of the Sun.

the mean energy distribution of solar type stars whose average $B - V$ color is ± 0.63 and of the Sun are compared. The 1959 curve was determined by the author (Code, 1960) based on earlier data. The 1972 curve is for the same stars using the mean of Hayes and the Oke-Schild calibration of Vega and the Arvesen *et al.* solar spectrum. The curves are normalized at 5556 Å and show remarkable agreement over the entire spectral region.

The determination of the ultraviolet energy distribution of stars is still in a relatively primitive state, particularly shortward of 1500 Å. Where we speak of differences of a few percent in the visual, discrepancies of a factor of 3 or 4 exist in the far UV. One of the basic difficulties has been the lack of a satisfactory fundamental radiation source. The intensity of a tungsten lamp or carbon arc falls off too rapidly to be of use below 2500 Å. A black body source must be operated at temperatures well above the melting point of any metals and hence the source must be in a plasma state. Boldt has succeeded in developing an optically thick arc running at a temperature of about 14000 K which in principle can yield calibrations to about 20% at 1000 Å.

Most energy measurements of stars in the ultraviolet have depended upon a more indirect calibration. In general a laboratory reference photomultiplier with a sodium salicylate coating has been calibrated longward of 2500 Å and the quantum efficiency of the sodium salicylate has been assumed to be uniform as a function of wavelength. This assumption has usually been checked by the use of nitric oxide ion chambers at 1216 Å. A review of methods of intensity calibration has been given by McWhirter (1971). Descriptions of the calibration of specific payloads are usually presented along with UV data (cf. Carruthers, 1969; Stuart, 1969; Evans, 1972). While accuracies of 15 to 30% are sometimes quoted, differences up to a factor of 5 exist. The history of ultraviolet spectrophotometry has been one of the large flux deficiencies relative to theoretical predictions, which have moved to shorter and shorter wavelengths as techniques have improved. It is now generally agreed that the observed fluxes are in reasonable agreement with theory longward of 2000 Å, while evidence is accumulating that there are no large flux deficiencies in the 1200–1500 Å region (Bless and Code, 1972).

One of the most promising fundamental sources of ultra-violet radiation is the synchrotron radiation from high energy electrons circulating in a storage ring. The University of Wisconsin operates a facility consisting of a synchrotron which accelerates electrons to about 50 MeV, injects them into a storage ring in which, after further acceleration to about 240 MeV, they can circulate for many hours. The source is stable in time, has a continuous energy distribution similar to a B5 star and can be calibrated absolutely. The fact that the energy distribution is similar to an early type star removes one serious source or error that has plagued ultraviolet calibrations in the past, namely scattered light problems and low rapidly changing intensities. The absolute intensity of the beam can be determined independently of any thermodynamic temperature scales or previous absolute radiation standards. This comes about because it is possible to measure the radiation from a single electron and use theory to determine the energy distribution, which is very insensitive to the electron energy longward of 1000 Å for 240 MeV electrons. We may check the theoretical calculations by measuring the angular distribution and polarization of the radiation.

We usually start with about 50 electrons and measure the step-wise decrease in intensity of the radiation each time an electron is ejected from the beam and thus one can determine the number of electrons producing a given signal. We have calibrated rocket payloads directly in the essentially collimated synchrotron radiation both before and after flight (Gaide, 1971; Bless et al., 1972). The calibration obtained by this method agrees well with our OAO-2 preflight calibration and is the basis of our present absolute energy determinations. Figure 3 shows the energy distribution of η UMa as measured by OAO-2 employing this calibration. The solid curve is from OAO-2 spectral scans. The solid circles are the results of filter measurements obtained with a rocket calibration payload. The open circles are the results of Schild et al. (1971), while the dotted curve results from the Hayes calibration. The circled cross is an OAO-2 filter measurement derived from solar type stars and the Arvesen solar spectrum. The energy distribution is similar to that of a blanketed model atmosphere with an effective temperature of the order of 17000 K. Evans (1972) has carried out a continuing program on absolute calibration in the ultraviolet which diverges from our results substantially shortward of 2000 Å. The Evans calibration yields an energy curve for η UMa shown by the dash-dot curve in Figure 3. The results of Moos *et al.* are shown by the dashed curve. The Evans calibration has been an

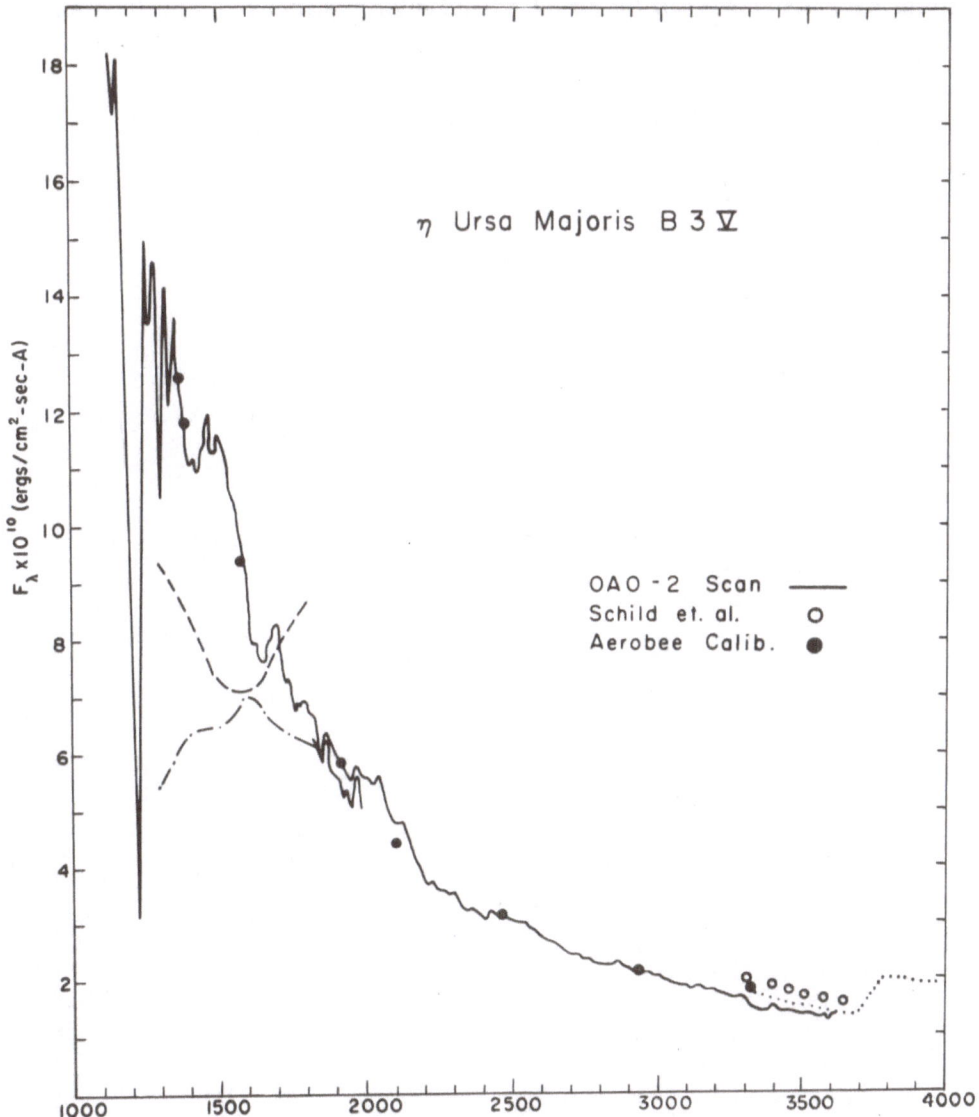

Fig. 3.   The energy distribution of ηUMa as measured by OAO-2 employing this calibration.

A. D. CODE

extensive and careful one and the source of the discrepancy is at present unknown.
The results of Carruthers are in agreement with the Wisconsin determinations and the
results of Stuart yield somewhat larger fluxes than the Wisconsin results. The ratio of
total integrated flux for ηUMa implied by the difference between the Wisconsin and
Goddard calibration is 2 which would indicate an effective temperature of the order of
15000 K for ηUMa for the Goddard result.

Figure 4 shows the energy distribution of S Mon based on the Wisconsin calibration. The filled circles are the Geneva balloon results (Navach, 1972). These observations have been corrected for an interstellar color excess of $E_{B-V} = 0^{m}.07$ using the

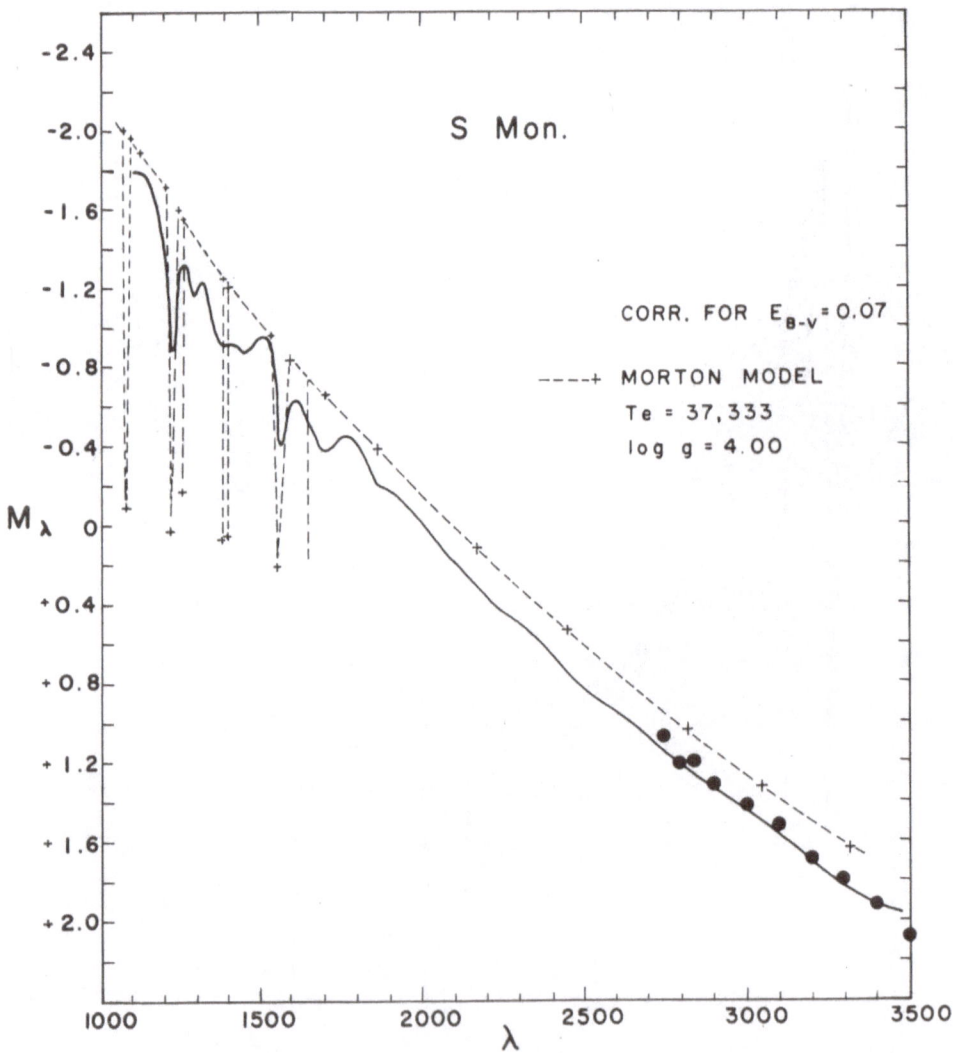

Fig. 4.   Energy distribution of S Mon based on the Wisconsin calibration.

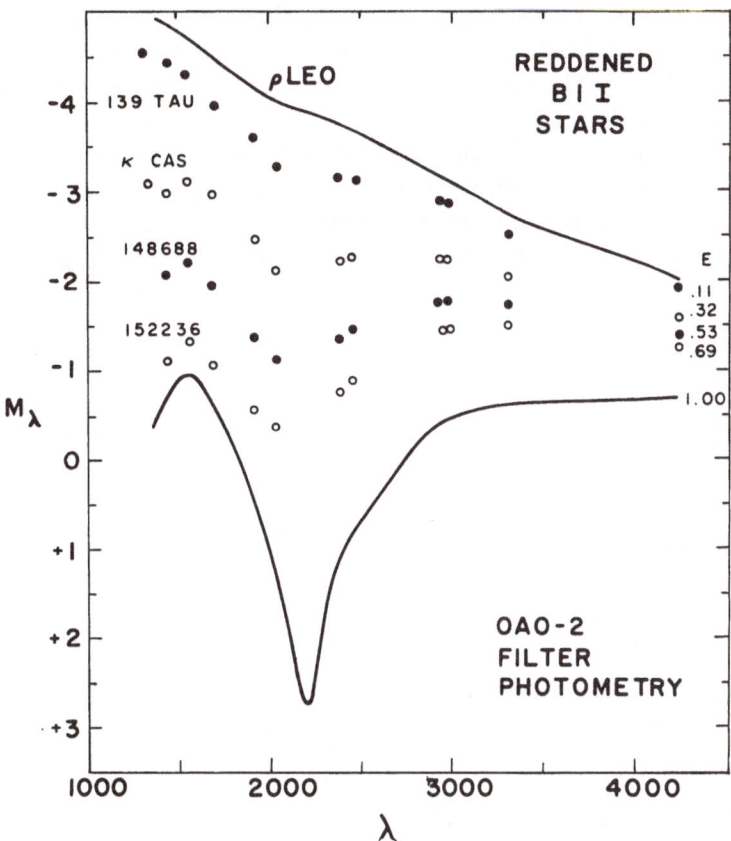

Fig. 5. Energy distribution of a number of reddened B1 I stars from OAO filter observations

mean extinction curve of Bless and Savage (1972) and compared with a Morton blanketed model for $T_{eff} = 37\,333$ K.

The Wisconsin absolute calibration has been well determined longward of 1300 Å but is not based on a fundamental calibration shortward of 1300 Å, although it agrees well with the Lyman alpha calibration by Blamont of his OGO-V instrument. It is clear, however, that much work remains to be done on the energy calibration shortward of 2000 Å.

4. Interstellar Extinction

The comparison of ultraviolet measurements of stars with theoretical predictions is limited by our knowledge of the interstellar extinction curve. The discussion by Bless and Savage (1972) shows that the extinction is large, variable and non-linear in the UV. A $(B-V)$ color excess of only $0\overset{m}{.}1$ could imply extinction at 1250 Å of as little as $0\overset{m}{.}3$ or as much as $0\overset{m}{.}7$. They have found, however, that stars of similar $B-V$ colors and spectral types do have the same spectral distribution in the ultraviolet and therefore the suggestion by Underhill (1972), that variations in reddening are due to line blocking, is not verified. Figure 5 shows the energy distribution of a number of reddened B1 I stars from OAO filter observations along with the computed curve for a $B-V$ color excess of 1 mag. The agreement for these stars with the mean extinction curve of Bless and Savage is good.

Figure 6 shows a spectral scan of ηUMa. The internal accuracy is very good (within 2%). The region in the neighborhood of 1700 Å shows a pronounced maximum. We have measured monochromatic colors at 1700 Å for many early type stars. Table I contains the results for 65 little reddened stars. The mean ratio of E_{1700-V}/E_{B-V} is 4.4.

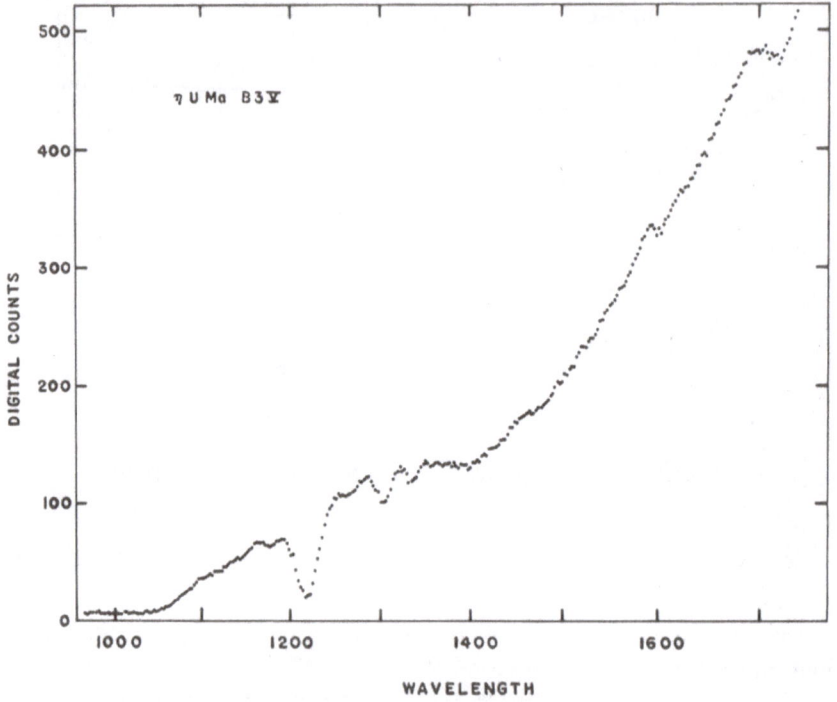

Fig. 6.   Spectral scan of η U Ma.

I have determined $1700 - V$ normal colors for these stars. A two color plot employing these data may be found in the paper of Bless and Savage (1972).

We have chosen 1700 Å for several reasons. It is a high point in the continuum and also relatively free of predicted lines. It occurs at a minimum in the interstellar extinction curve, and finally is at a wavelength in the Balmer continuum which is least sensitive to changes in gravity in model atmosphere calculations. The $1700 - V$ is of course an order of magnitude more sensitive to changes in temperature than $B - V$. The color index is based on the Wisconsin absolute calibration and agrees very well with Morton's blanketed models. Identification of the $(1700 - V)_0$ colors with model atmospheres provides an effective temperature scale similar to those recently proposed on the basis of other criteria.

TABLE I
$(1700-V)$ Fluxes

HR No.	Star	Sp. Type	V	$(B-V)$	$1700-V$	E_{1700}	$(B-V)_0$	$(1700-V)_0$
2456	S Mon	O7	4.66	−0.24	−3.84	0.35	−0.32	−4.19
3165	ζ Pup	O5 f	2.25	−0.29	−4.00	0.18	−0.33	−4.18
3207	γ Vel	WC8	1.82	−0.27	−3.97	0.22	−0.32	−4.19
1228	ξ Per	O7.5	4.06	+0.01	−2.80	$1.32^{R}_{4.00}$	−0.32	−4.12
1899	ι Ori	O9 III	2.76	−0.23	−3.77	0.35	−0.31	−4.12
8622	10 Lac	O9 V	4.88	−0.20	−3.64	0.48	−0.31	−4.12
1931	σ Ori	O9.5 V	3.83	−0.24	−3.89	0.26	−0.30	−4.15
6175	ζ Oph	O9.5 Vnn	2.57	+0.02	−2.51	$1.57^{R}_{4.90}$	−0.30	−4.08
1851-52	δ Ori	09.5 II-III + B2 V	2.20	−0.21	−3.58	0.35	−0.29	−3.93
1948-49	ζ Ori	09.5 Ib	1.74	−0.21	−3.45	0.26	−0.27	−3.71
1542	α Cam	09.5 Ia	4.29	+0.03	−2.23	$1.47^{R}_{4.90}$	−0.27	−3.70
1855	ν Ori	B0 V	4.63	−0.26	−4.07	0.18	−0.30	−4.25
6165	τ Sco	B0 V	2.82	−0.25	−3.93	0.22	−0.30	−4.15
5953	δ Sco	B0 V	2.33	−0.10	−3.12	$0.98^{R}_{4.90}$	−0.30	−4.10
1903	ε Ori	B0 Ia	1.70	−0.19	−3.22	0.22	−0.24	−3.44
1788	η Ori	B0.5 Vnn	3.35	−0.19	−3.29	0.40	−0.28	−3.69
1756	λ Lep	B0.5 IV	4.29	−0.25	−3.88	0.13	−0.28	−4.01
5984	β¹ Sco	B0.5V+B2 V	2.55	−0.08	−2.99	$0.70^{R}_{3.50}$	−0.28	−3.69
1220	ε Per	B0.5 III	2.89	−0.18	−3.34	0.44	−0.28	−3.78
4853	β Cru	B0.5 III	1.24	−0.24	−3.69	0.18	−0.28	−3.87
7446	κ Aql	B0.5 IIIn	4.95	0.00	−2.51	$1.37^{R}_{4.90}$	−0.28	−3.88
2004	κ Ori	B0.5 Ia	2.06	−0.18	−3.23	0.18	−0.22	−3.41
	42 Ori	B1 V	4.60	−0.19	−3.36	0.31	−0.26	−3.67
1789	25 Ori	B1 Vn	4.95	−0.21	−3.47	0.22	−0.26	−3.69
5056	α Vir	B1V+B3	0.96	−0.25	−3.54	0.04	−0.26	−3.58
5944	π Sco	B1V+B2	2.92	−0.19	−3.47	0.31	−0.26	−3.78
5993	ω¹ Sco	B1 V	3.99	−0.04	−2.89	$0.77^{R}_{3.50}$	−0.26	−3.66
2571	15 CMa	B1 III	4.82	−0.21	−3.39	0.22	−0.26	−3.61
5267	β Cen	B1 III	0.61	−0.23	−3.56	0.13	−0.26	−3.69
8238	β Cep	B1 III	3.20	−0.21	−3.40	0.22	−0.26	−3.62
6084	σ Sco	B1 III	2.89	+0.14	−2.33	$1.08^{R}_{2.70}$	−0.26	−3.41
2294	β CMa	B1 II-III	1.98	−0.24	−3.53	0.04	−0.25	−3.57
1203	ζ Per	B1 Ib	2.86	+0.10	−1.79	$1.16^{R}_{4.00}$	−0.19	−2.95
2084	139 Tau	B1 Ib	4.83	−0.07	−2.73	0.42	−0.19	−3.15
6247	μ¹ Sco	B1.5 IV	3.02	−0.23	−3.38	$0.09^{R}_{3.50}$	−0.25	−3.47
5695	δ Lup	B1.5 IV	3.21	−0.22	−3.46	0.13	−0.25	−3.59
6580	κ Sco	B1.5 III	2.41	−0.20	−3.50	0.22	−0.25	−3.72
5469	α Lup	B1.5 III	2.31	−0.21	−3.25	0.18	−0.25	−3.43
6028	13 Sco	B2 V	4.58	−0.15	−3.10	0.40	−0.24	−3.50
5708	ε Lup	B2 IV–V	3.36	−0.17	−3.14	0.31	−0.24	−3.45
153	ζ Cas	B2 IV	3.66	−0.18	−3.17	0.26	−0.24	−3.43
39	γ Peg	B2 IV	2.86	−0.21	−3.46	0.13	−0.24	−3.59
5248	φ Cent	B2 IV	3.82	−0.21	−3.35	0.13	−0.24	−3.48
5571	β Lup	B2 IH	2.67	−0.21	−3.46	0.13	−0.24	−3.59
1790	γ Ori	B2 III	1.66	−0.21	−3.32	0.13	−0.24	−3.45
2618	ε CMa	B2 II	1.50	−0.21	−3.17	0.09	−0.23	−3.26
5948	η Lup	B2.5 IV	3.40	−0.24	−3.49	0.00	(−0.24) −0.22	−3.49
2282	ζ CMa	B2.5 IV	3.02	−0.13	−2.91	0.40	−0.22	−3.01
5812	τ Lib	B2.5 V	3.65	−0.17	−2.95	0.22	−0.22	−3.17

Table I (Continued)

HR No.	Star	Sp. Type	V	$(B-V)$	$1700-V$	E_{1700}	$(B-V)_0$	$(1700-V)_0$
7623	θ^1 Sgr	B2.5 IV	1.24	-0.15	-2.81	0.31	-0.22	-3.12
2106	γ Col	B2.5 IV	4.35	-0.21	-2.66	0.04	-0.22	-2.70
1497	τ Tau	B3 V	4.29	-0.14	-2.62	0.26	-0.20	-2.88
1641	η Aur	B3 V	3.19	-0.18	-2.83	0.09	-0.20	-2.92
5191	η UMa	B3 V	1.86	-0.18	-2.76	0.09	-0.20	-2.85
5626	λ Lup	B3 V	4.04	-0.20	-2.83	0.00	-0.20	-2.83
2159	ν Ori	B3 IV	4.42	-0.15	-2.80	0.22	-0.20	-3.02
2199	ξ Ori	B3 IV	4.48	-0.17	-2.86	0.13	-0.20	-2.99
1934	ω Ori	B3 III$_e$	4.59	-0.11	-2.66	0.40	-0.20	-3.06
542	ε Cas	B3 V$_p$	3.38	-0.15	-2.53	0.22	-0.20	-2.75
472	α Eri	B3 V$_p$	0.49	-0.17	-2.60	0.13	-0.20	-2.73
5712	ϕ^2 Lup	B4 V	4.53	-0.16	-2.81	0.02	-0.18	-2.90
226	ν And	B5 V	4.52	-0.15	-2.38	0.04	-0.16	-2.42
1122	δ Per	B5 III	3.03	-0.12	-2.45	0.18	-0.16	-2.63
8773	β Psc	B6 V$_e$	4.52	-0.12	-2.34	0.09	-0.14	-2.43
8425	α Gru	B7 IV	1.73	-0.17	-2.18	0.00	(-0.17) -0.12	-2.18
1791	β Tau	B7 III	1.66	-0.13	-2.02	0.00	(-0.13) -0.12	-2.02
7039	ϕ Sgr	B8 III	3.17	-0.11	-1.72	0.00	(-0.11) -0.09	-1.72
1713	β Ori	B8 Ia	0.15	-0.03	-1.33	0.00	(-0.03) -0.02	-1.33
2095	θ Aur	B9.5 si	2.63	-0.08	$+0.01$	0.00	(-0.08) -0.03	$+0.01$
7011	α Lyr	A0 V	0.00	00.00	-0.66	0.00	0.00	-0.66
4905	ε UMa	A0 p	1.78	-0.03	-0.56	0.00	(-0.03) 0.00	-0.56
2491	α CMa	A1 V$_p$	1.45	-0.01	-0.62	0.00	(-0.01) $+0.03$	-0.62
2088	β Aur	A2 V	1.90	$+0.03$	$+0.27$	0.00	$(+0.03)$ $+0.06$	$+0.27$
1666	β Eri	A3 III	2.78	$+0.12$	$+0.20$	0.02	$+0.10$	$+0.11$
1702	μ Lep	A5 p	3.29	-0.11	-1.85	0.00	$-0.11)$	-1.85

In the process of determining effective temperatures, however, I was impressed by the sensitivity of the determinations to line blanketing and for the hottest stars to the effects of extended atmospheres. Figure 7 compares two of Cassinelli's (1971) model atmospheres with different curvatures with a 50000 K ATLAS model. Two effects should be noted. First, the decrease in the flux gradient in the Paschen and Balmer continua with increasing extension of the atmosphere which would therefore yield an artificially low effective temperature as has been pointed out by Heap (1972) and Conti (1972). Second, the absolute luminosity in the Lyman continuum is increased, providing more photons for ionization of interstellar clouds. In the case of extended atmospheres it is necessary to give up the concept of effective temperature and replace it by the total luminosity.

I believe that no essential improvement in the effective temperature scale for early

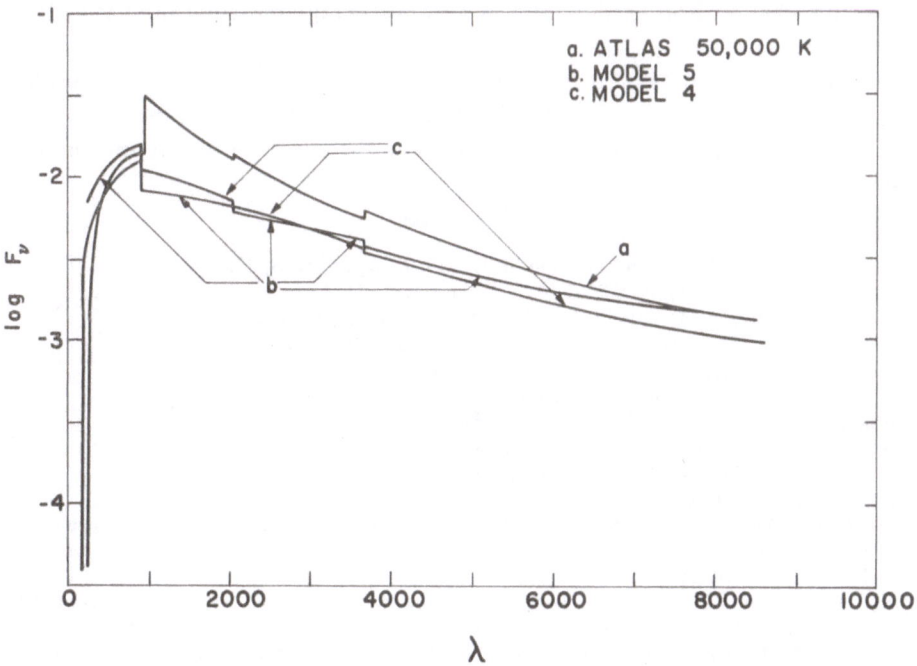

Fig. 7. Comparison of Casinelli's model atmospheres with 50000 K ATLAS model.

type stars can be achieved from measurements of stellar fluxes until such time as accurate fluxes are available in the ultraviolet down to the Lyman jump. It will then be possible to determine the integrated fluxes of early type stars and hence empirical effective temperatures for those stars for which angular diameters are available and empirical bolometric corrections for others. For the very hottest stars where uncertainties in the contribution of the Lyman continuum to the total flux is important, we require further studies of NLTE and extended atmospheres.

Doherty (1972) has investigated the ultraviolet energy distribution of late type stars. The agreement with model atmosphere calculations by Gingrich is remarkably good considering the large uncertainty in blanketing corrections required for these stars. High resolution studies will be required to derive satisfactory line identifications and blanketing coefficients.

5. Conclusion

The basic problems requiring further investigation that have been highlighted in this discussion are primarily observational.

The resolution of the remaining discrepancies in energy determinations must be resolved. Of particular importance is a redetermination of the Balmer jump of Vega or the slope and absolute flux in the Balmer continuum. The differences in the Brackett continuum also deserve investigation.

It is necessary to resolve the difference in calibration in the ultraviolet between

1300 Å and 3000 Å Shortward of 1300 Å absolute energy measures are basically non-existent.

Line blanketing presents many important problems requiring high resolution spectroscopy particularly in the ultraviolet along with laboratory work.

Finally, theoretical work must be continued on the effects of blanketing and on extended atmospheres.

References

Arvesen, J. C., Griffin, R. N., and Pearson, B. D.: 1969, *Appl. Opt.* **8**, 2215.
Bahner, K.: 1963, *Astrophys. J.* **138**, 1314.
Bless, R. C. and Code, A. D.: 1972, *Ann. Rev. Astron. Astrophys.* **10**, 197.
Bless, R. C., Fairchild, T., and Code, A. D.: 1972, NASA SP-310, 361.
Bless, R. C. and Savage, B. D.: 1972, *Astrophys. J.* **171**, 293.
Broadfoot, A. L.: 1972, *Astrophys. J.* **173**, 681.
Carruthers, G. R.: 1969, *Astrophys. J. Letters* **156**, 97.
Cassinelli, J. P.: 1971, *Astrophys. J. Letters* **8**, 105.
Code, A. D.: 1960, *Stars and Stellar Systems* **6**, 87.
Doherty, L.: 1972, *Astrophys. J.*, in press.
Duncan, C. H.: 1969, 'Radiation Scales and the Solar Constant', Goddard Space Flight Center Report No. X–713–69–382, Greenbelt, Md.
Evans, D. C.: 1972, NASA SP-310, 347.
Fernie, J. D., Hagen, J. P., Hagen, G. L., and McClure, L.: 1971, *Publ. Astron. Soc. Pacific* **83**, 79.
Furukawa, P. M., Haagenson, G. L., and Scharberg, M. J.: 1967, NCAR Technical Note, No. 26.
Gaide, A. E.: 1971, in F. Labuhn and R. Lüst (eds.), 'New Techniques in Space Astronomy', *IAU Symp.* **41**, 386.
Glushneve, I. N.: 1964, *Soviet Astron. AJ* **8**, 163.
Hayes, D. S.: 1967, unpublished dissertation, University of California, Los Angeles.
Hayes, D. S.: 1970, *Astrophys. J.* **159**, 165.
Heap, S. R.: 1972, *Astrophys. J. Letters* **10**, 49.
Johnson, F. S.: 1954, *Meteorol. J.* **11**, 431.
Johnson, H. L.: 1955, *Ann. Astrophys.* **18**, 292.
Johnson, H. L.: 1965, *Comm. Lunar Planetary Lab.* **3**, No. 53, 67.
Kharitonov, A. W.: 1963, *Soviet Astron. AJ* **7**, 258.
Labs, D. and Neckel, H.: 1968, *Z. Astrophys.* **69**, 1.
Martinov, D. J.: 1959, *Soviet Astron. AJ* **3**, 633.
Matthews, T. A. and Sandage, A. R.: 1963, *Astrophys. J.* **138**, 30.
McWhirter, R. W. P.: 1971, F. Labuhn and R. Lüst (eds.), 'New Techniques in Space Astronomy', *IAU Symp.* **41**, 369.
Navach, C.: 1972, private communication.
Nicolet, M.: 1951, *Ann. Astrophys.* **14**, 249.
Oke, J. B.: 1964, *Astrophys. J.* **140**, 689.
Oke, J. B. and Schild, R. E.: 1970, *Astrophys. J.* **161**, 1015.
Saidy, F.: 1960, *Monthly Notices Roy. Astron. Soc.* **115**, 493.
Schild, R. E., Peterson, D. M., and Oke, J. B.: 1971, *Astrophys. J.* **166**, 95.
Stebbins, J. and Kron, G. E.: 1957, *Astrophys. J.* **126**, 266.
Stuart, F. E.: 1969, *Astrophys. J.* **157**, 1255.
Tousey, R.: 1963, *Space Sci. Rev.* **2**, 3.
Wallace, L., Caldwell, J. J., and Savage, B. D.: 1972, *Astrophys. J.* **172**, 755.
Willstrop, R. V.: 1965, *Mem. Roy. Astron. Soc.* **69**, 83.

DISCUSSION

Pecker: I think that you cannot rule out the measured UV energy spectra on the basis of comparison with model atmospheres, as suggested by Underhill (I quote you!), as the models are extremely uncertain, and their spectra still more uncertain in the UV! A small change in their parameters can change indeed many things.

Hack: In your first slide you have shown the difference between the mean of 6 G2V stars and the Sun There is a sharp difference of $\sim +0.2m$ at $\lambda^{-1} \sim 2.4$ What is the reason for that difference?

Code: We don't know.

Kodaira: Even in the absolute calibration of the solar flux, we have difficulties for $\alpha < 2000$ Å. There has been a discrepancy of a factor of 3 between the values obtained by NRL (July 1966, August 1970) and those by HCO (September 1968). A Japanese group has started to make independent measurements and has obtained the first successful data with a rocket (September 1971). Our data at $\lambda = 1629, 1684, 1739$ Å $(\varDelta\lambda = 8.3$ Å$)$ are in very close agreement with HCO data (that is, lower values). The full results will be soon published by K. Nishi (Tokyo Observatory).

Van den Bergh: I should like to sound a very minor cautionary note! Although the agreement between the Sun and typical G2 V stars is very gratifying there is, I believe, one small but rather well established discrepancy. The ultraviolet cyanogen absorption in the Sun is too strong The observed solar CN strength corresponds to that in typical metal-rich main sequence stars with $B - V$ in the range 0.66 to 0.68.

LINE BLANKETING IN REAL STELLAR ATMOSPHERES

P. L. BERNACCA

Asiago Observatory, Italy

Abstract. Ultraviolet observations of line blanketing in the spectrum variables α Scl and α^2 CVn show that the usual procedure of allowing line blocking in model atmospheres has to be revised. It is also suggested that the region around 2900 Å could be safely used for determining effective temperatures.

Everybody knows that a blanketed model atmosphere is one where the line opacities are included in the computation of the emergent continuum flux. Blanketed model atmospheres exist that include the hydrogen lines (Klinglesmith, 1971) as well as the strongest ultraviolet lines (e.g. Bradley and Morton 1969, and references therein). The procedure followed in these calculations makes use of the radiative flux conservation relation

$$\int_{0}^{\infty} F(\lambda, \tau) \, \mathrm{d}\lambda = \text{const}. \tag{1}$$

The constancy of the integrated flux is maintained by adjusting the temperature structure of the atmospheres, that is the law $T(\tau)$, so that a blanketed model corresponds to an unblanketed one of higher effective temperature. In other words blanketing yields a higher continuum between the lines.

I think it is of interest to consider some recent observations made by OAO-II in the ultraviolet, which give us a hint as to how blocking and backwarming works in a real atmosphere. The cases of interest here are those of the stars α Scl and α^2 CVn.

The star α Scl is a helium-weak star of Population I and it was observed in the ultraviolet region by Bernacca and Molnar (1972) (BM). Most recently Norris (1971) has obtained scanner observations of its continuum down to λ 3400. More detailed information on the star can be found in the above two papers. In Figure 1 the photometry data from BM and the scanner observations from Norris are presented together with the run of the flux F_λ according model calculations by Klinglesmith (1971). The star α Scl has been given a $T_{\text{eff}} = 14400$ by using ultraviolet blanketed models (Norris, 1971) and one sees indeed from the figure that the ground based observations (crosses) fit reasonably well to a hydrogen line blanketed model of $T_{\text{eff}} = 14000$.

If we consider the ultraviolet observations made on January 23, 1971 (open circles) we can say that the effective temperature of α Scl is more close to 16000 K. On July 10, 1970 (filled circles), the energy output from the star at 2380 Å and 2460 Å is decreased and at the same time an increase of luminosity appears at 4250 Å. This behaviour has been explained by BM as a result of variable line blocking in the 2400 Å region which redistributes energy (backwarming) longward of the Balmer discontinuity. The blocking can be attributed to the strengthening of Ti II and Sr II lines which are known to be unusual in this star.

B. Hauck and B. E. Westerlund (eds.), Problems of Calibration of Absolute Magnitudes and Temperature of Stars, 146–148.

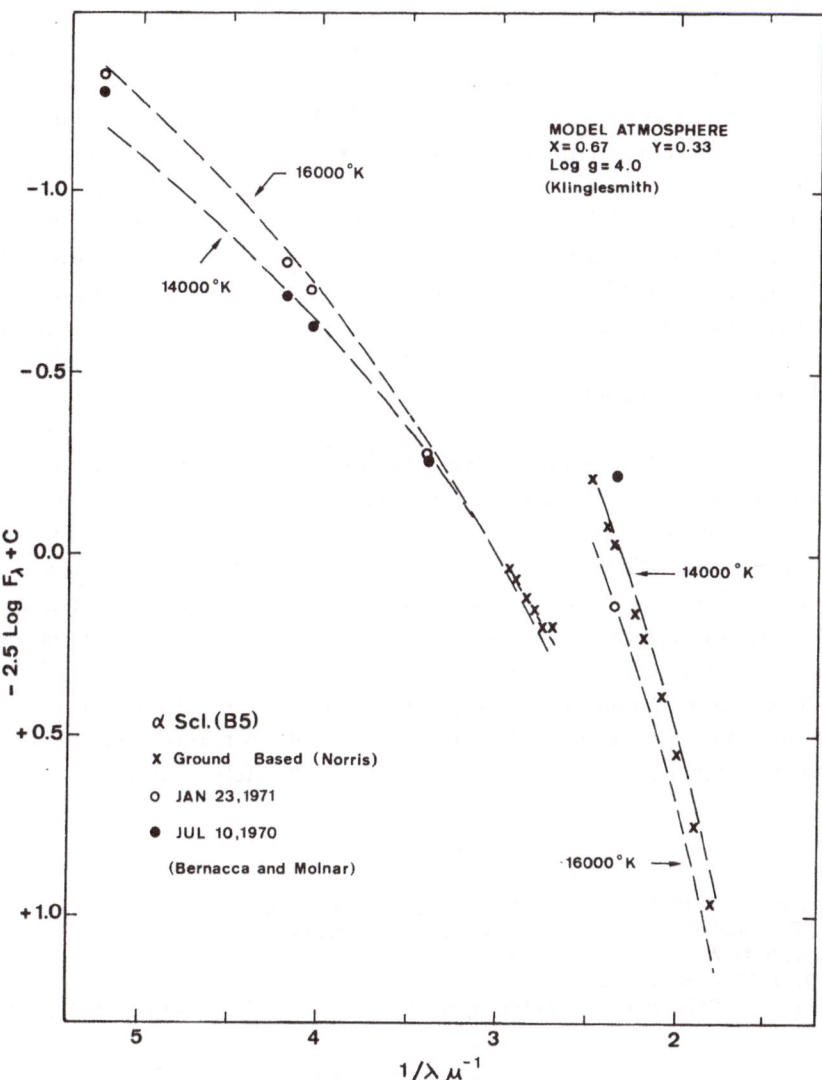

Fig. 1. Ultraviolet and ground based observations of α Scl are compared with hydrogen line blanketed models. The flux is normalized at $1/\lambda = 3.0\,\mu^{-1}$. The ground based observations by Norris (1971) from $1/\lambda = 1.8\,\mu^{-1}$ to $1/\lambda = 2.95\,\mu^{-1}$ have been easily extrapolated at $3.0\,\mu^{-1}$.

From the figure we may infer that it is possible that Norris' observations were made at a phase where the atmosphere of the star was similar to that observed on July 10, 1970.

The second remarkable object is the Ap variable α^2 CVn studied in the ultraviolet by Molnar (1972), to whom we refer for the relevant information concerning this star. It is sufficient here to point out that the star α^2 CVn is a light variable in U, B, V, and at 3317 Å, 2460 Å, 2380 Å, 1910 Å, 1550 Å, 1430 Å, and 1330 Å. The star has a constant luminosity at 2945 Å and 2985 Å. The intensity of the metal lines varies periodically with a period of 5.47 days.

All the light curves in the region $\lambda\lambda$ 1910–2460 have a minimum at phase 0.0 when the rare earths have maximum lin strength in the spectrum of the star. At the same phase the V light curve has a maximum. All the light curves below 1910 Å have a minimum at phase 0.1–0.2 when the U, B and λ 3317 light curves show a maximum.

This is interpreted as variable blanketing: blocking of radiation between 2000 and 2600 Å and consequent backwarming in the V band; blocking shortward of 2000 Å and backwarming in the U and B bands and in the near Balmer continuum. The behaviour of α^2 CVn is similar to that of α Scl as far as the line blocking between 2000 Å and 2600 Å is concerned.

Now, how does line blanketing behave in stars with normal spectra? Underhill (1971) has shown the existence of a large amount of line blocking in the ultraviolet spectrum of main-sequence B-type stars and we should worry about any fitting of the observations to classical blanketed models. There is no reason to exclude that α^2 CVn and α Scl indicate how line blanketing works actually in any stellar atmosphere. Instead of increasing the continuum between the lines, it is conceivable that the radiation substracted in a certain region of the spectrum is degraded longward. Where it is reemitted can be well indicated by the two peculiar stars presented above.

A minor conclusion is that the region around 2900 Å seems to be free from both blocking and backwarming effects so that it could be safely used to give a star a $T_{\rm eff}$.

References

Bernacca, P. L. and Molnar, M. R.: 1972, *Astrophys. J.* **178**, No. 1.
Bradley, P. T. and Morton, D. C.: 1969, *Astrophys. J.* **156**, 687.
Klinglesmith, D. A.: 1971, NASA SP-3065.
Norris, J.: 1971, *Astrophys. J. Suppl.* **23**, 213.
Underhill, A. B.: 1971, *Proceedings OAO-Symposium*, Amherst, Mass., NASA-SP-310, 367.

DISCUSSION

Jaschek: Your data provide the first reliable basis for calling α Scl a spectrum variable, confirming thus earlier suspicions.

Schild: In noting a 2000 K $T_{\rm eff}$ difference between a model atmosphere and OAO II flux of α Scl, you mention the models of Klinglesmith; can you tell how line blocking was treated in the Klinglesmith model?

Bernacca: Klinglesmith's models are hydrogen blanketed models and the line blocking is treated in the usual way, that is by imposing the constancy of the integrated flux at each optical depth.

ABSOLUTE SOLAR INTENSITIES AND THE SOLAR CONSTANT

H. NECKEL

Hamburg Observatory

and

D. LABS

Heidelberg Observatory, F.R.G.

Abstract. Absolute intensity measurements made in high altitudes on board of an aircraft are of less accuracy than ground based observations made at locations with favourable atmospheric conditions. As things stand, the rms-errors appear to be $\geqslant 3\%$ and $\leqslant 2\%$ respectively. Some of the reasons are sketched.

The talk of Dr Code this morning has pointed out the tendency to *over*estimate absolute measurements, which are made from aircrafts at high altitudes, in comparison to ground-based observations. As you may remember, Dr Code preferred the solar energy distribution obtained by Arvesen *et al.* (1969), the accuracy of which was given as $\pm 3\%$ in the visual region, as $\pm 6\%$ at 3200 Å, and as $\pm 25\%$ at 3000 Å. These errors are clearly larger than those errors which are caused by atmospheric extinction for ground-based observations, provided these are made at places with good atmospheric conditions.

In fact, the gain in stability of atmospheric extinction which one gets, if one ascends e.g. from the Jungfraujoch Scientific Station at 3.6 km to a height of 12 km, is not as large as to compensate the disadvantages, which are necessarily connected with aircraft flights.

As at the ground, the short wave region below 0.3 μ is cut off completely also at 12 km altitude by the ozone layer. The infrared region beyond 3–4 μ is generally blocked by at least one, mostly more glass or quartz windows in front of the equipment. That means, the advantages ascribed to high altitude observations can concern the spectral region from 0.3 to 3 μ only.

The extinction is – of course – less at an altitude of 12 km than at 3.6 km, but it is much more difficult to obtain its correct, instantaneous value. The flight-time is limited (2–4 h) and usually permits only 1–2 spectral scans per day. To evaluate approximately the extinction, data of different flights have to be combined. This procedure requires that the (zenith-) extinction at a given altitude (e.g. 12 km) does not vary with the seasons (Arvesen's observations were made in August and October!) and that the extinction is the same above a desert area in the middle of a continent as it is above the ocean. These assumptions may be very misleading.

The accuracy obtainable for ground-based observations is to be seen in Figure 1 ($\lambda > 0.3$ μ!). Here a comparison is made between observations which were obtained mainly at high mountain stations with good atmospheric conditions, and the continuum of the 'Harvard Smithsonian Reference Atmosphere' (HSRA).

For wavelengths above 0.6 μ the standard deviation of one observation relative to

B. Hauck and B.E. Westerlund (eds.), Problems of Calibration of Absolute Magnitudes and Temperature of Stars, 149–152.
All Rights Reserved. Copyright © 1973 by the IAU.

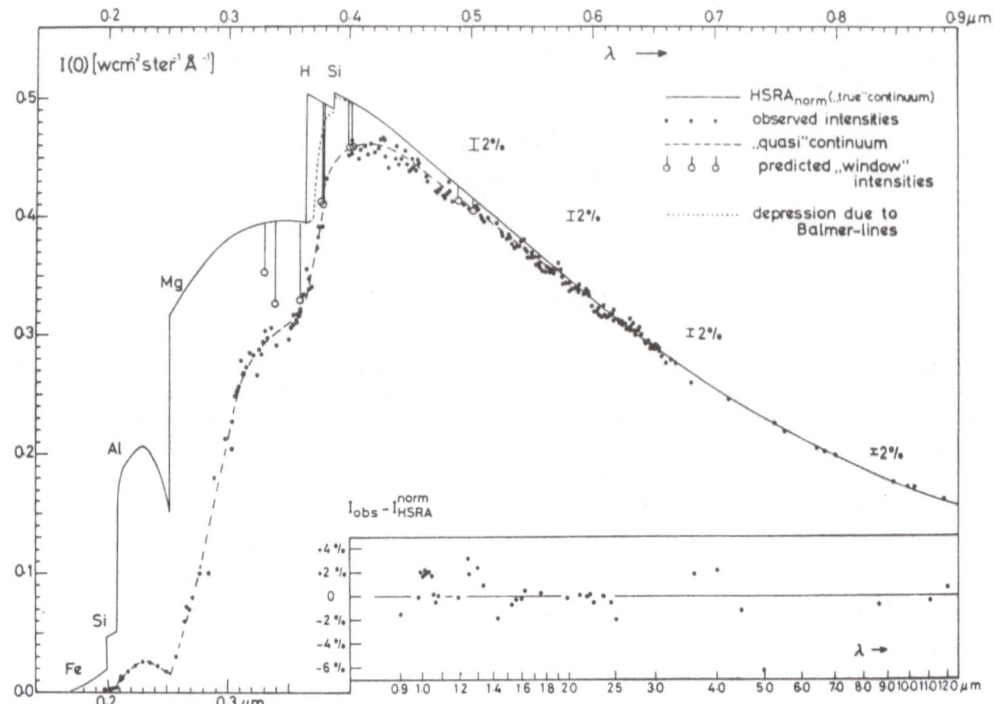

Fig. 1. Normalized HSRA continuum and the actual observations according to: Bonnet (1968; $\lambda < 0.29\ \mu$), Houtgast (1968; $\lambda\lambda$ 0.2977–0.3275 μ), Labs and Neckel (1968, 1970; $\lambda\lambda$ 0.3290–1.2470 μ), Pierce (1954; $\lambda\lambda$1.2986–2.5074 μ), Farmer and Todd (1964; $\lambda\lambda$3.6–5.0 μ), and Saiedy (1960; $\lambda\lambda$8.63–12.02 μ). Predicted window intensities and Balmer line depression, both being related to normalized HSRA continuum, according to Holweger (1970).

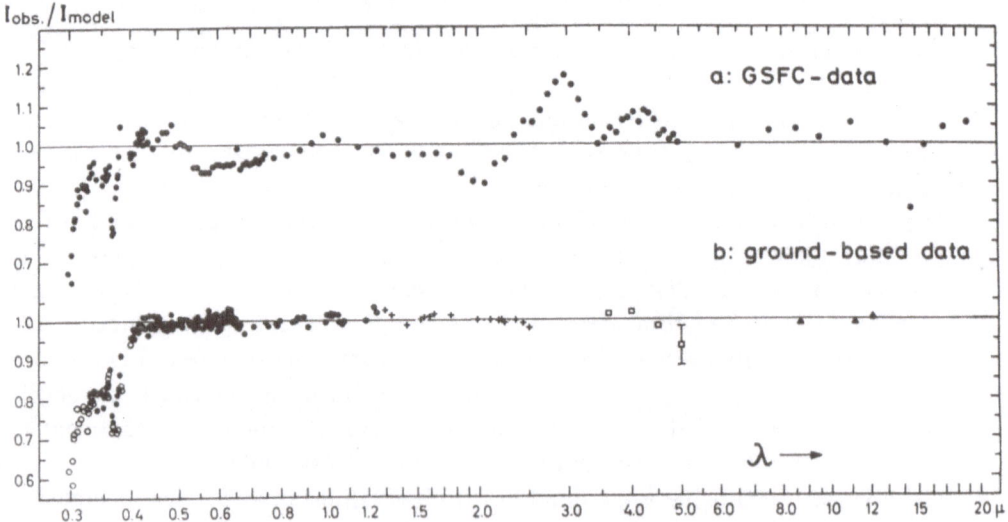

Fig. 2. Ratio of observed 'continuum' intensities to model-continuum. (Below 0.5 μ the observed intensities are those in the highest 'windows'. For details see Labs and Neckel, 1968, 1970.) (a) For the 'GSFC-continuum' corresponding to the irradiance observed within 'NASA 711 Galileo' flight experiment of GSFC. (b) For high mountain observations with careful determination of atmospheric extinction, according to Houtgast (○) 1970, Labs and Neckel (●) 1970, Pierce (+) 1954 Farmre, and Todd (□) 1964, and Saidy (▲) 1960.

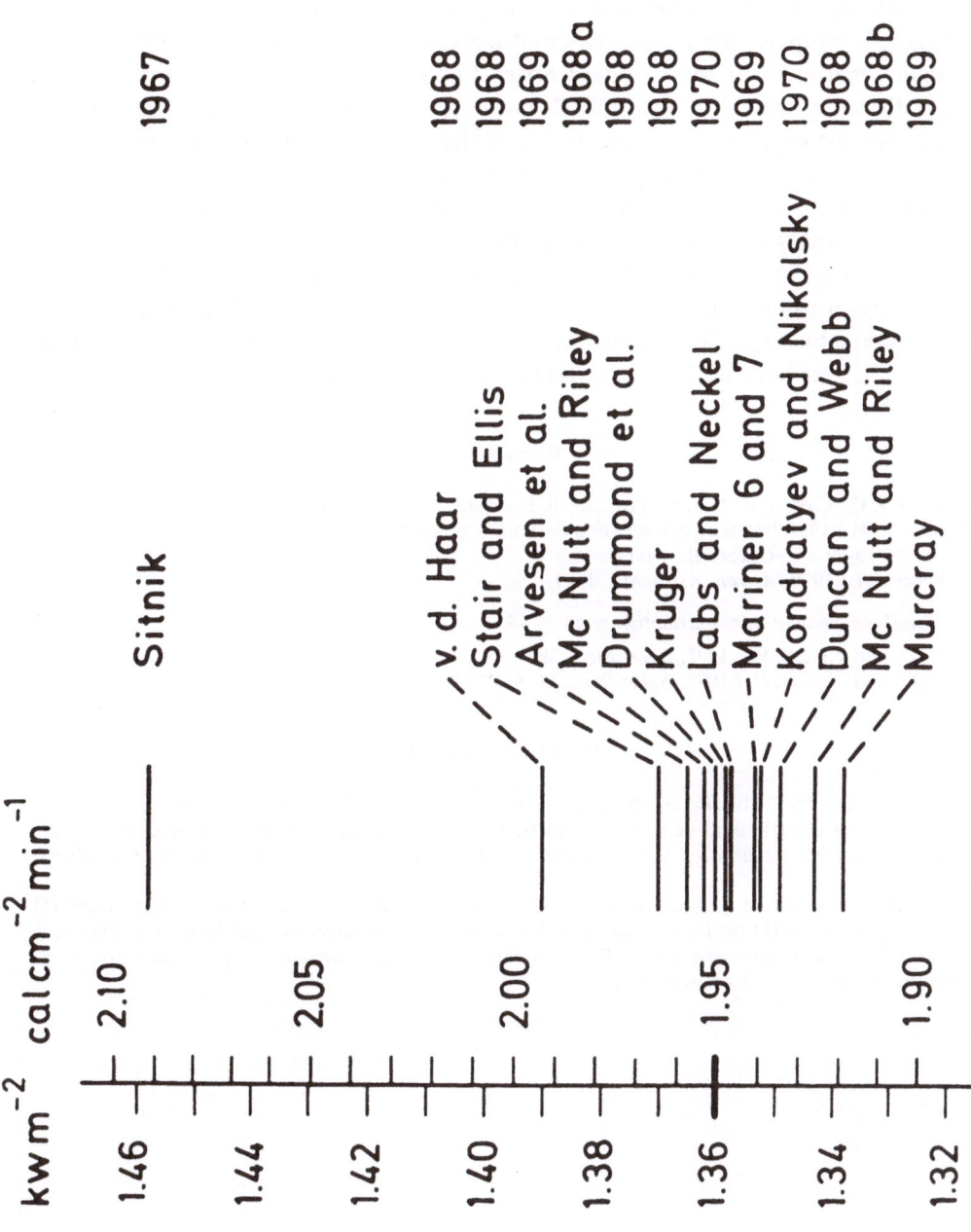

Fig. 3.   Recent measurements of the solar constant.

the HSRA continuum is $\pm 1.3\%$. Shortward of 0.6 μ the observed intensities are below the model continuum and form a 'quasi continuum'. Down to 0.36 μ the differences are fully explainable by the 'line haze' evaluated quantitatively by Holweger (1970). Also in this spectral region the mean error appears to be less than 2%.

Figure 2 compares the ground-based observations with the results obtained within a flight experiment of the Goddard Space Flight Center (GSFC). It appears very likely that the 'waves' of the 'GSFC-continuum' relative to the model continuum reflect just the experimental inaccuracies of the original irradiance data rather than intrinsic characteristics of the solar atmosphere.

Figure 3 may give some idea about the correctness of the absolute scale of the solar intensities discussed above. This figure compares all values of the solar constant obtained since 1967, including the value derived by us, which was mainly based on the ground-based irradiance data. The value of Arvesen (originally 1.390 kW cm^{-2}) has been lowered by 2.5%, due to a correction of the lamp calibration reported by Duncan (1969). The best value of the solar constant appears to be close to 1.95 cal cm^{-2} min^{-1} or 1.36 kw m^{-2}.

References

Arvesen, J. C., Griffin, R. N., and Pearson, B. D.: 1969, *Appl. Opt.* **8**, 2215.
Duncan, C. H.: 1969, 'Radiation Scales and the Solar Constant', Goddard Space Flight Center Report No. X-713-69-382, Greenbelt, Maryland.
Holweger, H.: 1970, *Astron. Astrophys.* **4**, 11.

For details concerning this contribution see:

Labs, D. and Neckel, H.: 1971, *Solar Phys.* **19**, 3.
Labs, D. and Neckel, H.: 1972, *Solar Phys.* **22**, 64.

DISCUSSION

Pecker: I agree with Dr Neckel on the great weight of good ground-based experiments. But the 'best fit' with HSRA model could be somewhat misleading, as Gingerich has precisely been using Labs and Neckel energy distribution in the spectrum (it is, so far as T is concerned, a purely empirical model).

Neckel: Figures 1 and 2 should demonstrate that it is possible to find a model, which obeys (1) the observations and (2) the fundamental physical laws such as hydrostatic equilibrium etc. There are many sets of observations, e.g. the GSFC-data and the Russian observations, which can not be represented by such a model atmosphere.

ABSOLUTE CALIBRATION OF STELLAR TEMPERATURES

H. KIENLE

Ege University Observatory Izmir, Turkey

and

D. LABS

Landessternwarte Heidelberg-Königstuhl, F.R.G.

Abstract. The scale of effective temperatures T_{eff} is based on observed absolute radiation temperatures T_r, which are defined by Planck's radiation law

$$i^*(\lambda)/i_{\text{Au}}(\lambda) = [\exp(c_2/\lambda T_{\text{Au}}) - 1]/[\exp(c_2/\lambda T^*) - 1],$$

where T_{Au} designs the absolute temperature of the gold point. A relative scale of radiation temperatures can be derived from spectrophotometric comparisons with a standard star. The absolute calibration of the standard star (α Lyr or Sun) demands a careful comparison with a standard radiation source of well known spectral energy distribution (Black Body or Synchrotron). With ground-based observations atmospheric extinction is to be taken into account; with extraterrestrial observations detectors may be used which are absolutely calibrated in a radiation laboratory under space conditions.

One of the relevant parameters which determine the physical state of a star is its *effective temperature* T_{eff}. It measures the total output of energy. With the Sun T_{eff} can be derived from measures of the solar constant, with stars from absolute bolometric magnitudes. From spectrophotometric observations of stellar spectra we can only derive *radiation temperatures* T_r based on the energy distribution of continuous spectra ('distribution temperature', 'gradient temperature'). The transition from observed T_r to T_{eff} is object of a theory of stellar atmospheres (*model atmospheres*).

The primary aim of observation must be the establishing of a *scale of absolute radiation temperatures* based on the fundamental definition

$$\frac{i_*(\lambda)}{i_{\text{Au}}(\lambda)} = \frac{\exp(c_2/\lambda T_{\text{Au}}) - 1}{\exp(c_2/\lambda T_*) - 1},$$

where T_{Au} denotes the temperature of melting gold on the thermodynamic scale ('gold point')

The problem of calibration. We have to compare the radiation of a star with a primary standard of radiation of known absolute temperature. The block scheme of Figure 1 shows how such a comparison may be performed. The radiation of star S after passing interstellar space meets a detector D (telescope + photometric device) either at a ground-based or an extraterrestrial orbiting observatory. If the *absolute response curve* of the whole detector arrangement is known from comparison with primary standard radiation we get from spectrophotometric observations the absolute intensity

$$J(\lambda)\,\Delta\lambda = \frac{1}{r^2} \int_{\Delta\lambda} J^*(\lambda)\, p_i(\lambda)\, p_a(\lambda)\, \mathrm{d}\lambda,$$

B. Hauck and B. E. Westerlund (eds.), Problems of Calibration of Absolute Magnitudes and Temperature of Stars, 153–155.
All Rights Reserved. Copyright © 1973 by the IAU.

where r = distance, $p_i(\lambda)$ interstellar absorption, and $p_a(\lambda)$ atmospheric extinction.

If the transmission coefficient of the terrestrial atmosphere is carefully determined (for the time of observation!) only interstellar absorption ('reddening') remains as a possible source of systematic error of observed temperatures $T(\lambda)$.

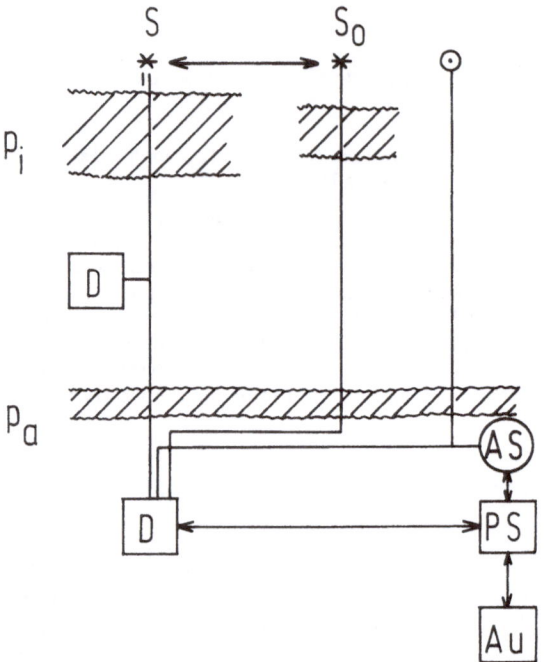

Fig. 1.   S = star, S_0 = standard star, $\bigcirc$ = sunlike star, p_i = interstellar absorption, p_a = atmospheric absorption, D = detector, AS = artificial star, PS = primary standard, Au = gold point.

Establishing the temperature scale can be split up into two parts:

(a) By intercomparison of stars of different spectral types with a standard star S_0 we get a scale of *relative temperatures* T/T_0.

(b) Absolute calibration of the standard star allows the transition from T/T_0 to *absolute temperature T*.

As standard stars spectral type A0 stars (especially α Lyr) have been used in the past. It would perhaps be preferable to choose early B stars (ηU Ma, S Mon) because of the better definition of the continuum, correction for lines being smaller.

The *absolute temperature scale* derived from comparison with a standard star can be checked by data from absolute calibration of the Sun which is to be regarded as a representative G2V star with known spectral energy distribution.

Primary standard radiation sources are the Black Body (BB) at about 3000 K for wavelengths larger than 2000 Å and the *Synchrotron* for the extreme UV below 3600 Å. Both must be calibrated by comparison with BB-radiation at fix points of the thermodynamic temperature scale (melting point of gold or copper). The overlapping part of the energy curves between 2000 Å and 3600 Å offer a useful checking of the absolute values.

Absolute calibration may be performed either by deriving the absolute *response curve* of the whole detector arrangement or by interference of an '*artificial star*' of well known spectral energy distribution. Often the response curve has been calculated from its different components (absorption by optical elements, sensitivity of photometric elements). No doubt it will be better to calibrate the whole detector in the laboratory with help of primary or secondary standards.

If a well calibrated artificial star is used for comparison, the whole detector arrangement will be eliminated ('zero method').

Calibration Work at Heidelberg Radiation Laboratory

(a) Absolute spectral energy of Sun's radiation from $\lambda = 0.3 \mu$ to 1.2μ. Standard value of solar constant.

(b) Blackbody radiation for $T = 3000$ K.

(c) Comparison Synchrotronradiation – Blackbody.

(d) Secondary standards for the region from 0.16 to 10 μ.

(e) Artificial star.

COMPARISON OF THEORETICAL AND OBSERVED INDICES
FOR THREE PHOTOELECTRIC PHOTOMETRIC SYSTEMS

F. RUFENER and A. MAEDER

Geneva Observatory, Switzerland

Abstract. A comparison of the theoretical and observed indices has been made for the UBV, $uvby$ and the Geneva's $UBVB_1 B_2 V_1 G$ systems. In the actual state of the photometric calibrations, coincidence between theoretical and observed indices better than 0^m1 are probably not significant.

1. Introduction

The comparison of theoretical fluxes obtained by means of model atmospheres with observed fluxes can be made in at least two manners. By direct comparison with spectrophotometric recordings carefully calibrated for example relative to the black body. Another comparison is possible by means of a wide or an intermediate band photometry. One must then know the actual shapes of the pass-bands of the photometry considered. Since it is difficult to determine these directly with a few percent accuracy, the majority of the pass-bands of photometric systems have been calibrated by means of spectrophotometric recordings. The methods of calibration are numerous. Two difficulties complicate this indirect calibration, on the one hand the spectrophotometric reference generally (α Lyr) posesses several absolute calibrations the respective qualities of which are not completely obvious; on the other hand, the instrument profile of the spectrophotometer is not always known explicitly and the recordings are published in the form of a limited number of fluxes measured in more or less 'clean' windows of the spectrum (the observed line blocking is often not mentioned).

This note intends to compare the observed sequences with those computed by means of theoretical model stellar atmospheres; this comparison is made with the three following photometries: Johnson-Morgan's UBV, Strömgrens $uvby$ and Geneva's $UBVB_1B_2V_1G$. In each case, the same four models will be used, i.e. those by Mihalas (1966) describing main sequence stars between B8 V and A2 V. These are model atmospheres calculated with blanketing by hydrogen lines, but without metallic lines for $\log g = 4$ and $\theta_{\mathrm{eff}} = 0.40$; 0.45; 0.50; 0.55. For each photometry, we use these models to compute the colour indices by following the procedures recommended by the authors who have calibrated the pass-bands of these systems. The computation program is due to E. Peytremann. The integrations are carried out by Simpson's method, except for the case of lines where the program automatically applies the trapezoidal method with an integration step that varies according to the region of the spectrum. The results obtained differ sometimes from those published by other authors. These differences must probably be attributed to different methods of integration. Here, the three comparisons are made by applying the same computational technique.

B. Hauck and B. E. Westerlund (eds.), Problems of Calibration of Absolute Magnitudes and Temperature of Stars, 156–162.

2. *UBV* Photometry

The determination of this photometry's pass-bands has been the object of numerous comments which we will not repeat. A difficulty arises from the fact that the observations of the indices $U - B$ are not completely reduced to outside the atmosphere. On the other hand, the existent calibrations use different absolute spectrophotometric references. Let us consider three recent descriptions:

A. PROCEDURE OF AZUSIENIS AND STRAIZYS (1969)

In Table I of the paper by these authors, four response curves restore the probable profiles of the original natural system. These are the functions $\varphi_U p$; $\varphi_B p$; φ_B; φ_V. We compute the natural indices by taking Mihalas (1966) models for F_λ

$$C_{U-B} = 2.5 \log \int F_\lambda \varphi_B p \, d\lambda - 2.5 \log \int F_\lambda \varphi_U p \, d\lambda$$

$$C_{B-V} = 2.5 \log \int F_\lambda \varphi_V \, d\lambda - 2.5 \log \int F_\lambda \varphi_B \, d\lambda.$$

The normalization proposed by the authors is

$$U - B = C_{U-B} - 1.33$$
$$B - V = C_{B-V} + 0.67.$$

This has been established by means of spectrophotometric recording for which the calibration of α Lyr adopted by Code (1960) is considered.

B. PROCEDURE OF MATTHEWS AND SANDAGE (1963)

One must adopt for the three filters UBV the response curves $S(\lambda)_1$ given in Table A1 by Matthews and Sandage (1963); these correspond to the natural response for the zenith of Mount-Wilson. The calculation of the natural indices of the models is obtained by means of the relations:

$$(u - b)_1 = 2.5 \log \int F_\lambda S_B(\lambda)_1 \, d\lambda - 2.5 \log \int F_\lambda S_U(\lambda)_1 \, d\lambda$$

$$(b - v)_1 = 2.5 \log \int F_\lambda S_V(\lambda)_1 \, d\lambda - 2.5 \log \int F_\lambda S_B(\lambda)_1 \, d\lambda.$$

The transformations recommended by Matthews and Sandage (1963) to obtain the standard indices are

$$U - B = 0.921 (u - b)_1 - 1.308$$
$$B - V = 1.024 (b - v)_1 + 0.81.$$

They have been obtained with spectrophotometric continua calibrated after the α Lyr description from Code (1960). No blocking correction is applied.

C. PROCEDURE OF MATSUSHIMA AND HALL (1969)

That is the Matthews and Sandage's method modified only by the change of coefficients in the transformations to standard

$$U - B = 0.896\,(u - b)_1 - 1.288$$
$$B - V = 0.982\,(b - v)_1 + 0.791\,.$$

These authors have computed them by filtering spectrophotometric recordings calibrated by Hayes' (1970) calibration of α Lyr. Moreover, they applied correction estimations for the effect of the hydrogen lines absent in the spectrophotometric continua (blocking correction).

Figure 1 represents with a dotted line the theoretical sequences A, B and C computed by means of the above mentioned procedures, for the four models of Mihalas (1966). The full line is the observed sequence defined by Johnson (1966).

3. *u v b y* Photometry

According to Crawford (1966) the optical transmission of the filters is sufficient to define the pass-bands of the *uvby* photometry. It suffices to know in addition a normalisation constant for each filter (this is equal to saying that over the width of each filter one considers the response of the cell, the reflections of the mirror and atmospheric transmission to be free of chromatic effects). Matsushima (1969) has published optical profiles as well as the normalisation constants which calibrate the system. These result from a filtration of spectrophotometric continua coherent with the calibration of α Lyr by Hayes (1970). A particular correction is estimated to compensate for the effect of the H δ line on the *v* filter. No correction for blocking is necessary for the indices *u-b* and *b-y*. We have applied this procedure to the four models by Mihalas. Figure 2 represents with a dotted line the resulting computed sequence. The observed sequence is the one established by Matsushima (1969) as being the best mean of observations for that region of the main sequence; this is shown with a full line.

4. $UBVB_1B_2V_1G$ Photometry

The pass-bands of the Geneva Observatory photometric system have been determined by Rufener and Maeder (1971). The method applies also spectrophotometric continua as means of calibration but is rather different from those used for the above mentioned photometries. Indeed, instead of establishing coherence between the spectrophotometric data and the actual observations by means of a linear transformation or by the choice of additive constants, this method seeks, by means of a representation by Fourier series, an optimum adjustment of the function $S(\lambda)=r(\lambda)s(\lambda)\varepsilon(\lambda)$ which, apart from giving the optical response of each filter also characterises the standard system. We have $r(\lambda)$ for the reflectivity of the telescope and the transmission of the Fabry lens. $s(\lambda)$ represents the electrical response of the photomultiplier. $\varepsilon(\lambda)$ corrects

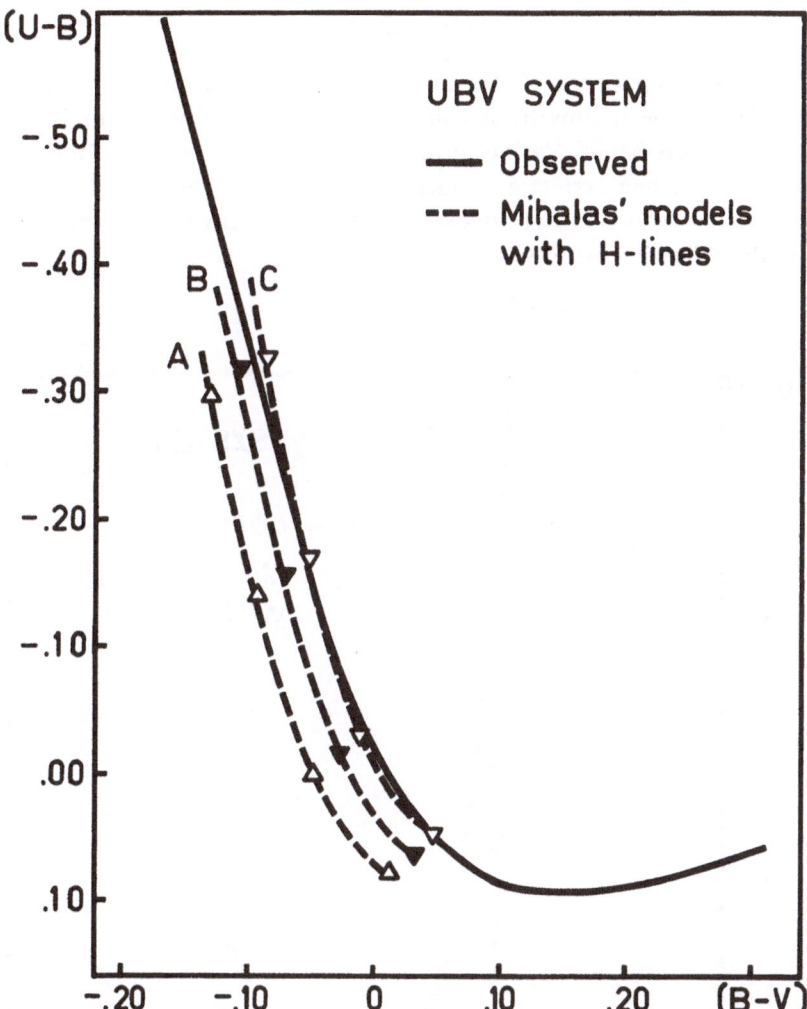

Fig. 1.   Comparison of the observed and computed sequences for the response curves: (A) Azusienis and Straizys (1969); (B) Matthews and Sandage (1963); (C) Matsushima and Hall (1969).

the residual imperfections of the reductions outside the atmosphere. This function is close to unity.

Two calibrations of the reference star (α Lyr) have been used, those by Code (1960) and by Hayes (1970). The spectrophotometric continua used were those of O stars and subdwarfs showing very few strong lines. 55 Cyg was also used but was weighted much less. No correction of blocking in the sense of Matsushima was made, Figure 3 shows Mihalas' models filtered with the Geneva photometry pass-bands; no normalisation is necessary. Dotted line A is according to Code's (1960) calibration, dotted line B according to Hayes (1970). The sequence with a full line is the observed sequence.

5. Conclusion

For these three comparisons of the same models with the observed sequences, the fluctuation of the deviations from coincidence is essentially due to the errors of definition of the pass-bands. Various processes used by photometrists for observations and calibration of their system are much too inaccurate, their use should cease i.e.

(1) The partial reduction out atmosphere. That is to say ignoring the terms K_2C and K_3 in the following expression

$$m_0 = m_Z - K_1F_Z - K_2CF_Z - K_3F_Z^2,$$

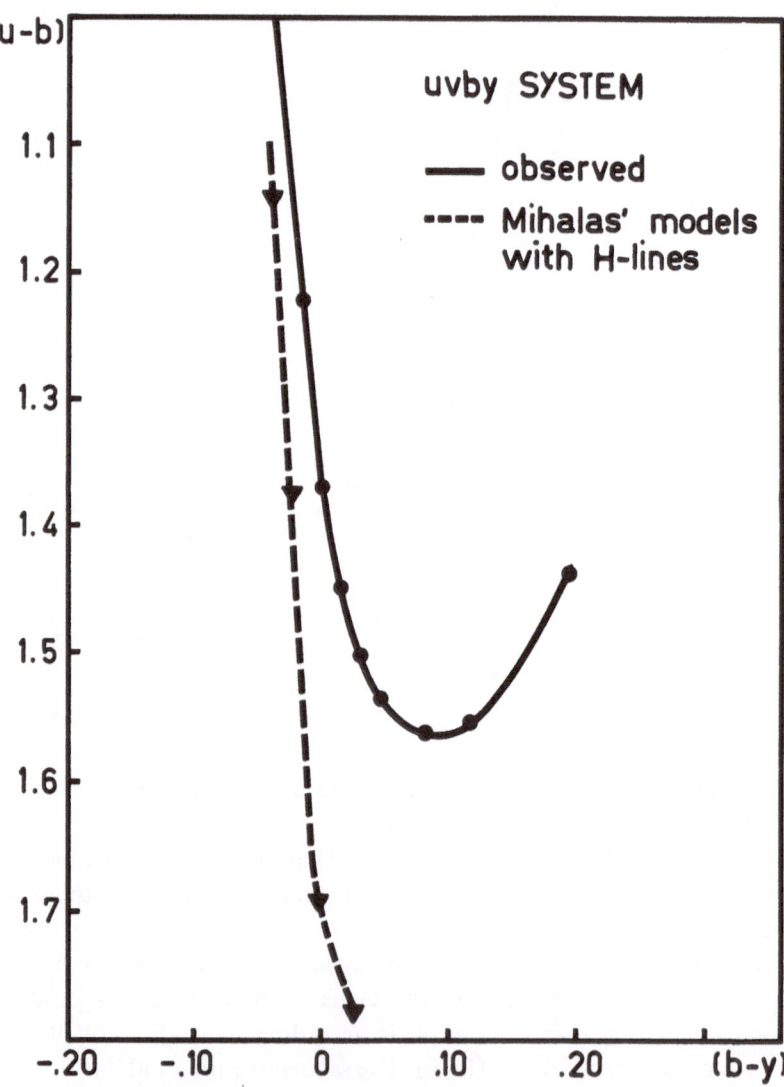

Fig. 2.   Comparison of the observed and computed sequences following Matsushima (1969).

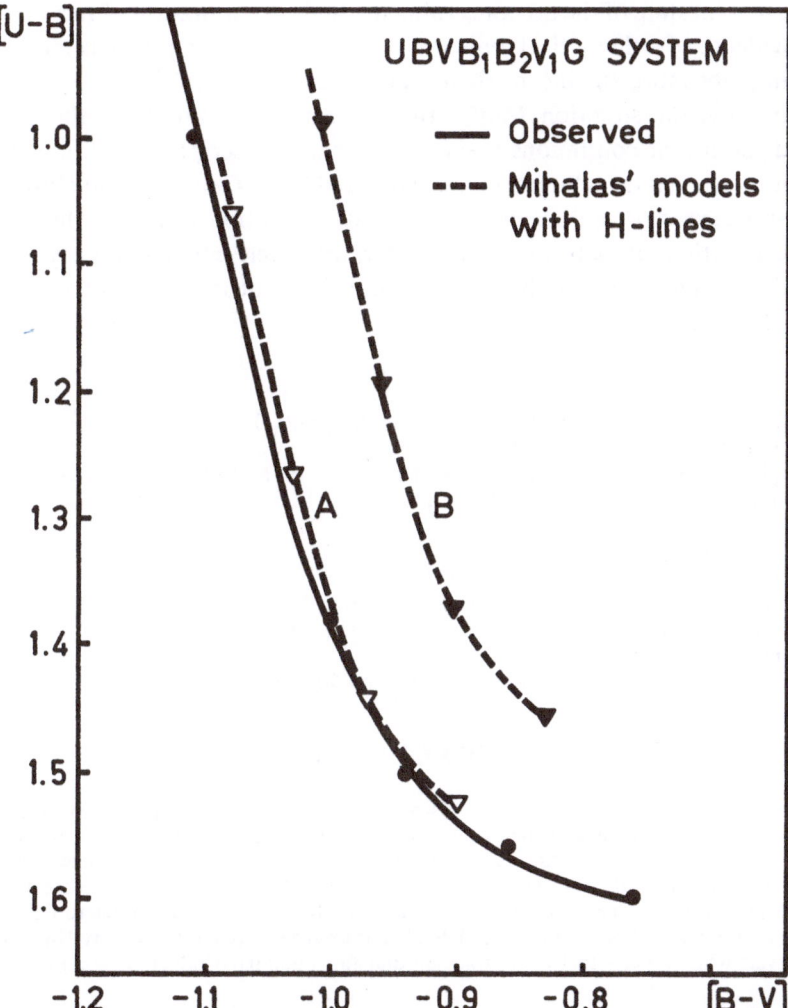

Fig. 3.   Comparison of the observed and computed sequences with: (A) Code's calibration;
(B) Hayes calibration.

where m_0, m_z are the magnitudes respectively out atmosphere and at ground, C is the colour index of the star, F_z the air mass and K_1, K_2 and K_3 the extinction coefficients.

(2) Identifying of the filter's optical response with response of the natural system. This implies, in the filter's spectral interval, to neglect the sensitivity variations of the PM as well as the mirrors reflecting power variations.

(3) The empirical estimate of blocking corrections applied to the calibrating spectrophotometric continua (frequently neglected correction).

(4) The fact to remedy to items 1 to 3 through a linear relation between the calculated and observed colours gives pass-bands only adapted to the stars considered for calibration (in general unreddened B and A stars).

In the present state of the art concerning methods of calibration of these pass-bands, the coincidence of a theoretical sequence with an observed sequence is misleading and is certainly not more significant than a deviation from coincidence $<0^m.1$. It is not easy to improve this situation. May be that a harmonisation of the methods of calibration and the use of continuous spectrophotometric descriptions calibrated properly for a selection of stars including in particular reddened O stars would allow to reduce these uncertainties of calibration. An alternative solution would be the very careful global calibration of natural system (filter, PM, telescope) relatively to physical standards of radiation and the quasi simultaneous observations of the considered photometry standard stars.

References

Azusienis, A. and Straizys, V.: 1969, *Soviet Astron. AJ* **13** (2), 316.
Code, A. D.: 1960, *Stars and Stellar Systems* **6**, 50.
Crawford, D. L.: 1966, in K. Lodén, L. O. Lodén, and U. Sinnerstad (eds.), 'Spectral Classification and Multicolour Photometry', *IAU Symp.* **24**, 170.
Hayes, D. S.: 1970, *Astrophys. J.* **159**, 165.
Johnson, H. L.: 1966, *Ann. Rev. Astron. Astrophys.* **5**, 571.
Matsushima, S.: 1969, *Astrophys. J.* **158**, 1137.
Matsushima, S. and Hall, D. L.: 1969, *Astrophys. J.* **156**, 779.
Matthews, T. A. and Sandage, A. R.: 1963, *Astrophys. J.* **138**, 30.
Mihalas, D.: 1966, *Astrophys. J. Suppl.* **13**, 1.
Rufener, F. and Maeder, A.: 1971, *Astron. Astrophys. Suppl.* **4**, 43.

DISCUSSION

Jaschek: I would just like to make a brief historical comment. At La Plata we have made a photometric catalogue (which is now at the printer) in which we have collected all photoelectric measurements made after those of Stebbins 1917. As a by-product we have tried to make out what color system each observer used. For the nearly two thousand references listed, we were uncapable of getting details for about one third, simply because the observers did not bother to tell if they used filters and what kind of photocells they used. Such a procedure is even today followed by some variable star observers who measure to 1% accuracy an undefined and irreproductible quantity.

FIRST RESULTS OF THE ABSOLUTE CALIBRATION OF
THE u, v, b, y SYSTEM OF FILTERS

S. TAPIA and H. L. JOHNSON

Steward Observatory, University of Arizona, Tucson, Ariz., U.S.A.

and

D. L. CRAWFORD

Kitt Peak National Observatory*, Tucson, Ariz., U.S.A.

Abstract. Absolute flux densities for a zero magnitude A0V star at the effective wavelengths of the u, v, b, y spectral bands are presented.

King (1952) has shown that a magnitude obtained in a given spectral band behaves, to a first approximation, like monochromatic magnitude at an effective wavelength given by

$$\lambda_{\text{eff}} = \frac{\int\limits_0^\infty \lambda \phi(\lambda)\,\mathrm{d}\lambda}{\int\limits_0^\infty \phi(\lambda)\,\mathrm{d}\lambda},$$

where $\phi(\lambda)$ is the sensitivity of the measuring instrument defining the spectral band. If the absolute flux density of a source is known at the effective wavelength of a given spectral band, the relation between instrumental and monochromatic magnitudes permits the absolute calibration of that spectral band. We have used the relation derived by King (1952) to obtain the absolute calibration of the u, v, b, y system of filters.

Several standard stars and a lamp of calibrated spectral distribution were observed with u, v, b, y filters at the Kitt Peak National Observatory. The observed magnitudes were combined with the relative sensitivity of the photometer to derive absolute flux densities. The first results have been used to test the calibration computing the absolute flux densities of a zero magnitude A0 V star at the effective wavelengths of the u, v, b, y spectral bands. The computed effective wavelengths and absolute flux densities are given in Table I. A comparison of our results with published absolute flux densities for A0 V stars is presented in Figure 1.

A check of the spectral distribution of the lamp used in the observations is in progress. The final results and directions for use of the calibration will be published elsewhere.

* Operated by the Association of Universities for Research in Astronomy, Inc., under contract with the National Science Foundation.

B. Hauck and B. E. Westerlund (eds.), Problems of Calibration of Absolute Magnitudes and Temperature of Stars, 163–164.

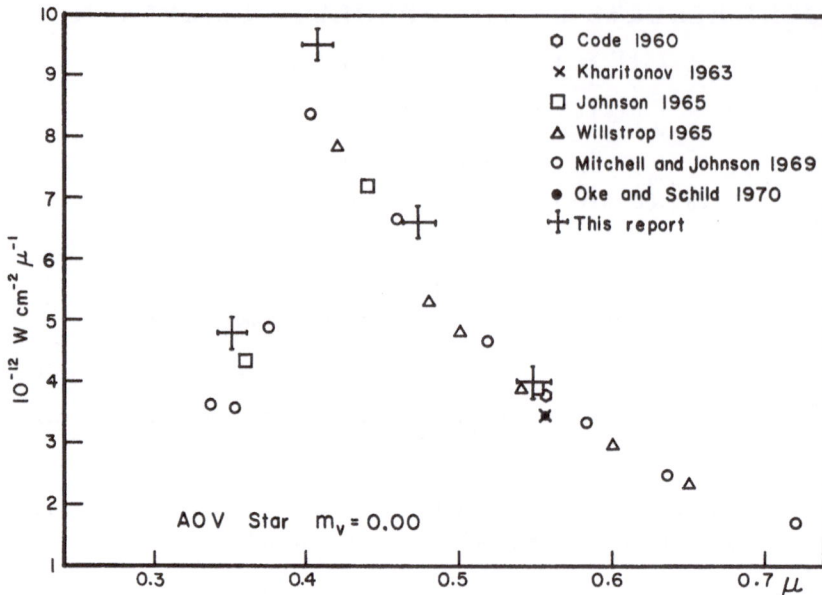

Fig. 1. Absolute flux densities, obtained from different calibrations, for a zero magnitude A0 V star. The abscissae are wavelengths in units of micron and the ordinates are flux densities in units of 10^{-12} W cm^{-1} μ^{-1}. The values of Table I are shown with crosses.

TABLE I

Absolute calibration of the
u, v, b, y system. First results

Filter	λ_{eff}	Absolute flux density (zero mag. A0 V star)
u	$0.348 \, \mu$	4.82×10^{-12} W cm^{-2} μ^{-1}
v	$0.401 \, \mu$	9.52×10^{-12} W cm^{-2} μ^{-1}
b	$0.462 \, \mu$	6.63×10^{-12} W cm^{-2} μ^{-1}
y	$0.545 \, \mu$	4.00×10^{-12} W cm^{-2} μ^{-1}

References

Code, A. D.: 1960, *Stars and Stellar Systems* **6**, 85.
Johnson, H. L.: 1965, *Commun. Lunar Planet. Lab.* **3**, No. 53, 73.
Kharitonov, A.: 1963, *Soviet Astron.* **7**, 258.
King, I.: 1952, *Astron. J.* **57**, 253.
Mitchell, R. I. and Johnson, H. L.: 1969, *Commun. Lunar Planetary Lab.* **8**, No. 132, 1.
Oke, J. B. and Schild, R. E.: 1970, *Astrophys. J.* **161**, 1022.
Willstrop, R. V.: 1965, *Mem. Roy. Astron. Soc.* **69**, 83.

RADII AND FLUXES OF LONG-PERIOD CEPHEIDS

A. OPOLSKI

Wrocław Astronomical Observatory, Wrocław, Poland

Abstract. The radii of 9 long-period Cepheids have been determined by means of the modified Wesselink method and some general conclusions have been formulated. The relations $R - P$ and $F_v - M_v$ for stars recognized as fundamental and first overtone pulsators have been established.

The determination of stellar fluxes F_v in the V system, relating to the solar flux F_v may be based on the fundamental relation:

$$\log F_v/F_{v\odot} = -0.4\,(M_v - M_{v\odot}) - 2\log R/R_{\odot} \tag{1}$$

In order to use this relation for long-period Cepheids we should secure the correct values of their radii. In this case they are more essential than the absolute magnitudes M_v because of the coefficients 0.4 and 2 respectively. The radii of pulsating stars are usually computed with the well known Wesselink method, which has been applied in our considerations but with some modifications. The proposed alterations should facilitate the practical use of this method and should increase the accuracy of the results.

In the Wesselink method we try to separate two factors acting simultaneously and defining the observed changes of star magnitude: the changes of fluxes and of radii

$$\Delta V = -5\log\,(1 + \Delta R/R) - 2.5\,\Delta\log F_v, \tag{2}$$

where ΔR is the displacement resulting from the integration of the radial velocities. In practice the component $5\log(1+\Delta R/R)$ is most frequently replaced by a more convenient but only approximatively equivalent value $2.17(\Delta R/R)$.

But we may proceed in the following way. Let us fix that R_{max} denotes the maximum of the radius, which for the Cepheids occur near the phases 0.3–0.4 and is rather flat. Now we introduce the correcting factor c in order to get the exact equality

$$5\log\,(1 + \Delta R/R_{max}) = 2.17\,(\Delta R/R_{max})\cdot c. \tag{3}$$

The value c depends in small degree on $\Delta R/R_{max}$:

$\Delta R/R_{max}$	c
0.00	1.000
−0.05	1.028
−0.10	1.055
−0.15	1.084
−0.20	1.116

and can be computed even by applying an approximate value of R_{max}. The Equation

B. Hauck and B. E. Westerlund (eds.), Problems of Calibration of Absolute Magnitudes and Temperature of Stars, 165–170.

(3) can be transformed as:

$$5 \log (1 + \Delta R/R_{\max}) = Nr, \qquad (4)$$

where $N = 2.17/R_{\max}$ determines the maximum value of the radius, and the 'rectified displacement' r:

$$r = \Delta R \cdot c \qquad (5)$$

assures the exact proportionality to the photometric effects expressed in the magnitude scale.

By this method we have for each phase three values determining the state of a star: V, $(B-V)$, and r. The farther procedure can be continued in two ways: graphically or numerically.

1. Graphical Method

According to Wesselink assumption for two phases with equal $(B-V)$ or for $\Delta(B-V)=0$, we have also the same values of F_v or $\Delta \log F_v = 0$. So the Equation (2) is reduced to

$$\Delta V = - N\Delta r \qquad (6)$$
$$\text{or} \quad N = - \Delta V/\Delta r; \quad \text{or} \quad R_{\max} = - 2.17(\Delta r/\Delta V).$$

Let us draw two graphs with the common axis $(B-V)$, Figure 1. Perpendicularly are V axis upwards and r axis downwards. By means of the values V, $(B-V)$, and r

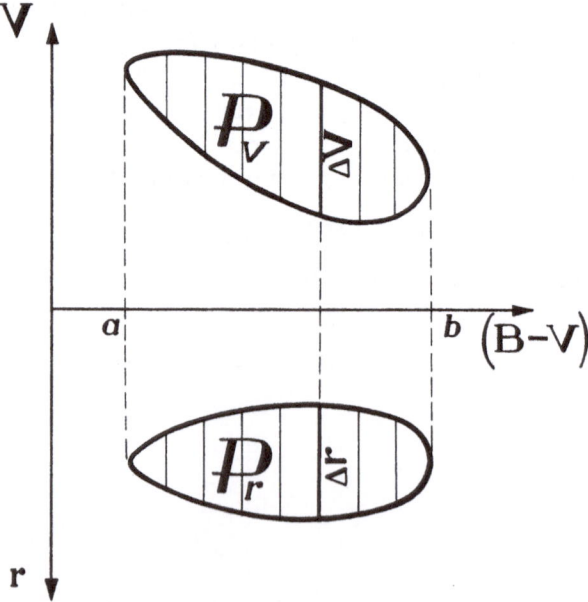

Fig. 1. Graphical method of N and $R_{\max}$ determination.

we trace two loops described by star during one period. Each line perpendicular to the $(B-V)$ axis and cutting both loops passes through the points with equal $(B-V)$. So the sections of the loops ΔV and Δr can be used as the values needed in the formula (6). But each such a cutting has this property. So it follows that N can be calculated as

$$N = -\frac{\int_{a}^{b} \Delta V \, \mathrm{d}(B-V)}{\int_{a}^{b} \Delta r \, \mathrm{d}(B-V)}, \tag{7}$$

where the integration includes the whole range of $(B-V)$ changes. In practice this integration can simply be executed by measuring the surfaces of the loops P_v and P_r.

$$N = P_v/P_r; \qquad R_{\max} = 2.17 P_r/P_v \tag{8}$$

2. Numerical Method

In this method we change the Wesselink basing assumption and enlarge the range of its applicability by introducing the coefficient k so that

$$-2.5\Delta \log F_v = k\Delta(B-V). \tag{9}$$

The coefficient k is regarded as a constant or as a mean value in the case, when it depends on the phases used in the formula (9). In terms of fluxes the Wesselink assumption: $\Delta \log F_v = 0$, when $\Delta(B-V)=0$ is equivalent to the statement that F_v is a monotonic function of F_B, but not of the shape $F_v = AF_B$, with constant A. In this last case we would have constant $(B-V)$ and it would not be possible to separate photometrically the changes of fluxes from the changes of radius. Whereas according to the assumption (9) we have

$$F_v = AF_B^a; \qquad a \neq 1; \qquad k = \frac{a}{1-a}. \tag{10}$$

This formula with two parameters A and a can represent, at least in approximation, a rather large class of real relations between both kinds of fluxes in the range of changes occuring during the star pulsation. Though it is to be remarked that in reality the fluxes as the functions of many atmospheric parameters – model, temperature, gravity – cannot be connected by only one function during the whole period.

By means of the formulas (4) and (9) the Equation (2) gains the simple shape:

$$\Delta V = -N\Delta r + k\Delta(B-V). \tag{11}$$

This type of relation was used by Opolski and Krawiecka (1956) and by Latyshev (1964). The differences ΔV, Δr, and $\Delta(B-V)$ can be calculated for each free chosen pair of phases. But in order to decrease the influence of the observational errors, it seems to be advantageous to calculate first the mean values $\bar{V}$, $\bar{r}$ and $\overline{(B-V)}$, and then to use the differences between the individual values and their mean: $\Delta V = V - \bar{V}$; $\Delta r = r - \bar{r}$; $\Delta(B-V) = (B-V) - \overline{(B-V)}$. So we can form the system of the Equations (11) with two unknown N and k. Taking into consideration the fact that $\Delta(B-V)$ are less accurate than ΔV, we can solve this system in the form

$$(B - V) = \frac{1}{k} V + \frac{N}{k} \Delta_r. \tag{12}$$

The individual values V, $B-V$ and r should be suitably distributed on the light-, colour- and displacement-curves. By this method we can get for each Cepheid two characteristic quantities: N – defining the radius of the star and k – which is a measure of the relation between the fluxes F_v and F_B. Also the accuracy of the values N and k can be calculated by standard methods.

We remark still, that the coefficient k may be obtained separately by the graphical method analogically to the N determination. For two phases with $\Delta r = 0$ we have $\Delta V = k \Delta(B-V)$ and the surface of two loops on the planes in the coordinate systems V, r and $(B-V), r$ determine the coefficient k.

As an example of the application of the above described methods we present in Table I the results obtained for 9 Cepheids. General remarks which can be deduced from these results are as follows:

(1) There are no systematic differences between N values obtained by graphical and numerical methods. But in Table I the results received by numerical methods are given.

TABLE I
Radii and fluxes of Cepheids

Star	$\log P$	$\langle M_V \rangle_{\text{int}}$	$\overline{\text{Sp}}$	$\log R_{\max}$	$\log \bar{R}$	$\log \dfrac{F_v}{F_{v\odot}}$	$\left(\log \dfrac{F_v}{F_{v\odot}}\right)$ F8
η Aql	0.855	-3.9	F 8.8	7.68	7.66	-0.22	-0.18
W Sgr	0.880	-4.0	F 7.7	7.69	7.67	-0.19	-0.21
S Sge	0.923	-4.1	F 8.7	7.75	7.73	-0.26	-0.22
β Dor	0.993	-4.3	F 8.1	7.79	7.77	-0.27	-0.27
ζ Gem	1.007	-4.3	F 9.0	7.79	7.78	-0.27	-0.22
X Cyg	1.215	-4.8	G 0.0	8.03	8.00	-0.56	-0.46
T Mon	1.432	-5.5	G 1.1	8.14	8.10	-0.50	-0.34
l Car	1.551	-5.9	F 9.1	8.18	8.15	-0.42	-0.36
SV Vul	1.655	-6.1	G 2.1	8.40	8.36	-0.77	-0.57

(2) When we try to improve the accuracy of N and k by shifting the phases of the 'rectified displacement' r, as it was suggested by Fernie and Hube (1967), it appears that the smallest errors of N and k occures at the diminishing of the phases r (and ΔR) by about $0\overset{\text{P}}{.}05$. This is particularly essential near the phases of sharp minimum of the radius. From the photometric data we should fix $R_{\min}$ in the phases near 0.87, whereas from the radial velocities $R_{\min}$ appears later, at the phases 0.92. This can be explained as due to the fact that, after common falling down of the whole atmosphere, the

movement upwards begins first in the photospheric layer, responsible for the radiation measured in the B and V systems, and then $0^{p}.05$ later, in the upper layers of the atmosphere, where the narrow absorption lines used for radial velocity determination are formed. This may be connected with the emission effects observed in some lines at these phases. Similar conclusion is to be found in the paper by Latyshev (1964).

(3) The best accuracy of the results gained by suitable diminishing of r phases is joint with the smallest value of N and consequently with the greatest value of R_{max}. The mean values $\bar{R} = R_{max} - (\frac{1}{2}$ amplitude of $\Delta R)$ given in Table I, for long-period Cepheids are systematically greater than values given by Fernie (1968). Following Fernie's considerations we can establish the period-radius relation with the division on the fundamental and first overtone pulsators, Figure 2. From the investigated

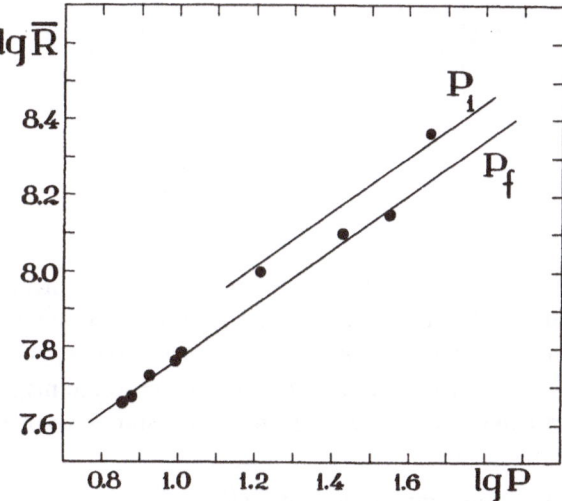

Fig. 2.   Radius-period relations for fundamental and first overtone pulsators.

stars only X Cyg and SV Vul can be regarded as pulsating in the first overtone. For fundamental pulsators we have the relation:

$$\log \bar{R} = 0.904 \log P + 6.88. \tag{13}$$

From the separation of two lines we can estimate the ratio of the periods P_1/P_f as about 0.7.

The values of R_{max} have been applied in the formula (1) in order to calculate the relative fluxes for Cepheids under consideration. As the absolute magnitudes the values $\langle M_v \rangle_{int}$ given by Fernie and Hube (1968) have been accepted, Table I. We can expect that for a homogeneous group of stars the fluxes depend chiefly on the absolute magnitude and on the spectral type. To remove the second dependence a correction for the differences of spectral types ΔSp have been used:

$$\Delta \log F_v = -0.05 \Delta \text{Sp}, \tag{14}$$

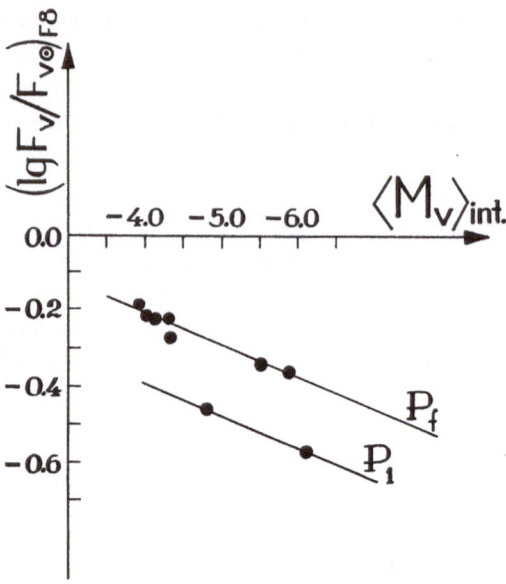

Fig. 3. Relative fluxes-absolute magnitudes relations for fundamental and first overtone pulsators.

where ΔSp is expressed in the tenth of spectral class. This formula results from the relation $(B-V)$-surface brightness given by Parsons and Bouw (1971) and from the relation $(B-V)$-spectral type. In this way we got the fluxes reduced to the mean spectral type F8. The dependance of these values on the absolute magnitudes is shown in Figure 3. Again we can observe two relations. For 7 stars recognized as fundamental pulsators we have

$$\log F_v/F_{v\odot} = 0.0822 \langle M_v \rangle_{\text{int}} + 0.12.$$ (15)

For the first overtone pulsators the values of $\log F_v/F_{v\odot}$ are smaller by 0.15. These fluxes concern the state when the star in the greatest extension reaches the mean magnitude.

References

Fernie, J. D.: 1968, *Astrophys. J.* **151**, 197.
Fernie, J. D. and Hube, J. O.: 1967, *Publ. Astron. Soc. Pacific* **79**, 95.
Fernie, J. D. and Hube, J. O.: 1968, *Astron. J.* **73**, 492.
Latyshev, I. N.: 1964, *Astron. Zh.* **41**, 644.
Opolski, A. and Krawiecka, J.: 1965, *Contr. Wrocław Astron. Obs.* No. 11.
Parsons, S. B. and Bouw, G. D.: 1971, *Monthly Notices Roy. Astron. Soc.* **152**, 133.

PART VI

THE USE OF MODEL ATMOSPHERES
FOR TEMPERATURE

THE USE OF MODEL ATMOSPHERES
FOR TEMPERATURE-GRAVITY CALIBRATION

J. C. PECKER

Collège de France, Institut d'Astrophysique du C.N.R.S., France

Abstract. Regardless of the degree of elaboration of series of models, just *how* can they be used for calibration purposes? And how much is this calibration *sensitive to the* quality of the model theory?

These two questions are the basis of our discussion, which covers: I – The general principles of the use of model atmospheres in stellar calibration (1 – The two dimensional classifications; 2 – The use of the total luminosity; 3 – The cases of Vega and Sirius; 4 – The calibration of ST $- T_{\text{eff}}$ relation); II – The failures of the two parameters model atmospheres (1 – The observational need for more-than-two-parameters classification; 2 – The abundance of elements, the line formation, and the model atmospheres; 3 – Various sources of unadequacy of models; 4 – Envelopes or shell features; their influence on model-building; 5 – The case of HD 45677. Diagnostic of early-type stars; 6 – Various unexplained spectral features); III – The present state of the model factory (1 – The classical models; 2 – New concepts in the description of a stellar atmosphere; 3 – New approaches in model making; 4 – Conclusions).

Résumé. Indépendamment du degré d'élaboration des réseaux de modèles, *comment* peut-on s'en servir pour la calibration des spectres? Et à quel degré une telle calibration est-elle *sensible à la qualité* des modèles utilisés?

Ces deux questions forment la base de notre discussion, qui couvre les points suivants: I – Principes généraux de l'utilisation des modèles d'atmosphère dans la calibration stellaire (1 – Les classifications à deux dimensions; 2 – L'utilisation de la luminosité totale; 3 – Le cas de Véga et Sirius; 4 – La calibration de la relation TS $- T_{\text{eff}}$; II – Les échecs des classifications utilisant des modèles à 2 paramètres (1 – Le besoin observationnel pour une classification à plus de deux paramètres; 2 – L'abondance des éléments, la formation des raies, et les modèles; 3 – Différentes sources d'inadaptation des modèles à la réalité; 4 – Caractéristiques d'enveloppes ou de couches; leur influence sur la construction des modèles; 5 – Le cas de HD 45677; le diagnostic des étoiles chaudes; 6 – Diverses observations spectroscopiques non expliquées); III – L'état actuel de l'usine à modèles (1 – Les modèles classiques; 2 – Nouveaux concepts dans la description d'une atmosphère stellaire; 3 – Nouvelles approches en vue de la construction des modèles; 4 – Conclusions).

The concept of model atmosphere has indeed made its first appearance in astrophysics as early as at the time of Secchi, when colors and temperatures were associated in a one-to-one relation.... Actually, the first models ever used, much later, were 'black-bodies'.

Very slowly, the development of the classification of stellar line spectra (independently of their colors), by the Harvard School, under the leadership of Pickering, A. Cannon, A. Maury..., in a sequence A, B, C... brought a considerable amount of new information. The discovery of the Saha ionization law, the identification of spectral lines it allowed, has helped to achieve this transformation of the Harvard initial sequence (completely artificial, based on a simple labelling), into the well known, and still in use, sequence: OBAF, etc....: this important step, to which such scientists as R. N. Russell, C. Payne... have contributed in an essential way, allowed to identify the spectral sequence with a *temperature* sequence; the stars were no more classed according the Balmer lines intensities, but according their temperature. The new classification, together with its calibration in T, was not in a too bad agreement with the temperature scale based on the colors or on the energy distribution in the visual range.

We shall not go into a detailed study of the historical developments.*

But we should now mention that, as early as the first determinations of radii from interferometric measurements, in cases where parallaxes were known from trigonometric measurements, temperatures were determined from an estimate of the total flux – without much need of a reference to the black-body model, or to any other model (see Section 1.3).

We shall find, in the following, the simultaneous use of a similar 'double' methodology in the framework of the very sophisticated models and detailed measurements we can found now in the literature.

Let us come back to the Saha-law calibration of the Harvard classification.

It allowed immediately not only the removal of some ambiguities, but essentially the possibility of making easier the interpolation of any newly measured star in the grid of the comparison stars used for the calibration of the classification. It allowed therefore immediately the interpretation of any stellar spectrum in term of 'temperature'.

Since then, much has been done in using always more elaborate models to calibrate stellar classifications. This sentence essentially defines what will be the topics I intend to cover:

(a) Regardless of the degree of elaboration – physical or numerical – of a series of models, just HOW can they be used for calibration purposes? What are the PRINCIPLES involved in this use... when facing the stellar spectra, gathered in the large piles of tracings on our desk? and:

(b) Just how much are the results of this calibration technique SENSITIVE to the degree of elaboration of the models, to their quality? Nowadays, just how far have we proceeded in this elaboration? And what do we have to do in order to make a better progress, independently of the progress of the observations?

As I feel we have indeed MUCH TO DO, I want immediately warn seriously the observers that, still for a long time, they should be very cautious in using models, implicitely or explicitely, whomever they may come from; and I certainly do not intend to give them recipes they could blindly apply....

1. General Principles of the Use of Model Atmospheres in Stellar Calibration

They have been described often (see Strömgren, 1963; De Jager, 1955; Pecker, 1955; Van Regemorter, 1959, for example). But I feel that, at least shortly, I should try to reproduce once more the basic organigrams, with very few modifications... (Figure 1).

* See, in the bibliography, articles of books marked with the sign*, which give some historical view or some bibliographical study, on this question.

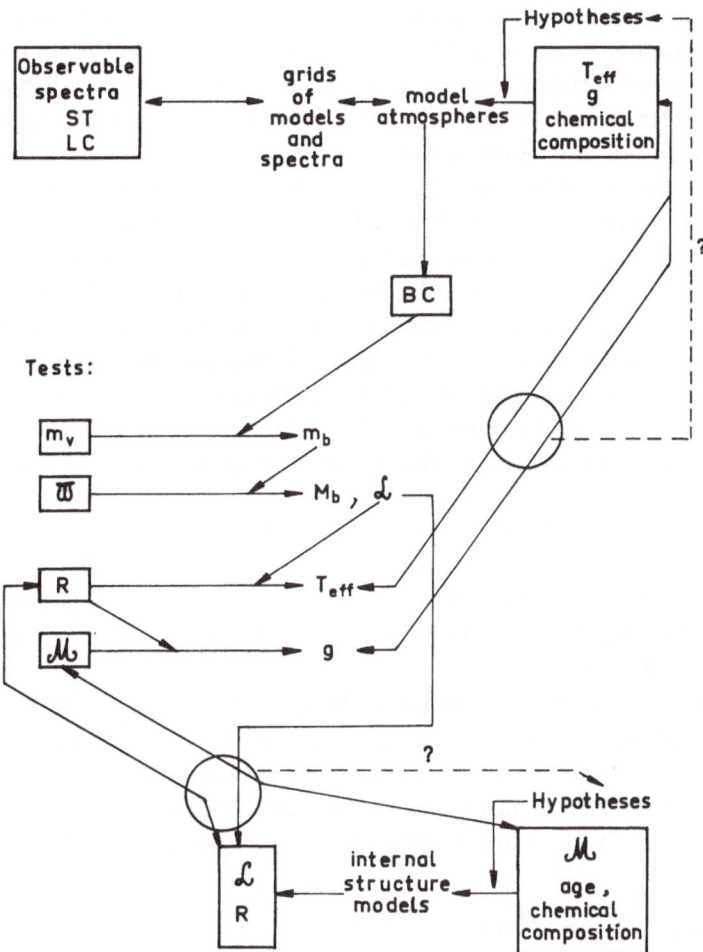

Fig. 1. The organigram of $T_{eff}-g$ calibration procedures. On the left side, the observed data; on the right, the hypothesis of the theory, and its parameters. The question-marks refer to the tests: Are the R, $\mathcal{M}$ values directly deduced from observations compatible with the T_{eff}, g values deduced from spectral data and atmospheres theory? Are the $\mathcal{L}$, $\mathcal{M}$, R values deduced from observations compatible with those deduced from internal structure theory? In other terms, are the observations compatible with either one or the other of the two sets of hypotheses and theories?

1.1. THE TWO-DIMENSIONAL CLASSIFICATIONS

On one side, we have the observed data. For all stars, they are essentially *spectral data*, and they give only relative values of the variation of intensity across the spectrum. The spectral data are indeed of a great complexity. All classifications (from the rougher ones, purely qualitative, well adapted to faint objects, to the precise quantitative classifications, which can be used only for the brightest stars) are selecting spectral data as *spectral 'criteria'*, as sensitive as possible to the physical quantities

that seem to be relevant to the authors of the classifications.* Some are good indicators of temperature, some others of pressure, – for example, the width of the metallic
lines, function of collisional broadening, is decreasing when the pressure or the density
decreases; for example, the ratio of He II to He I lines, when observable (early-type
stars), is increasing with temperature. But already now, we feel some ambiguity:
other parameters are influencing the line width, such as magnetic fields, microfields
of non-thermal velocities; etc.... We shall have to come back later on that kind of
difficulties, which may lead, when ignored, to misleading conclusions.

At least, one can say that it can be, and it has been, reasonably asserted that (a-
part from some irregularities or misfits, on which we shall come back) all spectra can
be fitted into a *two-dimensional classification*, all spectra can be defined by the given
datum of the *spectral type* (ST) and of *the luminosity class* (LC), with more or less
accuracy, according the ways they have been measured, and according their brightness. This is a purely empirical result, at least as it appeared twenty or even ten years
ago, and is the basis of the well deserved success of the MK classification, for example.
It means that the models that are intended to fit the observations should also be
defined by two parameters, and that, in general, will be sufficient.... This idea
has been, for years, of a great help; nowadays, it looks more like a severe and artificial
limitation, which has been often acting as a brake against any progress in the interpretation of stellar spectra.

Of course, the first parameter to use is the *effective temperature* T_{eff}, which defines
the total flux of energy which crosses each square centimeter of the atmosphere during
each second of time, assuming implicitely (this assumption is essential: think of H
and K lines, think also of fast rotators, such as those described by Maeder at this
meeting!) the fact that this flux is the same at all points of the stellar surface:

$$F = \int_0^\infty F_\nu \, d\nu = \frac{\sigma}{\pi} T_{eff}^4 \,. \tag{1}$$

This quantity has not the meaning of the temperature of any given well defined region
in the atmosphere; it just reminds its user of the paramount importance taken in the
early days of spectral classification by the black-body representation of the spectrum.
Here, it is related essentially to the flux per centimeter square of the stellar surface
and does not need to be linked with the total luminosity.

The other parameter is, naturally, the *gravity*, g, which defines the pressure p in the
atmosphere, provided hydrostatic equilibrium is assumed to be present. Then, we have:

$$dp = - g\varrho \cdot dh \,. \tag{2}$$

---

* We do not want to give here any exhaustive list, and we send back the reader to the classical
papers of Fehrenbach (1958), Keenan (1963), Strömgren (1963), to the article by Schmidt-Kaler
(1965); some more recent systems include the Spite system (1966), the Barbier-Morguleff-Gerbaldi
(1972) system, oriented towards metallicity indices, and the IR systems (see Pecker, 1971), oriented
towards envelopes criteria.

As the optical depth (at a frequency v) is defined by:

$$d\tau_v = - \varkappa_v \varrho \, dh,\tag{3}$$

one sees easily that

$$dp = (g/\varkappa_v) \, d\tau_v.\tag{4}$$

If the opacity is nearly constant, then p is nearly equal to $(g/\bar{\varkappa}) \, \bar{\tau}$, and, – everything being the same (a dangerous sentence: 'toutes choses égales d'ailleurs' en français!),– well defined by g, at an optical depth $\bar{\tau}$ near unity.

Building models with a given T_{eff} and a given g, assuming some more or less natural assumptions, of physical nature, about the nature of the stationary equilibrium reached in a stellar atmosphere, and their analytical expressions, about the symmetries of the geometry,... but assuming that nothing else differed from star to star, one is thus able to build artificial spectra depending upon two parameters. Then it is possible either to assign to a given star (used as a calibration star) a value of T_{eff} and g, or to assign, to each set of values of T_{eff} and g, a combination ST-LC, using for building these correspondences *any* set of two criteria used by the observers to define ST or LC, or measured in the spectrum of the particular calibration star under study.

Let us assume now that this operation is easy and unambiguous* (!!!). Let us note essentially here that, apart from the physical uncertainties on the physics of the atmosphere itself, our theoretician has been led (by the insistance of the observer) to admit *a priori* that nothing else but T_{eff} and g varies from star to star. In other terms, he has forced himself to admit a unique chemical composition, no variation of the magnetic fields (as they cannot be computed from T_{eff} and g), the unicity of rotation (for the same reason), perfect similitudes in the envelopes or shells, etc..... One sees now that this classical methodology might be seriously misleading; this, at least for two reasons:

(a) the 'criteria' might be depending upon additional parameters, even if carefully selected;

(b) the models themselves might, through some physical coupling, depend upon neglected phenomena, and might therefore be completely inadequate.

1.2. THE USE OF THE TOTAL LUMINOSITY

The use of models, to determine the T_{eff} and g of a given calibration star, or to determine the relation ST, LC vs T_{eff}, g, is possible if the spectrum is known with a sufficient resolution. But it does not need the use of any other measured data than the spectrum.

Another well-known approach is possible, using not the spectrum, but other observable quantities – parallax, magnitude, eventually mass.

The effective temperature can indeed be linked with the total *luminosity* of the star, if its *radius* R is known:

$$\mathscr{L} = 4\pi R^2 \sigma T_{eff}^4.\tag{5}$$

---

* A severe difficulty comes in particular from the fact that many much-used criteria are not so well defined in terms of the profile of the filter utilized for the measurements.

But let us note that this relation is ambiguous, as T_{eff} can indeed, when treated as an atmospheric parameter, depend upon the point on the stellar surface.... So (5) *defines* a T_{eff} which may be different from the T_{eff} in (1).

Let us note also that this $\mathscr{L}$ is only *radiative* energy; mechanical energy however may be important, and is generated as well in the central parts of the stars.

This quantity $\mathscr{L}$ is linked naturally with the stellar absolute magnitude M, which depends not only upon the flux per cm^2 of the stellar surface, but also upon the size of the star:

$$M_b = -2.5 \log \mathscr{L} + \text{constant}. \tag{6}$$

The subscript b is used for 'bolometric'; but we can hereafter delete this subscript, M being clearly the magnitude corresponding to the total flux of the star. The constant in formula (6) is determined by the reference to the solar values. For the Sun, one has (Table I):

$$M_{b\odot} = 4.62$$
$$\mathscr{L}_{\odot} = 3.866 \times 10^{33}.$$

TABLE I

Basic data used in this paper

Sun	ST. LC	G2 V		
	$R_{\odot}$	$= 6.960 \times 10^{10}$ cm		
	$\mathscr{L}_{\odot}$	$= 3.866 \times 10^{33}$	$M_b = 4.62$	Allen (1963)
	$\mathscr{M}_{\odot}$	$= 1.989 \times 10^{33}$ g		
	T_{eff}	$= 5.785$ K		
	$\log g$	$= 4.44$		
Sirius A	ST, LC	A1 V		
	δ	$= 0''00612$	$\pm\ 0''0001$	Brown *et al.* (1967)
	ϖ	$= 0''375$	$\pm\ 0''006$	Jenkins (1952)
	m_v	$= -1.43$	$\pm\ 0.02$	Allen (1963)
	$R/R_{\odot}$	$= 1.75$		from δ,
	R	$= 1.22 \times 10^{11}$ cm		
	$\mathscr{M}/\mathscr{M}_{\odot}$	$= 2.28$	$\pm\ 10\%$	Allen (1963)
	$\mathscr{M}$	$= 4.54 \times 10^{33}$ g		
	$B - V$	$= 0.01$		Allen (1963)
Vega	ST, LC	A0 V		
	δ	$= 0''0037$	$\pm\ 0''0005$	Brown and Twiss (1956, 1964)
	ϖ	$= 0''123$	$\pm\ 0''008$	Jenkins (1952)
	$R/R_{\odot}$	$= 3.23$		
	R	$= 2.25 \times 10^{11}$ cm		
	m_v	$= +0.05$	$\pm\ 0.02$	Allen (1963)
	$B - V$	$= 0.00$		Allen (1963)

Bolometric correction, in the A0 region
      BC $= -42.5 + 10 \log T + 29000/T$ Allen (1963)
      (general interpolation formula)

Therefore the constant is equal to 88.587 ± 0.01, the error being essentially due to the internal dispersion of the whole magnitude system, involving the comparison of the Sun's magnitude to the stellar standards.

The radius R and the gravity g are also related by a relation which brings in the *mass* $\mathcal{M}$:

$$g = G\mathcal{M}/R^2. \tag{7}$$

Clearly the knowledge of M (i.e. of $\mathcal{L}$) would then bring the additional knowledge of the radius, and of the mass $\mathcal{M}$ of the star, if T_{eff} and g are known from the spectrum, – even if one does know directly these quantities, as possible in a few cases (binaries...). If we know $\mathcal{M}$ through the study of a binary system, then we have the possibility of an interesting check of the consistency of the whole analysis.

But in order to proceed in such a way, again, we need some additional use of the model, and a very essential one indeed. For the observed quantity is not the *total absolute magnitude M*, but the *visual absolute magnitude M_v*, if we know, through the measurement of the parallax ϖ, how to deduce it from the *visual apparent magnitude* m_v.

To extrapolate the visible spectrum to unobservable parts, i.e. to compute the bolometric correction $(-BC = M_v - M)$, we need the models. The well-known relations involving magnitudes are:

$$M = m + 5 + 5 \log \varpi \tag{8}$$

$$m - m_v = M - M_v = BC. \tag{9}$$

There is here some ambiguity. The absolute magnitude is often coming not from trigonometric parallaxes, but from '*absolute magnitude spectral criteria*'. This is the case, for example, for the O-stars: in their case, the number of stars of known parallax, that could be used for calibration, is so small as to almost forbid their use. The absolute magnitude criteria are often used to assign to a star a LC: but indeed the latter is still purely a quantity characterizing only the atmosphere, and quite independent of the radius. If it found that the LC is also a total luminosity criterion, it means essentially that there is, at least in the samples in use, a strong statistical correlation between the three quantities T_{eff}, g, and $\mathcal{L}$. Two of them seem to be sufficient to define the third one, in practice.

This implies that there exists an $\mathcal{M} - \mathcal{L}$ relation, independent of R. We know from empirical evidence that indeed this is actually the case.

This could imply also that there is a singly T_{eff}, g relation, independent of the luminosity: but this is not true: we have several sequences in the HR diagram. The fact that actually, we have such sequences implies thus another relation. On the surface of Figure 2, which is the locus of stars in the $(T_{\text{eff}}, g, \mathcal{L})$ space, stars do not appear everywhere: a second additional relation therefore must exist; and this could be a $\mathcal{M} - R$ relation.

Actually, these two additional relations are given by the theory of stellar interiors, and they both depend upon the three quantities $\mathscr{L}$, $\mathscr{M}$, R. Empirical evidence however gives some additional weight to their existence, as it is well-known.

The model atmospheres, together with the use of measured spectra, are thus allowing some determination of T_{eff}, g, BC, (which may depend, and indeed does, of the

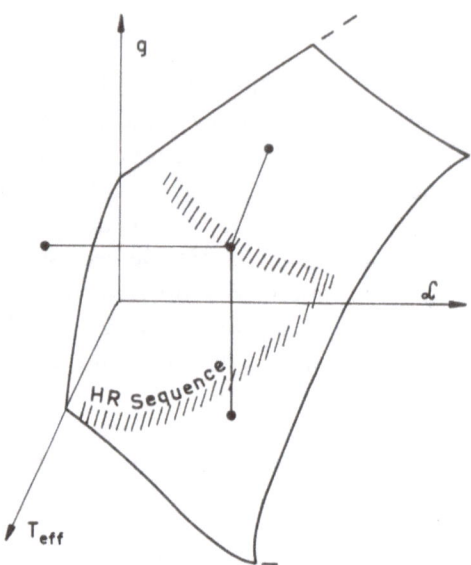

Fig. 2.   Two-dimensional sequences in the T_{eff}-g-$\mathscr{L}$ space. The empirical possibility of two-dimensional classification forces the representative point of a given star to fall on a single surface; the existence of sequences on the HR diagram forces the representative point to fall on some limited bands of that surface.

models in use). They do not tell anything else. If we use any observational luminosity criteria, in addition, to claim we know $\mathscr{L}$, it means therefore that we *assume* some internal equilibrium relation without saying so. We assume such relations are valid at least for both the stars used for the sampling of the classification and for the star under study.

The trouble is of course the statistical character of the various phases of the procedure. A given star might always not satisfy the internal equilibrium equations (star not yet on the main sequence, already having left the main sequence, etc....) – or the atmospheric traditional equilibrium equations....

1.3. THE CASE OF SIRIUS AND VEGA

In both cases, accurate knowledge, and particularly, direct determinations of the radius R, and a relative adequacy of the model atmospheres should allow checks.

We can essentially compute T_{eff} from the measured data of magnitudes, of radius, of parallax, and from the interpolated statistical values of BC (which is relatively small, by convention, in this ST region, near A0–A1).

The values we have used are collected in Table I, where is also included an interpolation formula for the BC.

Then we can write the two Equations (5) and (9)+(8)+(6), as follows:

$$\log \mathcal{L}_1 = -3.147 + 4 \log T_{\text{eff}} + 2 \log R \tag{10}$$

$$\log \mathcal{L}_2 = -0.4m_v + 0.4A - 2(1 + \log \varpi) +$$
$$+ 4 \log T_{\text{eff}} + 11.600/T_{\text{eff}} + 18.435, \tag{11}$$

where A represents the interstellar extinction expressed in magnitudes. Assuming no interstellar absorption (B-V is very small indeed, see Table I), we get for each star a set of two curves $\mathcal{L}(T_{\text{eff}})$. Their intersection gives the values of T_{eff} we need – without intervention of the model atmosphereexcept for the evaluation of the BC – obviously a one of first approximation... (see Figure 3). The estimation of the errors on the measured data allows to delineate the margin of the determinations.

The first surprise is to find for Vega a relatively low temperature – lower indeed than most of the determinations made from the detailed use of models (Table II;

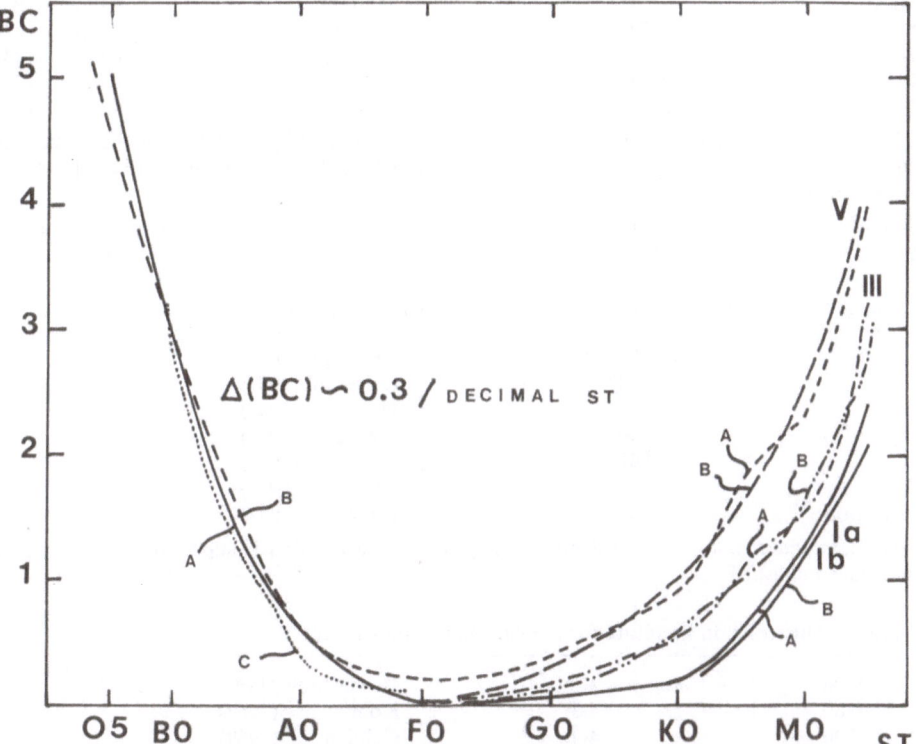

Fig. 3. The bolometric correction generally in use. Full line: LC I; mixed line; LC III; interrupted line: LC V (on the right part of the diagram). On the left part of the diagram, all stars are falling on the same curves. A: according Schmidt-Kaler (1965); B: according Allen (1963); C: according Harris (1963). One has estimated the error made on the BC for one tenth of spectral type to be of the order of 0.3 mag., in the B star area, where this error is maximum.

TABLE II

Some determinations of effective temperatures, either directly, or from model atmospheres

(a) Determinations of T_{eff} of an A0 star (after Keenan, 1963)

T_{eff}	method	authors
9 500	Model Atmosphere	Hunger (1955)
10 700	Eclipsing binary	Kopal (1955)
10 500	Eclipsing binary	Popper (1959)
10 900	Model Atmosphere	Melbourne (1960)
9 400	Model Atmosphere	Bless (1960)

(b) Fundamental determinations of T_{eff} (after Harris, 1963)

(stars with observed angular diameters)

μ Sco	B1.5 V	27 500 K	Harris (1963)
α CMa	A0 V	9 350 K	Popper (1959)
β Aur	A2 V	10 500 K	Popper (1959)
Sun	G2 V	5 784 K	Allen (1950)
$\gamma\gamma$ Gem	M1 V	3 650 K	Popper (1959)
α Boo	K2 IIIp	4 090 K	Kuiper (1938)
α Tau	K5 III	3 780 K	Kuiper (1938)
α Sco	M1 Ib	3 230 K	Kuiper (1938)
α Ori	M2 Iab	3 460 K	Kuiper (1938)
β Peg	M2 II-III	3 080 K	Kuiper (1938)
o Cet	M6 (var)	2 360($\pm$) K	Kuiper (1938)

This list is based on angular diameters from Pease, 1931; more and better data should now be computed from the lists of measurements by Brown and Twiss. In the present publication, we obtain from these new measurements (see Figure 4):

α CMa	A1 V	10 100 K	(this publication)
α Lyr	A0 V	8 200 K	(this publication)

(c) Effective temperatures from model atmospheres (after Harris, 1963)

O5	44 600°	Underhill (1957)
O9	36 800°	Underhill (1957)
B2 V	27 800°	Underhill (1957)
A0 V (α Lyr)	9 500°	Hunger (1955)
A3 V	8 900°	Osawa (1956)
A9 V	7 560°	Osawa (1956)
F2 V (σ Boo)	6 800°	Code (1954)

These ancient determinations are indicated on Figure 4 and are obviously superseded by many more models since published.

(d) Various values used in literature for a Sirius model atmosphere

$T_{eff} = $	$\log g = $		
9 700	4.3	Boyartchuk (1962)
9 700	4.3	Kohl (1964)
9 700	4.3	Gehlich (1969)
10 000	4.0	Strom *et al.* (1966)
10 080	4.44	Warner (1966)
10 200	4.35	Schild *et al.* (1971)
10 290	4.3	Latham (1971)
10 500	4.0	Gros (1972)

Figure 4). According to the usual $ST - T_{eff}$ scale, it corresponds to A0–A1 (Sirius), but to A4–A5 (Vega) instead of A0 ...Vega appears as colder than Sirius!...

Can we discuss the errors in any reasonable way? The Table I gives the values, from the original sources, of the errors that affect each measurement. Therefore, we can estimate:

$$\Delta \mathscr{L}_1/\mathscr{L}_1 = 2\Delta R/R \cong 0.03 \quad \text{(Sirius)} \tag{12}$$
$$\cong 0.27 \quad \text{(Vega)}$$

or

$$\Delta \log \mathscr{L}_1 \cong 0.012 \quad \text{(Sirius)}$$
$$\cong 0.12 \quad \text{(Vega)}$$
$$\Delta \mathscr{L}_2/\mathscr{L}_2 = 0.4\Delta m + 0.4\Delta A + 2\Delta \varpi/\varpi + 0.4\Delta \, (BC), \tag{13}$$
$$\cong 0.056 \quad \text{(Sirius)}$$
$$\cong 0.154 \quad \text{(Vega)}$$

or

$$\Delta \log \mathscr{L}_2 \cong 0.02 \quad \text{(Sirius)}$$
$$\cong 0.07 \quad \text{(Vega)}.$$

We have estimated the error on BC to be of 0.04 (see Figure 4), which *seems* to be sheer pessimism, in the present state of *usual* model atmospheres. From (12) and (13) – see Figure 5 –, one obtains the following values:

Sirius: $9500 < T_{eff} < 10700$

Vega: $7700 < T_{eff} < 9900$.

A source of error has been noted by Popper (1959): the interferometric measurements

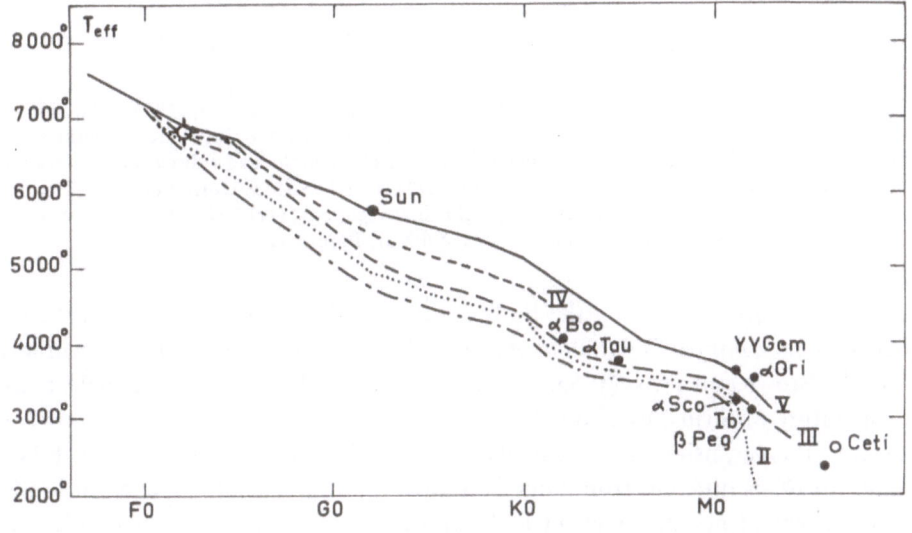

Fig. 4a.

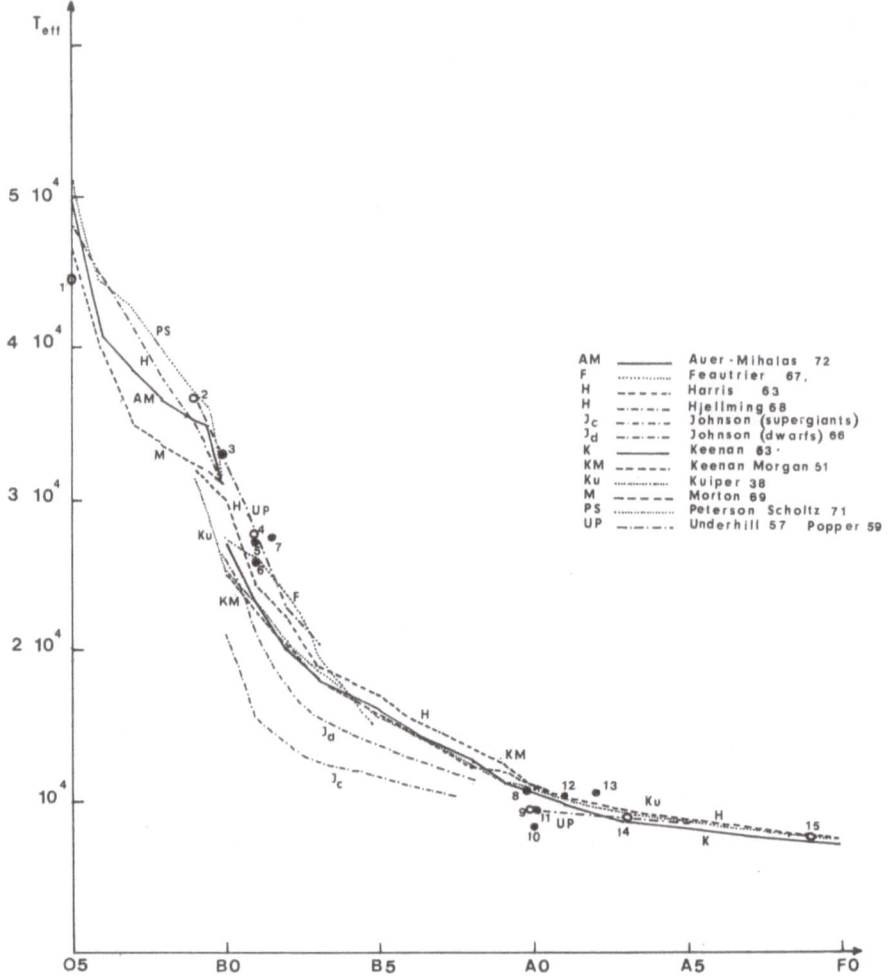

Fig. 4b. The T_{eff} − ST calibration. (a) from F0 to M stars, according Keenan (1963). The different curves refer to different LC. (b) from O5 to F0, according various authors. Open circles denote model atmospheres used by Harris (1963) (see Table 3). Black circles denote a few direct determinations: (3) MD 164402, Dufton (1972); (5) ζ Per, Cayrel (1958); (6) HD 96248, Dufton (1972); (7) μ Sco, Harris (1963); (8) η Leo, Wolff (1971); (10) α Lyr (this investigation); (11) Sirius, Popper (1959); (12) Sirius, this investigation; (13) β Aur, Popper (1959).

of the diameter are affected by the limb darkening. By assuming a limb-darkening coefficient of 0.45, Popper finds a diameter of 0.0072 instead of the 0.0068 measured at that time by Brown and Twiss (1956, 1964). This increases $\log \mathcal{L}_1$ by 0.05 and decreases the temperature of Sirius, by about 100°.

However, Popper, using the same analysis as we made, finds for Sirius 9350° ± 340°. His value for *BC* is different from ours by about 0.07: this would increase our $\log \mathcal{L}_2$ by 0.03 and would not be sufficient to bring our values to an agreement. We shall come back on the *BC*, – but at this stage we do not understand the difference between our calculations and Popper's. The m_v is -1.43 for us, -1.47 in his computation. But the diameter value is definitely smaller in the more recent measurements by Brown *et al.* (1967) than the measurements used by Popper.

Can we understand why we obtain temperatures that seem too high for Sirius, too

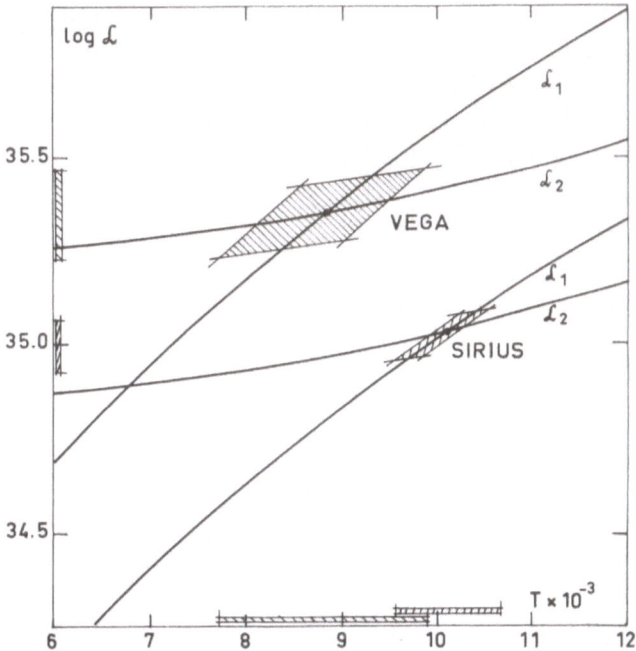

Fig. 5. Determination of T_{eff} and $\mathscr{L}$ for Sirius and Vega. The curves represent the functions de-scribed in the text (Equations (10) and (11)).

low for Vega? If, instead of having put $A=0$ in the equations, we would have taken some positive value for the absorption, we would have obtained higher temperatures, more compatible for Vega with the usual calibrations: A value of $A\cong0.7$ mag. (Vega) is indeed fitting the data. But it seems really high, and difficult to reconcile with the very low value of the color index $B-C$, unless we assume a practically grey circum-stellar additional absorption of that order of magnitude.

We have so far not determined the value of g; we can do it for Sirius, without reference to the model atmosphere, as Sirius is a well-known double system. We then obtain:

$$\log g = 4.309$$

The error is then

$$\Delta g/g = \Delta\mathscr{M}/\mathscr{M} + 2\Delta R/R \cong 0.13$$
$$\text{hence} \quad \Delta \log g \cong 0.06.$$

Let us discuss the two cases separately, using the various available data.

1.3.1. *Case of Sirius*

The small measured values of R do not fit either the statistical relation for the statistical value of T_{eff} deduced from the spectral type A1 V (or the above value of T_{eff} (Figure 6). It means that we have to reduce the radius R, thus push Sirius under the MS; g has to be accordingly increased, and $\mathcal{M}$ decreased. Can we accept such an error in the analysis of a binary system so well known? It is doubtful; we are tempted to class Sirius as an A1 VI, or to class it as a A4–A5 V with an abnormally high T_{eff} ... but why? And anyway, is this linked with the fact that Sirius seems to have properties similar to metallic-line stars?

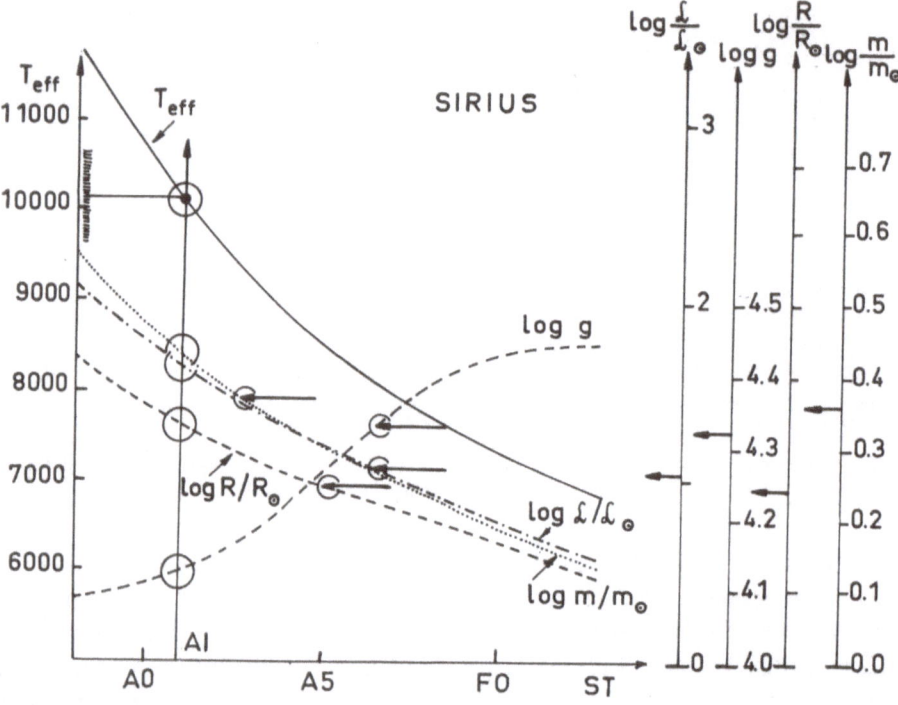

Fig. 6.   The case of Sirius. The arrows on the right indicate direct determinations. The heavy line on the left, parallel to the T_{eff} axis, is taken from Figure 5. The curves correspond to the usual calibration for class V stars.

1.3.2. *Case of Vega*

The measured value of R is too high for the generally admitted statistical value of T_{eff} – and much too high for the new one. A factor of 1.5 to 3 has to be applied. We think therefore that whatever is the reason for the T_{eff} anomaly, the star Vega is probably well above the main sequence and can be considered as of LC IV, (as suggested also by BCD classification). The luminosity, according the usual calibration, is allright. But according the new measurements of T_{eff}, the star is twice of three times

overluminous confirming that the star might be colder and bigger than its usual classification, – let us say A 4–5 IV (Figure 7).

A high value for the extinction may be the cause of the trouble; but it would still
keep Vega above the MS. An error in the other measurements is unlikely larger than

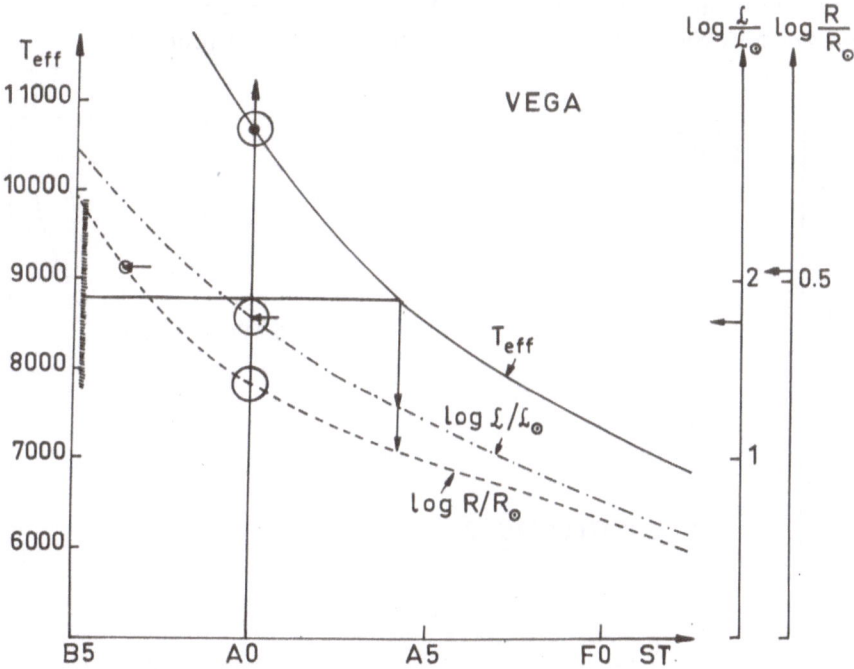

Fig. 7.   The case of Vega. Conventions as in Figure 6.

the margins we have adopted. Therefore, one is forced to admit that, at least in the
case of Vega, possibly in the case of Sirius too, something else is badly wrong... It
brings thus the question: *'how valid indeed is the usual* $ST - T_{eff}$ *calibration?'* As this
relation depends very little upon the LC, the possible changes in LC do not affect this
basic question.

Of course, we know that none of the two stars is a 'bona fide' star; even for Sirius,
which seems the less abnormal, we know that the use of D_B to fix the ST gives A0–A4;
but the gradients φ_b, φ_{uv} of the Paris classification give B8–B9. Whenever we happen
to know better a star, it seems indeed an universal rule that we put it progressively
out of power as a standard, ... isn't it true? After an extensive use of stellar demog-
raphy, we found now that the more classical stars, as Sirius, Vega, or S Monocerotis
are indeed psychotic individuals! – But they are not... We just happen to know them
better now. And we should not forget that, according Freud's views on psychosis, the
only thing wrong about such individuals is the way we tackle them, ... i.e. the theory
of model atmospheres.

In the particular case of Sirius or Vega, obviously no model, in the long series of available tables, replies in any satisfactorily way to the observed difficulties. In the case of Sirius, recent attempts by M. Gros (1972) to introduce a chromosphere, as suggested by the UV observations of Stecher (1969) and Carruthers (1968, 1969), do not reach any completely satisfying solution. Her model is represented in Figure 8; the remaining discrepancies are clear on her computed UV spectrum (Figure 9).

We may note on this figure that the UV spectrum is generally much smaller than that predicted by the models – due to line absorption, most probably.

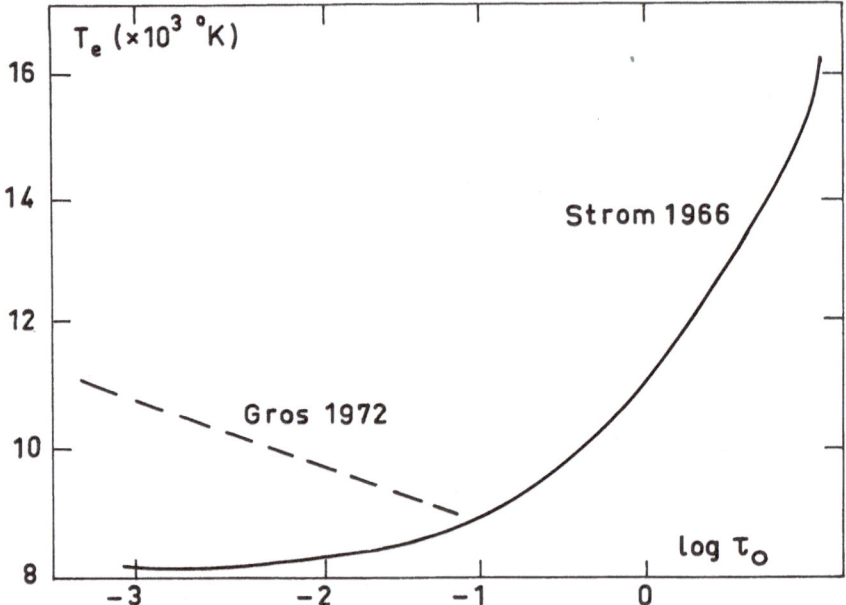

Fig. 8.   The models of Sirius, according Strom *et al.* (1966) and Gros (1972).

What kind of models should be used to reconcile computed spectra with observed spectra, in both cases of Vega and Sirius, – admitting that they cannot be the statistically valid models representing roughly well stars of the main sequence – if any such model does exist?...

(a) In the case of Sirius, the radius is probably smaller than in the main sequence; the temperature is probably correct. Therefore, the model is sufficiently well defined by the usual values of g and T_{eff}. However, the calibration obviously does not work too well.... Can we think the BC is in error? If so, the model is in error itself, the energy is distributed in the spectrum in a different way than in the case of a normal 'classical' model. Much flux is missing in the UV. Therefore, the BC is probably smaller, and $\mathscr{L}_2$ also. We shall then have to lower the temperature; which fits a number of things (g, R, $\mathscr{M}$), as well as the measurements of D_B.... The distribution of the radiation in the spectrum, strongly affected by an abnormal blanketing effect, might lead to an abnormal temperature gradient, and then to an excess of color, as

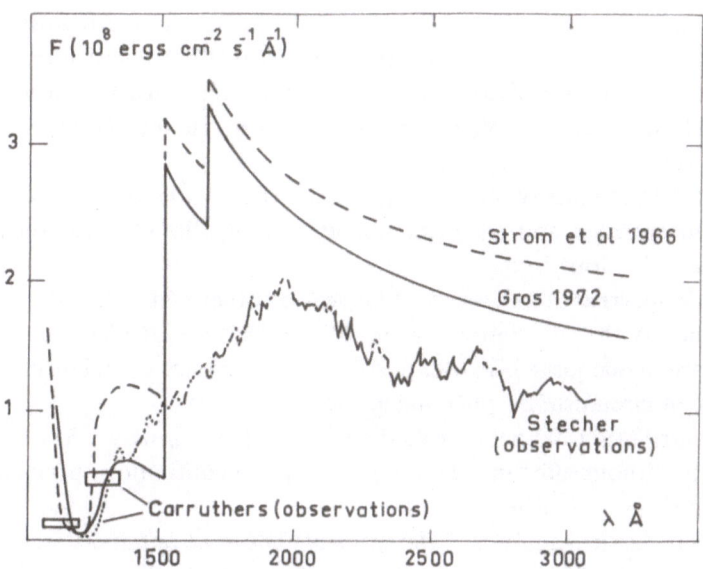

Fig. 9. The UV spectrum of Sirius. According models by Strom *et al.* (1966); Gros (1972), and measurements by Stecher (1969, 1970), and Carruthers (1968, 1969).

shown by the measurements of the φ_b and φ_{uv}.... In that case, the use of series of 'homologous' models derived from two-dimensional grids would be wrong and misleading. Let us note that the various authors who have computed T_{eff} for Sirius – independently of other differences in the data – have used different values for the *BC*. Our value, with the adopted temperature, was 0.40; Popper (who has been using larger radii, coming from previous series of measurements by Hanbury Brown and Twiss, 1956, 1964) used 0.27; Pease has used the value 0.72. However, this does not satisfy us: a change in the *BC* changes $\mathscr{L}_2$: but, according the fact that the representative curve for $\mathscr{L}_2$ is almost flat, the total obtained luminosity would be decreased, which pushes the representative point in diagram 6 still further away from where it should be.

Our only solution is indeed to accept a smaller radius, a larger g – and to hope that such models will fit better the visible spectra as do the usual models; of course the *BC* will have to be revised; but this will not be sufficient!... Undoubtedly, the building of models richer in their physical value as the ones now in use, will greatly help; but it has still to be done!

(b) The case of Vega is entirely different, as already seen. What kind of model could put the picture more coherent? Here some circumstellar *grey* absorption might do the trick. What would do some change in the *BC*? In order to improve the temperature fit, we should not, again, change the luminosity too much; therefore, a change in the radius would fit the classical T_{eff} value, provided we diminish the 1956 Hanbury Brown and Twiss value, by a factor 2.7. It is highly unlikely that we can do that. Taking a A4 ST value for T_{eff} would not fit the line spectrum, unless some strong abundance anomaly exists. Here, we feel that assuming a circumstellar absorption is a better choice – but a small value for g will still be necessary. And we stay unhappy... Indeed similar discouraging conclusions are reached in a very thorough analysis by Hardorp and Scholz (1968), as well as by Heintze (1968) who feels also that a small value of g is necessary.

Coming back to a sentence already expressed earlier ... the wrong thing is the theory of stellar atmospheres; and we must not use blindly the classical theory. Drastic changes have to be done in it.

(a) The UV spectrum (and, as we shall see for other stars, the IR spectrum) will force us to modify the calculations of the BC. Models are unadequate.

(b) They are unadequate because, possibly, of abnormal abundances, of extended atmospheres, of circumstellar phenomena (see Section 3).

(c) The contribution of chromospheric layers to the T_{eff} may affect the T_{eff} direct measurements without affecting the layers where the continuum spectrum is formed – or affecting them very little.

(d) The LTE classical analysis of the spectrum has to be put under strong suspicion.

(e) The treatment of blanketing in the usually available models is primitive.

What is this study of the difficulties encountered in the case of Sirius and Vega teaching to us? After all, both stars are amongst the standards that have been used to establish the average calibration of spectral sequences. So, the fit SHOULD be excellent ... And it is not!

Even if the models were systematically wrong, one could say they should be wrong in the same way. We have just shown it is not true. Therefore, the use of DIFFERENTIAL spectrum analysis (essentially based on the fact that if models are wrong they are wrong in the same way for the two stars to be compared) should be abandoned. Incidently, we shall see later that, for similar reasons, we should abandon the differential curve of growth analysis as well, an opinion we have expressed several times without being enough convincing so far – at least, we feel so...!

1.4. CALIBRATION OF ST $- T_{eff}$ RELATION

To calibrate the relation used in our preceeding discussion of the cases of Vega and Sirius, use has been done of well-known (let us say 'well-known' instead, with quotation marks!...) stars. The Table II taken from Harris and Keenan, gives the basic data used in such calibrations. The curves of Figure 4 show the relation, as finally established, according the same sources. The internal dispersion reveals the same kind of effect as discussed for Vega and Sirius and can be considered as real and physical, not as spurious.

Clearly, the two-dimensional models are only a historical step, but we may foresee that, for obvious reasons of convenience, it will still be applied, in practice, for many years.

At least, when doing so, an important question will have to be answered (and, whenever feasible, i.e. rarely, replied): let us assume (Figure 10) that the use of two criteria k_1 and k_2 gives, using a set of iso-k_1 and iso-k_2 curves, in the $T_{eff} - \log g$ plane, allows the determination of one point in this diagram. The use of k_3 and k_4 gives another

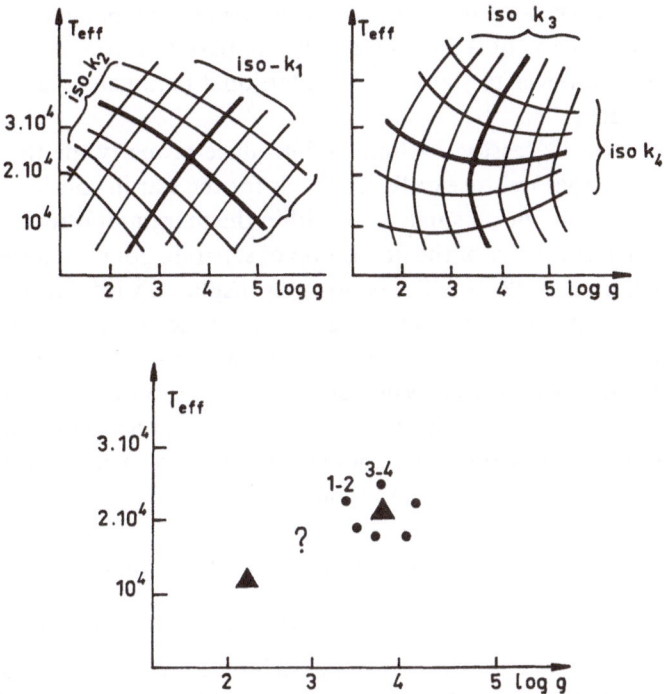

Fig. 10. The basic difficulty in using two-dimensional determinations of $T_{\rm eff}$ and g. The problem is: by using two different sets of two criteria, one find different solutions. The heavy triangle is the average of such determinations, represented by dots. With better models, the dispersion may be reduced... But will the average triangle be strongly displaced or not, in the twodimensional diagram?

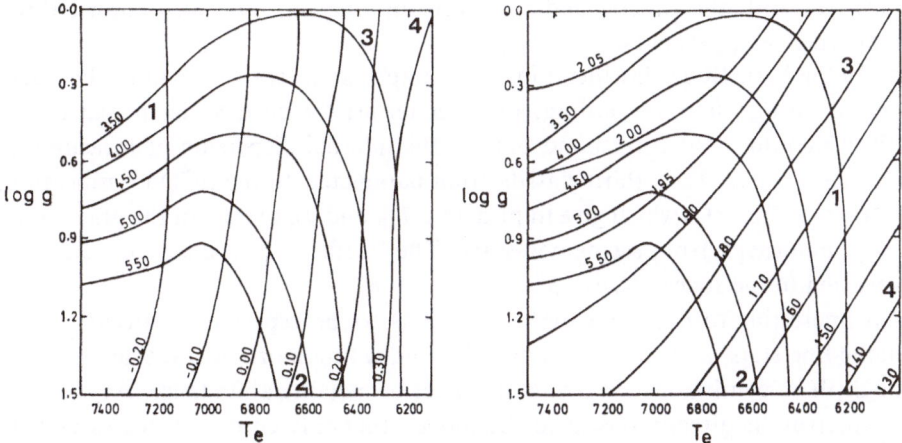

Fig. 11. (according Osmer) – Two-dimensional analysis; an actual case. The two-criteria method applied to two sets of criteria (left: Hγ, and C I; right: Hγ and D_B) and to some stars: 1: φ Cas; 2: α Per; 3: HD 10 494: 4: $+60°$ 2532. One can see how abnormal is especially the case of stars 1 and 4.

point. Any set of two criteria will allow us the determination of a new set of values for
T_{eff} and g. The dispersion of these points is due, in part, to errors in measurements, in
the essential part however to the need for additional parameters, and to the unadequacy of the models.

The question now is: 'assuming that we introduce properly the third, fourth, etc.
... parameters, and that we use excellent models for the star in consideration, we shall
find, instead of a few points, a single one, within the margin of errors. Will this point
fall near the center of gravity of the first group of separate points determinate with the
first approximation theory? Or will it be strongly displaced with respect to this group?

In the latter case, the two-parameters classical approach is not even giving a correct
order of magnitude, even when all possible sets of two criteria are used, and, although,
this discussion is rarely done in a systematical way, we have a hint that this is indeed
very often the case, and that an extremely strong warning, once more, should be made,
against the classical two-parameters model approach. A recent example is clearly
given by Figure 11, drawn according Osmer (1972).

2. Failures of the Two-Parameters Model Atmospheres

2.1. THE OBSERVATIONAL NEED FOR MORE-THAN-TWO-PARAMETERS CLASSIFICATIONS

We have seen clearly that even the best stars available for calibrating the classification,
whenever they could be approached both by direct methods and by model atmospheres
methods, gave results not so easy to interpret. They do not fit very well the statistical
relations established precisely by their use – which proves not so much an unadequacy of the samples which have been used, than the basic unadequacy of the
very idea of the feasibility of a two-parameters sampling.

Actually, independently of that kind of considerations, if became soon apparent,
whenever any two-dimensional system of spectral classification was used, that it was
inconsistent with the others, and therefore unsufficient. This was shown through
several types of approaches.

(a) In the Paris BCD classification, for example, some stars fall out of the 'surface'
where all normal stars are falling; in other terms, the surface has a real non-zero
width, which does not correspond only to the internal dispersion of measurements,
but to the real need for a third classification parameter. In the BCD classification, in
addition to λ_1 and D_B (which give in principle T_{eff} and g), the use of the gradient φ_b of
the continuous spectrum seems necessary. The 'surface' mentioned above is indeed
constructed in the space D_B, λ_1, φ_b.

It is interesting to note that the effect seems to be especially noticeable for the
'metallic-line stars' and for the subdwarfs – indicating more or less that the effect is
through their continuous spectrum, an effect of metallic abundance. As the continuous spectrum is directly linked to the model, this only comment makes us believe
that the influence of metallic lines and metallic continuum, not only on the spectrum,
but also, and essentially, on the model itself, is of a noticeable importance, as we shall
see later.

(b) Several stars are denoted in the usual classifications, by some additional symbol, such as the letters *e* (for emission), *p* (for peculiar) or still *f*, or others, to indicate abnormal spectral features, often appearing in emission. This is generally thought to imply envelopes, or extension of the atmosphere, which the models rarely predict when used in the classical simple-minded way.

(c) Several stars display 'symbiotic' spectra, where are juxtaposed, simultaneously, features corresponding to very different types of stars. They often could be interpreted not in terms of binary spectrum, but by using detached envelopes and the like, where physical conditions strongly differ from those of the classical photosphere.

The 'shell' type spectra belong more or less to this class; often observers, although they agree on the 'type' of peculiarity displayed by the spectrum, disagree on the interpretation, and naturally, as a way of consequence, on how various strange stars are related, physically, each to the other.

Accordingly, theoreticians should indeed certainly consider the symbiotic character of a spectrum as very general, in all stellar atmospheres.

(d) Several stars, classed in a certain way, according one type of classification, are classed in another way according another type of classification. Typical of this case are the metallic line stars. But we should note here that 'metals' (indeed, other elements than H and He, according some authors) cannot be considered as a whole. The ratio of abundances of two given elements seem to vary very much, from a star to another one, and this variation differs from the two elements selected to a group of two other elements. There are 'metallic-line stars'; but there are also 'baryum stars', 'europium stars', 'helium-poor stars', 'helium-rich stars', etc. ...

A particularly difficult region of the HR diagram is that of the cold stars. Three additional parameters (at least?) are necessary. The figure in Schmidt-Kaler (1965) shows this quite clearly. So does Figure 6.1 and Table 6.3 in Pagel (1971), who quotes Fawell and Greene.

Several cases illustrating that type of difficulties have been quoted in the literature. The following list of stars is only a partial list of examples of the non-suitability of a two-dimensional classification, according to Schmidt-Kaler.

(i) Emission line stars of early types: Oe, Be, Ae
    such as φ Per, $\varkappa$ Dra, HD 45677, etc. ...

(ii) Stars with strongly broadened lines (Bnn, Ann, Fnn)
    such as η UMa.

(iii) A-stars with peculiar spectra: Ap, Fp
    such as α And, β CrB.
    Generally, the abundances of ions Mn II, Si II, Eu II, Cr II, Sr II seem higher than normal.

(iv) Metallic-line stars: Am, Fm
    Type: α Gem B.
    The ratio Ca II (K-line) to metallic lines is abnormally weak.

(v) Wolf-Rayet stars WN5 to 8, WC5 to 8
    Types: HD 192163, 192103.

These stars have broad emission lines of ions of either C or N.

(vi) Population II stars

> Types: F to M.

Such as δ Lep, HD 140283.

Their spectra are characterized by abnormal intensities of CH and/or CN bands (possibly due to abnormal abundance ratios C/metals).

(vii) Carbon stars

> Types C0–C9 (R and N)
> Type: UX Dra

The bands of C_2 and CN are abnormally enhanced.

(viii) S-Stars

> Type S

such as: R And

The oxide bands (ZrO, YO, LaO, TiO) and the technetium lines are over-intense.

These are very clearcut cases. But let us remember the case of Sirius and of Vega, which, although often used as standards for the classifications, cannot be labelled unambiguously! ... We can therefore consider that the success of the two-dimensional classification has been indeed hiding its basic unadequacy ....

2.2. THE ABUNDANCES OF ELEMENTS, THE LINE FORMATION, AND THE MODEL ATMOSPHERES

Some of the difficulties described hereabove are obviously due to the dispersion in the various abundance-ratios.

Of course the first check to do, which would be quite convincing (could it be done without ambiguity), is the direct determination of abundances.

In a series of papers, qualified by some authors, in a regrettable way, as 'regrettable' (no quotation is necessary), several authors, including myself, have shown that, in the solar case, using a model deduced from the empirical analysis of the continuum data, the abundances of several metals are difficult to determine, a priori, from the classical LTE analysis of the curve of growth, or of individual lines. Important errors were the consequence of the LTE analysis. Similar results have been obtain since for the Sun, by De Jager and Neven (1967, 1968), and by Wijbenga and Zwaan (1972), for B stars, and helium, by Hearn (1970, 1971), for O–B stars and Mg, by Mihalas (1972).

We shall come back on some aspects of this discussion. Clearly, whatever is the interpretation of line intensities, abundances differences affect primarily the lines, and through the lines, spectral broad- or narrow-band analysis. Indices of metallicity can be deduced from different types of classification, used for different degrees of stellar brightness. The φ_b of the BCD classification is an index of metallicity; so is the m_1 index of the Strömgren classification; or again the index of the Barbier's method, recently developed by Morguleff and Gerbaldi, which is a combination of the GB (blue gradient 4010–4070 Å), D_B, Hβ photometric measures (Barbier, 1960; Barbier, Morguleff, 1964; Morguleff and Véron, 1970; Gerbaldi, 1972). It is not useful here to list the indices of metallicity that exist in the literature; but they are many... .

Some authors consider that absolute abundances are not necessary in classification purposes, so long as we can determine (through such methods as differential curve of growth analysis) relative abundances from star to star. The comment we made about the BCD classification shows that differential analysis, as we have already said, can, at the most, be qualitative, as structures of models are affected by abundance determination. Moreover, other things can affect the models, and give place to spurious abundance differences, that are deduced only from some idea a priori about the similitude of the convenient models....

Therefore, we can as well consider the effects we have just listed as indicators of the fact that models of the usual way are not universal, and that we should worry about the two-dimensional classical grids of models. And of course, the first thing is to look for the effect of the abundances not only on the spectra, but also on the models themselves.

How the existence of a variety of abundances do influence the model atmospheres and, hence, the T_{eff}, g calibrations?

The elements other than hydrogen (and helium, in early-type stars) are more or less 'impurities'. They can influence the models through:

(i) the continuum opacity;

(ii) the line opacity ('blanketing effect').

2.2.1. *The Continuum of Elements Other than Hydrogen*

The discussion of the continuum opacity is possible in various parts of the HR diagram, where different studies have looked for the influence of the most significant continua, particularly in the UV part of the spectrum. As precisely most of the metallic continua affect only the UV spectrum, observational tests are difficult; the only possible tests are the somewhat doubtful line intensities analyses.

The following elements have been considered in some detail: He, He^+, C, Ne, Ne^+,

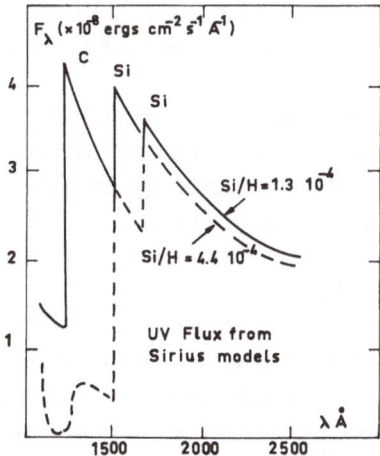

Fig. 12. (after Gros, 1972): The influence of Silicon abundance on the computed spectrum of Sirius in the UV.

Na, Mg, Al, Ca, Si. Other elements have been also introduced in the computations, with less accurate quantum-mechanical data. The Figure 12 shows a particular case where abundance differences has led to differences of spectrum in the wave-lengths where the responsible ions are not absorbing, according Gros.

We should note here that the models are to be criticized for many reasons (possibly different in each case), but, as they are more or less homogeneous series of models, the influence of metal continua on the spectral criteria is of significance.

Obviously, nothing more realistic can be said at this stage – except a strong warning: only a few elements have been properly considered, in many respects in a quite primitive way, and the only result we get is an order of magnitude of the effects that have to be added to effects of another nature (NLTE, for example), which may be much larger....

Another point to make at this point is the importance of space observations, in order to clear up the opacity sources that affect the UV, and the structure on the outer layers of the atmosphere which also affect the UV strongly.

2.2.2. *The Effect of Spectral Lines (Blanketing) on Model Atmospheres*

Irrespective of the abundance problem, the problem of blanketing is in itself a formidable problem.

Studies of the blanketing effect have been done several times in the case of the solar atmosphere, using the measured intensities of the lines in the observed spectrum as boundary conditions to the problem.

This question has been often obscured by a confused terminology. Actually, the effect of lines of all sorts is manifold. An historical semantical note is not, at this point, completely out of place. We shall at least outline such a note.

The first studies made about the blanketing effect considered only the 'pure absorption' lines, and in a still more restricted way, only those, amongst such lines, which were not linked with the adjacent continuum by some atomic transitions (lines which we shall call 'impurity lines' were thus the only ones to be considered really).

They conclude that the effect of introducing such lines in the spectrum has several effects (Figure 13):

(a) to *lower* the 'surface temperature' – (*the superficial cooling*), – and, correlatively (because the radiative flux has to be kept constant), to increase the temperature in the deeper layers, around the depths where the continuum is formed (*backwarming* effect).

(b) to *modify* the relation between any mean opacity coefficient $\bar{\varkappa}$ and $\varkappa_0$ (at $\lambda = 5000\,\text{Å}$), in the sense that $\bar{\varkappa}$ is increased, and correlatively $\bar{\tau}$ also, at a given depth τ_0, or at a given geometrical height h.

(c) to *block* the energy in the lines (*blocking effect*): this is a consequence of the effect (b): the necessary, observed, decrease of the intensity in the lines has to be compensated, the total flux being well defined for a given model, by an increase somewhere in the continuum; this effect has to be expressed with much caution; the only possible base of reference for this increase being the continuum of a model atmosphere of the same T_{eff} as the model with lines, but this one computed without the lines....

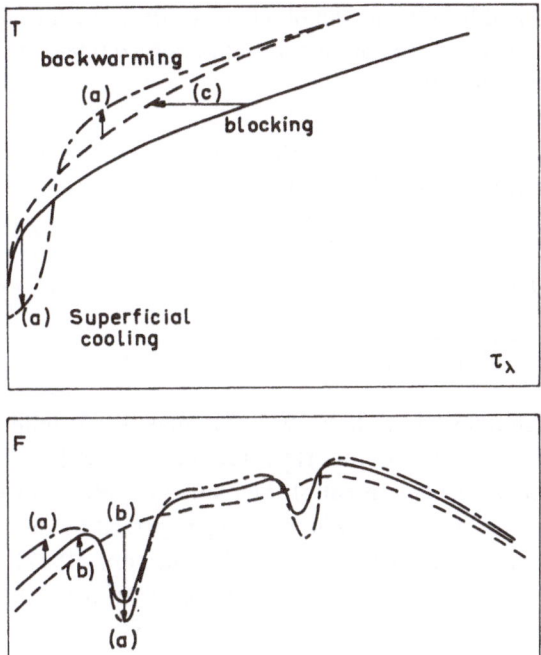

Fig. 13. The various aspects of the blanketing effects (schematic). Top full line: unblanketed model; dotted line: model blanketed by 'scattering' lines; mixed line: effect of 'pure absorption' lines. – Bottom: dotted line: spectrum computed without lines from unblanketed model; full line: actual spectrum, computed from unblanketed model; mixed line: flux computed from blanketed model.

The 'pure scattering' processes, in line formation (we continue here to use the classical terminology, how obsolete as it may appear now), do not take any energy from the radiation field; they just modify its angular distribution. Accordingly, the effect (a) is absent; no modification of the value of T_0, no backwarming either, is present. But the effect (b) (and as well the blocking effect) is present. It means that the relation $T(\bar{\tau})$ is not modified, but the relation $T(\tau_0)$ is modified, and therefore any expression $T(\tau_\lambda)$ of the model is indeed also modified; so is the spectrum $F(\lambda)$.

This has been discussed several times. Let us refer to the discussion by the author of one of the first exhaustive discussions along these lines. But neither has it been the first, nor the only one. Many authors have computed blanketed models of all sorts. Not only the discussion of the source-function in use is of importance; but also the way to schematize by a few typical-lines, the dozens of thousands lines that are to be taken into account: some authors use representations such as the symbolic 'picket-fence'; some use instead more realistic typical lines.

We shall not attempt here to review all these works. Let us quote only the early and classical result of Münch (1946) essentially identical to our schematical description, and the fine recent examples of blocking effect by Mihalas. Let us quote also the Strom-Kurucz results for Procyon where 30000 lines have been taken into account.

In the modern way of looking at the blanketing effect, we must clearly distinguish: (a) the impurity-lines cooling; (b) the continuum-coupled lines (or population cooling, according Thomas's terminology).

2.2.2.1. *Impurity lines.* We forget about the now obsolete use of 'absorption' vs 'scattering' lines. The general source-function is

$$S_\nu = \frac{\int_0^\infty I_\nu \varphi_\nu \, d\nu + \varepsilon B_\nu + \eta B_\nu^*}{1 + \varepsilon + \eta} \tag{14}$$

as often discussed (Thomas and Athay, 1961; Thomas, 1965; Jefferies, 1968; Mihalas, 1970, to quote some of the more important reference-books).

Using such a formula, the treatment of the blanketing effect, insuring the conservation of the radiative flux, is possible and can be performed by modern methods of the transfer solution, such as Feautriers's – used by Dumont (1972) –, or Gebbie's and Thomas's (1971) – the TCB method, now in the process of being applied by its authors and F. Praderie. It gives, quite naturally, results that are somewhat intermediary between the two classical extreme cases.

The problems which faced the earlier workers in the field face now the present ones. It seems to us that the method of 'typical lines' (such as developed, in the earlier context, by Labs, 1951, or Pecker, 1951) is the only possible one to achieve realistic results. But now, it will be necessary to assign to each line of the actual spectrum a certain range of the parameters ε, η of the source-function, and of the intensity parameter ($gfAb_L \lambda$). Clearly, it is far from easy!...

2.2.2.2. *The continuum-coupled lines (or population effects).* More important still is the importance (heavily emphasized by Thomas, in the principles, and used in computations by Auer and Mihalas (1972)) of the effect of the lines associated with the neighbouring continuum. We can understand easily that ionizations (by radiation or collision) are indeed coupling the electron temperature and the radiation field.

To describe the effect, let us schematize the opacity by the two levels + continuum atom. Then the spectrum has two continua and one line. One can compute a model M_1 by forgetting the opacity in the line. One can, at the contrary, take it into account: the effect of the line will be commanded by N_1, number of atoms in the level l, and is strongly coupled with both continua. The statistical equilibrium equations can be written, in the case with the line:

$$\frac{dN_1}{dt} = 0 = -N_1 \int_0^{\nu_1} J_\nu \alpha_\nu \, d\nu - \underline{N_1 N_e C_{12}} - N_1 N_e C_{1K} + \tag{15}$$

$$+ N^+ N_e C_{K1} + \underline{N_2 A_{21}} + \underline{N_2 N_e C_{21}}$$

$$\frac{dN_2}{dt} = 0 = -N_2 \int_0^{v_2} J_v \alpha_v \, dv - N_2 N_e C_{2K} - \underline{N_2 A_{21}} - \underline{N_2 N_e C_{21}} + \tag{16}$$

$$+ N^+ N_e C_{K2} + \underline{N_1 N_e C_{12}}$$
$$N^+ = N_e. \tag{17}$$

In the simplified case (without the line), the underlined terms, corresponding to collision – and radiation – induced transitions between levels 1 and 2 are vanishing, the line being absent. Without entering in the details, the three equations above give the ratios N_1/N_e, N_2/N_e and the value of N_e; they give very different results in both cases: clearly, in the case without-line, the level 2 is strongly overpopulated, the dominant term $A_{21}N_2$ being uneffective. Then the $L\alpha$ line, when taken into account, has for direct effect to decrease the opacity in the adjacent continuum (Balmer continuum), and therefore to increase the flux in the same Balmer continuum, for a given value of T_{eff}, as well as the temperature in the layers of formation of this continuum.

The various examples given by Auer and Mihalas (1972) or Mihalas (1971) show that the effect of lines is great on a model; it influences the model in rather deep layers: in deeper layers if $H\alpha$ is considered alone than if $L\alpha$ is considered alone; and in opposite direction (in NLTE) as easily understood when looking for the effect of various levels ($n=2$ is primarily a heating continuum, whenever $n=3$ cools – see Mihalas, 1971). Would all lines be considered, it is likely that even deeper layers would be influenced and spectral criteria calculated from the model would be strongly affected. A detailed study, which is still in the beginning, of the spectral criteria that are the less sensitive to these effects would be very valuable.

2.2.2.3. *The effect of different abundances; abundance determinations.* According the two preceding paragraphs, it appears clearly that, both through the influence of continuum absorption, and through the influence of spectral lines, the chemical composition affects definitely the model atmospheres, and leads to different computed spectra, according the composition. At this point, we should warn again both observers and theoreticians against the common error of using differential curve-of-growth analysis to have a hint on to which kind of chemical composition should be used in the 'modeling' of a given stellar atmosphere.

We should here state that the differential curve of growth method is highly qualitative when we have to compare stars that are considered as normal, to stars that are considered as abnormal, and precisely because of their 'abnormality'. The abnormality may be, or may not be, a matter of chemical composition; when it is the case, the abnormal composition may have distorted completely the model and led to absurd determinations, the temperature gradients coming explicitly in the curve of growth analysis. In addition, NLTE effects, as we have said, may be very important.

A very good example of this is given in the case of the hot stars, by Mihalas, for several lines, H-lines, or Mg II lines ... or for features of the continuum such as those noted by Mihalas or by Maeder (1971). Let us note the fact that a correct NLTE treatment forces the apparent variation of Mg abundance with ST to appear as spurious (Mihalas, 1972)!

Another effect of various abundances is that of the ratio He/H, which affects strongly the continuum. The Figure 14, derived from the tables of Klinglesmith (1971) (quite criticable actually in many ways, but at least sufficiently extensive and homogeneous for tests of that sort) shows that the models with different H/He ratio would give place to different evaluations of T_{eff} and g.

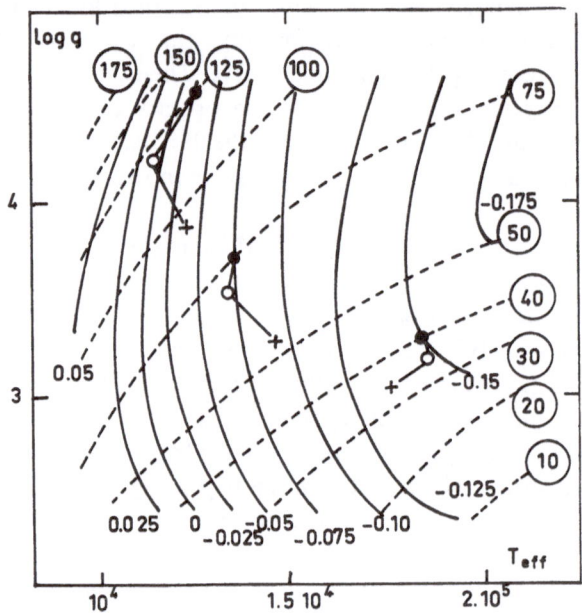

Fig. 14. The two-dimensional analysis as a function of helium content (after the computed spectra of Klinglesmith, 1971). The curves represent, for the abundances $X=1$, $Y=0$, the iso-Hγ (dotted lines, labeled in tenths of ångströms) and the iso-$(B-V)$ (full lines) labelled in magnitudes. The open circles correspond to the same determination made with $X=0.67$; $Y=0.33$; and the crosses to $X=0.143$; $Y=0.857$. One sees the influence of the assumed abundances on the $T_{eff}-g$ determination (compare with Figures 10 and 11).

But, a long time ago already, Underhill has insisted on the fact that apparent chemical composition differences (such as between WC and WN stars) are spurious and due only, or at least mostly, to structure differences. Similar views have been expressed often in the case of Am and Ap stars, as well as in the case of some late-type stars.

At this stage, we do not need to say more, although the subject is immensely large and still far from being explored in any satisfactory way.

2.3. VARIOUS SOURCES OF UNADEQUACY OF MODELS

Evidence for unadequacy of models has to come, in any particular case, from the fact that a model atmosphere made to fit some, or most, of the observed features, does not fit them all.

Apart from the abundance problem (for which we can at least adjust the abundance parameters so that to influence properly the models and the computed spectrum, lines included, and to fit the observations for a few lines of each significant element) we must mention that many kinds of uncertainties in the line source-function are leading to severe difficulties. Even if we treat properly the various levels of a given element, a Fe abundance (for example) made to fit the Fe I lines, does not fit the Fe II lines, or even does not fit all Fe I lines. A few examples are given by Auer and Mihalas (1972) and Mihalas (1971), and concern He I and He II lines.

Without being able here to do anything else but to list a few additional difficulties related to line formation, we can mention the following facts, more or less well-known, and often, if not always, neglected in the model building.

2.3.1. *Fluorescence Phenomena*

Fortuitous wavelength coincidence in the spectrum of two elements couples strongly their spectrum. When elements of small abundance are involved, it modifies their spectra, but does not affect too much the models. But if the lines in question are used as criteria to adjust a model to an observed spectrum, this may lead to severe errors.

The Figure 15 shows the case of He lines when one does take into account the overlap between $n_1 - n_2$ lines of hydrogen, and $2n_1 - 2n_2$ lines of ionized helium He II. For

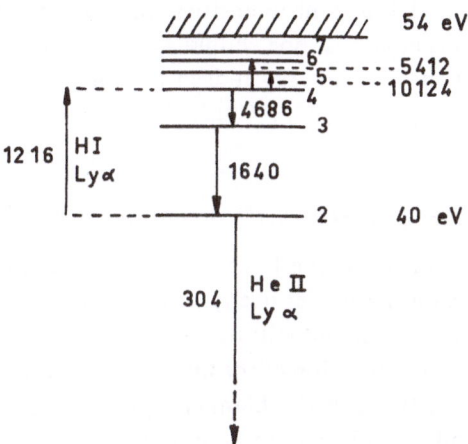

Fig. 15. The pumping of the He II transition 2–4 by the Lα line.

example λ1640 He II (2–3) weakens because Lα (H, 1–2), pumps He II (2–4) and the cascade 4–3–2 brings extra emission in the He line. The Pickering lines, at the contrary ($n=4$) strengthen, because of the pumping to the 4th level. Of particular interest is the behaviour of the near-IR line 10124.

Let us note here that the use of He II 4686 in stellar classification (it is actually used for O and Wolf-Rayet stars) might be misleading, if models do not take into account in a proper way the fluorescent effect we mention here, an effect which, although not enormous, is far from negligible.

2.3.2. *Magnetic Fields*

In lines of high Landé factor, a magnetic field induces a strong separation of the Zeeman components, and therefore reduces the saturation effect, giving place to the enhancement of total intensities of the involved lines.

Once such effect is suspected, it is of course possible to determine the magnetic field, although, even in the solar case, this operation is far from accurate, unless we can use the polarized-profiles analysis. And anyway, this will be quite unsufficient: we must indeed take the line splitting into account when solving the transfer problem – which we hardly know how to do well, except by very elaborate methods. The solution of the problem is made still more difficult by the fact that magnetic fields can vary on the stellar surface.

2.3.3. *Large Scale Doppler Broadening of Lines*

Just as the magnetic broadening enhances some lines, large scale velocity fields (rotation, convective motions, expansion) do affect spectral lines. They not affect in the first order the total intensity of the line: the splitting occurs after the transfer in the atmosphere. But they do affect definitely the profiles – hence the transfer solution, hence the model. So when a model does fit intensities, but cannot fit the profiles unless a large-scale velocity field is assumed, one should start to worry about the unadequacy of the model itself, which does not say indeed anything about how to predict the velocity field, but which depends upon it (this effect has been discussed in Pecker and Thomas, 1961).

2.3.4. *Small Scale Doppler (Non-Thermal) Broadening of the Lines*

There, even the total intensity of the lines is affected; but the profile also, as well known. The errors made on the model itself, by assuming non-thermal velocity field at all, or some unrealistic one, might still be large, and intervene, as in the three preceding case, through the modification of the first term of the numerator of the source-function S_v (formula 14) through the blanketing effect calculations. It is too easy (in both cases (2.3.3.) and (2.3.4.)) to fit observations by 'ad hoc' velocities fields which are just a 'deus ex machina', which no physical theory can predict in a correct way.

2.3.5. *Envelopes and Shells*

This problem seems to me so important as to need a complete paragraph of this section of our report, mostly because its intervention in the thinking about models is recent (not because the effect will be larger than the other sources of uncertainty above quoted in 2.3./1.–4.!).

2.4. Envelopes or Shell Features; Their Influence on Model Building

Whenever a shell is suspected, through the appearance in the spectrum of emission lines – or, at an higher dispersion, of strangely shaped profiles – the model maker should start to seriously worry.

The first idea is that some of the features of the spectra are, in a way, superimposed to those of a *normal* spectrum. Therefore, one is temptated to correct for the emission features (emission cores of lines) and to give as 'undisturbed spectrum' a substracted spectrum more or less interpolated and supposedly well represented by a classical model atmosphere.

The fact that such an attitude is generally not badly justified has been proved in the case of several Be star analyses, such as those from Burbidge and Burbidge (1953), amongst the pioneers, till, more recently, those from Delplace, Doazan and Briot, without mentioning the many other similar analyses (Figure 16).

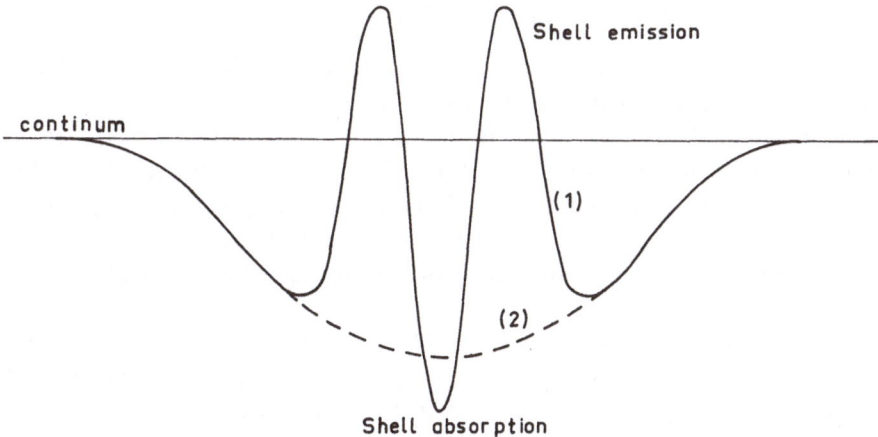

Fig. 16.   A typical shell-feature in an early-type star spectrum.

Unfortunately, this happy circumstance is not the happy end.

On one hand, some apparent 'dilution' effects are due only to a bad thermodynamical treatment, and more precisely to NLTE over populating ground levels. We have to keep in mind that the spectrum is an image of the variation of the source-function S with some τ, not with h; the relation $\tau(h)$ comes finaly from theoretical considerations except in very rare cases (some eclipsing binaries, for example).

On the other side, these doubts do not hide the real fact that the stars (the Be stars are a good case) are indeed surrounded by a complex envelope of dust and gas, of which the observable effects are manifold:

(a) in the visible spectrum, not only the emission features are visible, but the star is underluminous, when it is possible to determine its luminosity.

(b) this under luminosity may exist, even though there is no visible emission lines.

The emission features may appear in the UV and this just means that the chromo-spheric regions may have different behaviours (according the chromosphere thickness, possibly).

(c) in the IR, the envelope may strongly radiate, either through dust black-body radiation, or through free-free emission of the ionized gas of the H I and the H II region surrounding the star (Pecker, 1971; Dyck and Milkey, 1972).

The physical properties of the envelope will be discussed later on. At this stage, we may wonder how the diagnostic of such stars may be affected especially when keeping in mind that, after all, Sirius and Vega, although they are not showing conspicuous emission profiles, *may* have (at least Vega) extended absorbing atmospheres, and the fact that the Sun itself has an extended atmosphere which cannot be readily forgotten, as it includes our Earth itself...!

A very interesting particular case (quite similar, seemingly, to S Monocerotis), is that of HD 45677, which has been observed very extensively, and to which we shall devote the forthcoming paragraph.

2.5. The case of HD 45677. diagnostic of early-type stars

The star is well known as a variable Be star, with a strong IR excess, as observed by Low *et al.* (1970), or by Allen and Swings (1971).

We shall assume that the IR excess is due to a thin spherical dust shell; arguments in favor of such an interpretation have been given elsewhere (Pecker, 1971), but we should keep in mind that arguments in favor of the interpretation by the free-free radiation of H^- have been recently given in similar cases (Dyck and Milkey, 1972) – a controversy which shows us immediately the feeling that the diagnostic is far from obvious.

Anyway, the IR observations of HD 45677 can be interpreted as the radiation of a shell of $T_2 = 750°$; $R_2 = 600 R_*$, $\tau_{2, \text{IR}} = 0.34$; $\tau_{2, \text{vis}} = 4$ (which corresponds roughly to $A = 2.5 \log e^4 = 4.3$ mag. of visible extinction; we have assumed a $1/\lambda$ extinction law).

The bolometric correction, thus reaching several magnitudes, is enormous, and would not have been guessed, knowing only the visible spectrum.

The many consequences of this conclusion are not to be described here; let us only remember that, obviously, absolute magnitude criteria have not much meaning in this context....

But we cannot escape the conclusions that many early-type stars are surrounded by shells, envelopes, and the like. When we have only under hands the visible information, some important data are lacking.

The absolute magnitude of O stars have been calibrated by several authors, and quite recently, by Burnichon, (1972) who has reviewed other work on this question (Figure 17). The results are indeed distressing (see also Blaauw, 1963). The use of different samplings, of different criteria, is obviously the main cause of the dispersion; and no method is really protected against systematic errors.

The method used by Burnichon is certainly amongst the most reliable. She considers only O stars belonging to multiple systems in which stars of other types are present.

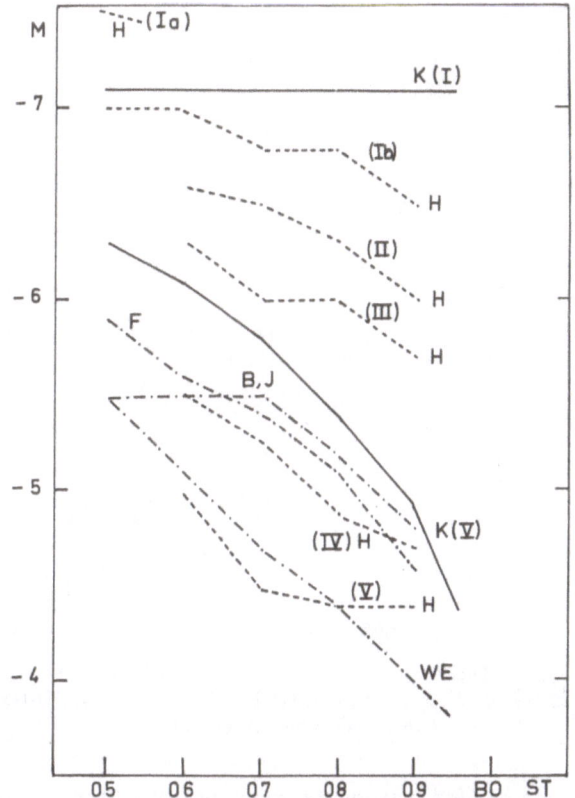

Fig. 17. (after Burnichon, 1972) – (a) The scale of absolute magnitude, according various authors, for O–B stars. The curves refer to: (K) Kopylov (1955); (H) Hack (1963); (F) Fitzgerald (1969); (B) Blaauw (1963); (J) Johnson and Iriarte (1958); (WE) Weaver-Ebert (1964); they are also labeled by the LC, in roman numbers. – (b) Comparison of the absolute magnitudes obtained by Burnichon (1972), with those obtained by other authors: (K) Kopylov (1958); (CA), Conti and Altschuler, (1972); (U) Underhill (1955).

The latter are calibrated, using well established absolute magnitude criteria (from the BCD classification of stars of type AFG). However a strong doubt still exists: are the AFG used in the calibration of the O stars under study really well calibrated themselves? Their belonging to a multiple system (which is rather young, as proved by the existence in it of an O star) might have affected the M_v–ST relation used for their study.

Apart from the M_v–ST relation, another source of undetermination of the luminosities is the bolometric correction BC. Clearly, it is always to be strongly put under question in the case of Be and Oe stars, – as shown by the example of HD 45677.

This at least assumed that the relation $T_{\rm eff}$–ST is well determined; but it is indeed just as bad as the M_v–ST calibration, as shown by the use of several sets of models (Figure 18).

Truly enough, this state of affairs is usually considered as peculiar to either the

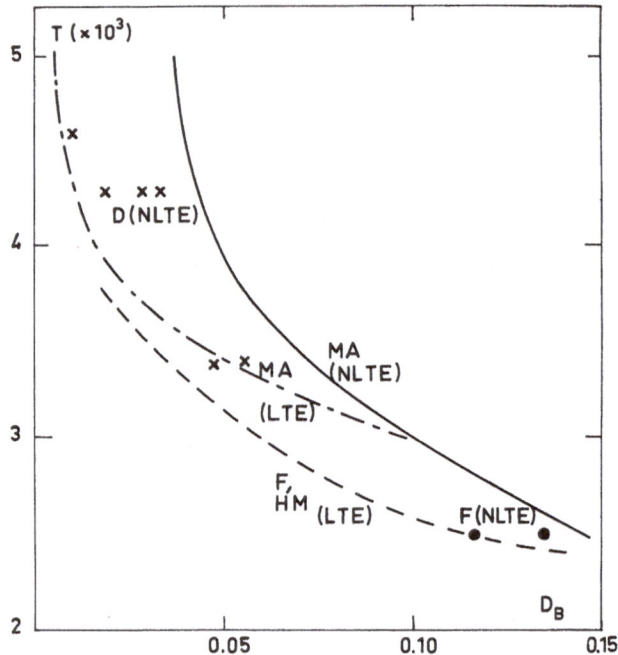

Fig. 18. (after Burnichon) – The $T_{eff} - D_B$ relation, from model atmospheres, in the O–B region. (MA): Mihalas and Auer (1970); (F) Feautrier (1967 (LTE), 1968 (NLTE)); (HM) Hickok and Morton (1968); (D) Dumont (1972).

early-type stars, or, for similar reasons, to some late-type stars, such as T Tau stars. But the case of Vega, and possibly Sirius, has shown us that indeed we should carry our worries all along the spectral sequence.

2.6. VARIOUS UNEXPLAINED SPECTRAL FEATURES

We have quoted the IR excesses observed for early-type stars; we have quoted the UV features observed for Sirius.... There are stars with UV excesses, in number... All of these features are generally unexplained, at least quantitatively (and we cannot be satisfied by any qualitative interpretation, which may be quite unconsistent...) The case of the star BD$^+$ 39° 4926, which has a very large $D_B(=0.71)$ as shown by Divan (1963), and confirmed later by several authors, certainly does not fit the usual calibration pattern.

Of a different nature are the discrepancies due to the unidentified absorption features of many spectra. Amongst the most conspicuous of them is the discontinuity at 4800 Å, discovered by Berger *et al.* (1956), which seems to prove the existence of some free-bound absorption increasingly intense for increasingly hot stars. This feature may be the origin of some difficulties encountered in fitting the continuum spectrum by model atmospheres (Krikorian, 1972; Figure 19). Some unsuccessful attempts to reproduce this 4800 Å discontinuity have been made (Van Regemorter, 1959), but there is no need to study them in detail now.

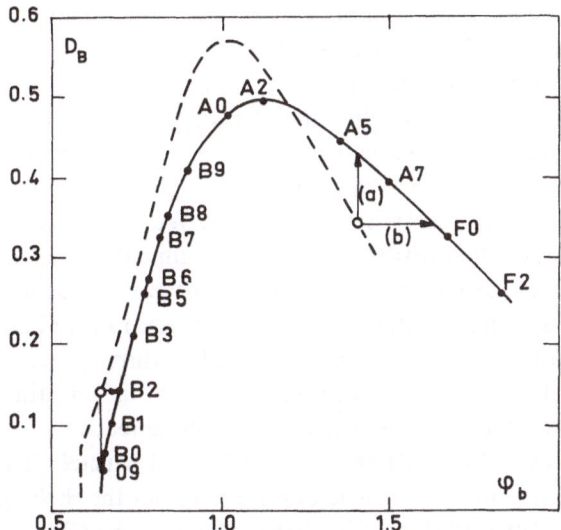

Fig. 19. (after Krikorian, 1972). The $D_B - \varphi b$ relation for LC V stars, according model atmospheres (dotted line) and the BCD calibration (full line). One sees that, whether one used one or the other of the two parameters for the ST determination, errors of a few tenths of spectral type can be made.

All this section is meant to say that models, arranged in two-dimensional grids, how well made they may be, are indeed unadequate. We must now look, in the present state of the use of models for calibration purposes, on the prospective aspects of this question. This will occupy the last section of this report.

3. The Present State of the Model Factory

3.1. THE 'CLASSICAL' MODELS

We shall not list here the various numerical aspects of the model making business. I shall send back the reader to my paper of 1965, in *Annual Reviews*, in which is described the essential of the methodology of non-grey models, and to Mihalas's book: essentially these references concern practically only RE models, which is a severe restriction.

Neither shall I try to review all models that have been built since that time: many have been done; some are excellent... I shall only refer to the papers of Mihalas and associates, mostly for early-type stars, of Vardya for late-type stars, where some references can be found. I would like to attract the attention, at this occasion, to the fact that a complete *tabulation* of models is a somewhat obsolete procedure (may I say 'unfortunately'?), the reasons being (a) the great number of parameters that can be introduced, (b) the fact that computers are available in many places, enabling scientists to use not tabulated models but borrowed programs. Therefore, one refers only to models 'built with the Feautrier's program' or 'with the Mihalas's program', without never seing these models actually published. It does not help the one who is supposed to give a clear account of the use of models and their validity!...

Instead, I shall try to describe, more or less according a paper now under completion by Pecker *et al.* (1973), the necessary evolution of point of view which should dominate, in the years to come, the whole approach to stellar atmospheres (see Thomas, 1970), going probably as far as to eliminate the concept of grid of models, the number of parameters becoming really excessively large for allowing us to apply simple interpolation methods in order to derive a model for a given observed star.

Let us first agree on one semantical definition. We shall call *classical* a model atmosphere which essentially satisfies a few equilibrium equations, and depends upon a few parameters. Needless to say, there is no more question to use black-bodies, or even grey models, although this is still often done by observers who have not paid enough attention to the evolution of the problem in the last thirty years.

Actually, the first non-grey models, very rough indeed, but rather stimulating, still now, in the sense that their methodology is still able to teach us a lot about the atmospheric structures, were due to Rudkjøbing (1947) and Mustel (1940). Their methods could not been applied to any realistic opacity law; but the Strömgren's method was on its way (Pecker, 1950, 1951) and its first applications was indeed the beginning of a new era: grid of grey models are not any more acceptable, since we know how to build non-grey ones. Let us here note that grey models have a tendency to give always a value of T_{eff} too low for a given spectral type, compared to the non-grey models. The new era, however, could not be effective (in view of the numerical complexity of the problem) before the availability of large computers; nowadays, non-grey models incorporate blanketing effect – at least in a simplified way –, and are the basis of the interpretation of stellar spectra, and of the calibration in T_{eff} and g of the ST and LC scales.

The difficulties are still however far from being removed; but, essentially, they are no more of a numerical nature, but much more on a physical, or even conceptual nature.

The models that can be now (September 1972) considered as 'classical' are:

– *In radiative equilibrium (RE)*. It implies that T_{eff} is given, and $F=(\sigma/\pi) \, T_{eff}^4$ is constant throughout the atmosphere. The chemical composition can be modified, according the wishes of the author; the opacity introduced in the models can be any opacity thought to be suitable; it can take line opacity into account. The only restriction is that we have to treat the opacity in accordance with the other hypotheses of the models.

– *In hydrostatic equilibrium (HE)*. It implies a given value of g_{eff}, constant in the atmosphere, and the validity of equation (2). Note that g_{eff} may be different from g, because of turbulence, or other mechanisms able to transfer momentum: we mean only that p varies in the atmosphere according the Equation (2), where g_{eff} is a constant.

– *In plane parallel geometry (PP)*

– *In local thermodynamical equilibrium (LTE)*, where the populations of all atomic levels – and hence opacities – strictly follow the laws of Boltzmann and Saha.

– *Without any magnetic field, or velocity fields of systematic nature* (i.e. macro-velocity fields: expansion, rotation, convection).

Fortunately, the meaning of the word 'classical' is quickly changing.

Nowadays, the hypothesis LTE, at least in the continuous opacity computations, is not any more necessary, since the work of Feautrier (1968), Auer and Mihalas (1972), and others. The blanketing account has even been taken into account in model building – but generally not in a complete way and almost always in LTE: it is either the impurity blanketing (such as the models of Bradley and Morton (1969), or Van Citters and Morton (1970), who take into account UV lines) – or the blanketing by lines coupled to the continuum (such as the models of Auer and Mihalas, 1972), where NLTE are essential. Sometimes, models include both NLTE effects and lines. The papers by Auer and Mihalas (1972) and Mihalas (1971) show the influence of these improvements in the models, and the quality of the new models when compared to the observations. Particularly illustrative is the case of Mg II lines in early-type stars. It is remarkable to note that NLTE, which was still heresy, less than ten years ago, seems now of such a quantitative importance...

The PP hypothesis has been also more or less overcome: Hummer and Rybicki, Cassinelli (1971), Cassinelli and Hummer (1971), Lucy and others, have introduced ways of computing models in spherical symmetry, and even with laws different from HE (laws such as $\varrho \propto r^{-n}$, applicable to some extended atmospheres). Similarly, the transfert in stochastic media is reaching a state where it will soon be possible to use it in atmosphere making. Maeder and Peytremann (1970) have considered rotating models where T_{eff} is a function of the latitude. Etc....

Unfortunately, such improvements bring in so many numerical complexities, that it seems so far impossible to take them into account and, at the same time, to keep realistic opacity laws.

Very little effort has been done to compute models departing from RE. Convective zones of first approximation (mixing-length theory) have been used for intermediate or late-type stars. But a really hydrodynamically consistent solution, although not far, possibly, from completion (Spiegel, Lacour, Zahn, see Souffrin (1971)), has not been performed so as to obtain realistic models. Leibacher's approach can be also taken into serious consideration whenever propagating waves may change significantly the structure of the atmosphere. But again, we are at the beginning of a long road...

Similarly, the introduction of magnetic fields, or velocity fields, has given place to interesting solutions of the transfer problems (let us quote the recent works by Simonneau and by Magnan, amongst many others); but this does not go far enough as to affect the models.

On the other hand, even the 'classical' models are not, within this very limited concept, very satisfactory in many parts of the HR diagram, because of an inadequate treatment of the opacity sources. The molecular lines and continua, the stable ions, etc...., are hardly going out of the shadow put on them by the uncertainties in molecular, atomic, ionic energy transitions physics, or by the formidable numerical work associated with a correct treatment, even in the classical frame (see Tsuji, Vardya, etc. ....).

3.2. NEW CONCEPTS IN THE DESCRIPTION OF A STELLAR ATMOSPHERE

In spite, or because, of the large drawbacks and small improvements of the classical theory of model atmospheres, there is nowadays a strong movement towards a new look at a stellar atmosphere. Whenever the classical atmosphere is more or less a boundary (rather sharp) between the star and the vacuum outside, we have now a tendency to consider it, as, *in essence*, a transition region between the internal parts of the star and the interstellar medium, a region which may be as extended indeed as to cover the space between the deepest part in the star from where photons can escape directly to the observer, to the place where atoms (or grains) do not know anymore to which star they belong.

In this sense, the atmosphere is the region which covers not only the classical atmosphere ('photosphere') but also the various kinds of extensions or envelopes.

Many things have to be said about the new concepts. We shall only try to abstract them here, as they will be described in more details in the series of papers to come (Gebbie, Pecker, Praderie, Thomas and others) and in the report of the President of Commission 36 of the IAU (Thomas, 1973).

The interstellar medium is characterized by the existence, locally, of a great many degrees of freedom, a state indeed of complete degeneracy. That is: we cannot describe the state of the medium by only its temperature, or by only the temperature and the radiation field.... Many other parameters (local, or not) are indeed necessary!

At the contrary, the internal parts of a star are completely degenerated; the temperature fixes every other property of the medium, except the density (they are not coupled each to the other): We have a black body in TE.

When going from inside to outside, the degeneracies progressively disappear. Radiation escapes, and is no more in equilibrium with the local properties of the medium, being somewhat diluted; this however is still in TE, as collisions dominate, and as, therefore, the local temperature commands the population of atomic energy levels; the Boltzmann, Saha, Maxwell laws are still valid – not any more the Planck law. We shall call this region the *lower photosphere*. The escape of photons is the major physical phenomenon which creates the model.

When proceeding outside the photosphere, the Boltzmann-Saha laws starts to be no more valid. NLTE effects start to prevail. But, still, the escape of photons is the way the thermodynamical structure of the atmosphere is built out. We shall call this region the *'upper photosphere'* essentially in RE (see Pecker and Thomas (1961) as one of the first papers where is described this behaviour).

When going still outside, the flux of photons is not any more the essential phenomena. Most of the photons truly enough do escape, and are only partly decoupled from the transparent medium (the opacity has decreased, the photons travel with less interaction with the medium). But we start to depart from RE; the mechanical energy plays an important part, but the density is still high enough for the flux of mechanical energy (waves) to interact with the medium. We have the *'chromosphere'*.

And we meet next the '*corona*' where definition of the local conditions is requesting the solution of energy equations taking solar wind into account.

There the magnetic conditions become predominantly important and command the flow of matter and the energy balance. Various regions can be found in the corona, according the variation of temperature; the extension of this region is leading to a decrease of the photon density, according a law in r^{-2}, of mass density in the solar wind, according a law in $r^{-2n} (n < 1)$, but to a still smaller decrease in the magnetic field (in $r^{-2n} (n < 1)$), the difference between n and 1 being essentially due to the geometry of magnetic lines of forces.

We have so far made a distinction of various regions in term of energy balance and mean free path. It affects the distribution of T and ϱ with some optical depth in the deep region, with geometrical depth in the outer regions.

We must also introduce a distinction between regions in term of momentum balance. There the pressure terms come in the picture, and so the HE hypothesis. In the deepest parts of the atmosphere, the momentum transfer is local, and the HE is prevailing, commanding the density structure, whenever the RE commands the temperature structure. When going outside, the momentum transfer changes its character, due to the mean free path of material particles. About where the NLTE or NRE effects starts to prevail, the HE starts to fail (this coincidence being actually fortuitous and certainly not general). According the case, and the physical processes which dominates in the hydrodynamical (and still outside, hydromagnetical) equations, the density structure may have different behaviours. Whereas the spectrum (made out of photons) is an image of the variation of the structure (through various source-functions) with optical depth, no diagnosis (except for objects of non-negligible apparent size, using eventually interferometric Michelson-type devices, or for such objects as eclipsed stars – either by a companion, or by the Moon's limb, or else) gives the variation of density with h, or of any optical depth with h, something that in general, only a purely theoretical (i.e. through a theoretical diagnosis of the observations) treatment can obtain. We have therefore appearances ('symbiotic' spectra, shell spectra, extended envelopes spectra) of which the spectral characteristics are often well described, but do not necessarily fit a conceptual definition of the words used for this description. For commodity purposes, with no certainty of fitting the usual meaning of the words in their purely spectroscopical definition, we shall use the word '*extended atmosphere*', whenever HE does not hold anymore, '*shell*', whenever there is even an increase of the total density outwards, somewhere, '*dust-shell*', or 'H I-*shell*', or else, whenever only one of the various components of the circumstellar matter undergoes a neat increase at some distance of the star; '*envelope*' whenever the density is very near the interstellar density, but the physical conditions are nevertheless still affected by the vicinity of the star (examples: the H II region, the C II region, are essentially envelopes).

The Figure 20 gives a schematical description of our terminology. We send back the reader to Pecker-Thomas (1961), to the Münich Colloquium (papers by Thomas (1969) and Pecker (1969)), to the Menzel jubilee symposium (Thomas), and to the forthcoming series of papers by Gebbie, Praderie, Pecker and Thomas, for further

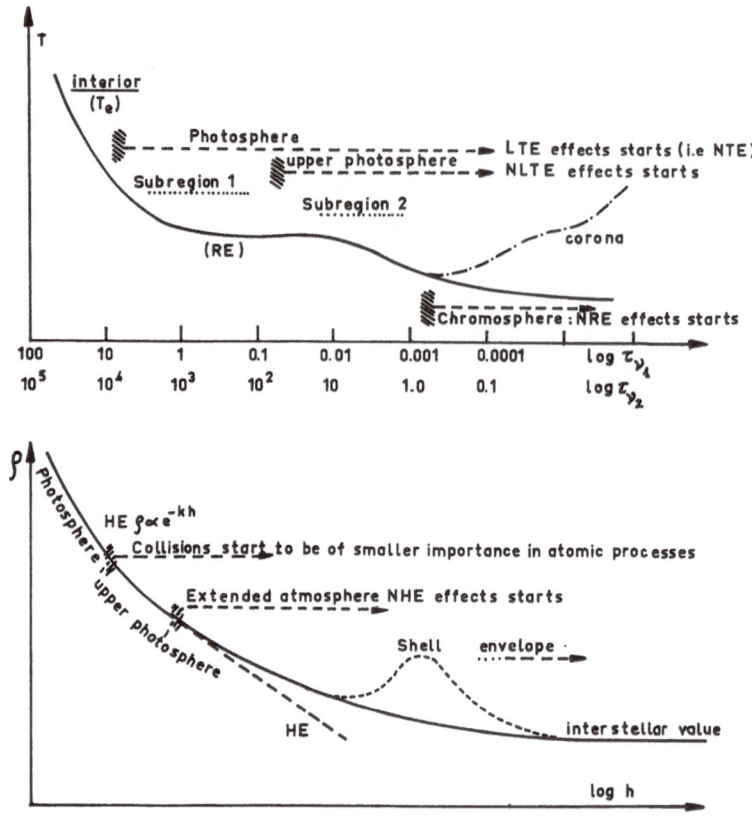

Fig. 20.   Terminology of the theory of stellar atmospheres (according concepts developed by Pecker *et al.*, 1973). The frequency ν_1 corresponds to the more transparent part of the continuum spectrum, ν_2 to the more opaque.

details and a better physical insight concerning the phenomena affecting each part of the atmosphere – each 'region' – and even the 'subregions' (defined, within each region, for example the photosphere, by the various sources of opacity which affect predominantly a part – subregion – or another of the photosphere: for example there is a photospheric subregion defined by the region affected by the escape of Lyman continuum photons, another one defined by the escape of the Balmer continuum photons, – and the like).

3.3. NEW APPROACHES TO MODEL MAKING

This new type of concepts obviously influences the methods to apply to build models. And we cannot, at the present time, conceive of a completely self-consistent method of solution; obviously, we can treat some problems such as the transfer of radiation, whenever we are at the infancy for hydrodynamical transfer, and still at the foetal state for the hydromagnetical transfer...

So far as the radiation equilibrium condition is the only one concerned, we need, to start with, a geometry, a chemical composition, a value for the flux of radiative energy, and a law for the momentum transfer.

To express the RE, it is customary to impose to the source-function the classical conservation energy:

$$\int_0^\infty \varkappa_v [J_v - S_v]\, dv = 0 \quad \text{at all depths}. \tag{18}$$

As S_v is related to T_e, we can write this in some explicit way – although this is not quite easy; essentially it can be written as follows:

$$T_e = \phi\,(T_e, \text{etc.}\ldots) \tag{19}$$

an expression which implies some kind of iterative procedure, the first member corresponding to the nth approximation, the second to the $(n-1)$th approximation.

It is essentially what has been done early by Strömgren, in a very slowly converging method, more recently by Gebbie and Thomas, in the very far reaching so-called TCB method, where the various sources of opacities intervene in successive terms, and can be introduced progressively when proceeding from the state of degeneracy of the internal layers to the outer layers; the Gebbie-Thomas procedure has the advantage of giving a physical way to improve any provisional solution taking into account only some components of the opacity, and to, accordingly, reach better solutions. For example, the introduction of the blanketing effect can be more easy in their method than in any other. Essentially, the method groups the terms in isolating in each of them a 'TCB' (temperature controlled bracket).

We do not need here, and do not want, to enter in the details of the physical meaning of this procedure. We want only to note that the failure of the Strömgren's method was partly due to the fact that in the solution of Equation (19), the second member is strongly depending upon the values of T_e and ϱ; whenever, essentially, in the TCB method the interaction works much more quickly, as the second member depends much more slightly on these parameters. This seems essentially due to the fact that the interaction procedure operates on a smaller part of the integrals of Equation (18). It actually subtracts out from Equation (19) large terms, equal in both members, which therefore do not affect the constancy of the flux, concentrating on the small, unequal terms, which, at a given depth, fix T_e.

The method converges so quickly, and is so easy to apply, that it can be handled with a reasonable accuracy by desk computers or even slide rules at least on simple problems.

But this does not yet solve either the HD problems, or the MHD...

At least, we should now comment on the fact that all parts of the atmosphere, all regions and subregions, are indeed intimately coupled each to the other, and therefore one is badly in error when treating a star as isolated from the outer regions, shells or envelopes, or extended regions, which necessarily surrounds it.

This coupling, in a way, has been known for a very long time. Two sub-regions influence each other, and give non-grey models different from grey models; the blanketing effect affects deep layers where line absorption may be very small, etc....

But we must note also that an envelope, or a shell, often acts as a reflecting wall, or semi-reflecting – semi-transparent wall. An H II – H I transition region, for example, is an H I wall; and the diffuse Ly C radiation inside an H II region is similar to that of a black-body of about 10^4 K. This radiation falls back to the star, and the usual boundary condition $I(\cos\theta<0)=0$ cannot be readily applied to a stellar photosphere. The whole thing, photosphere and envelope, has to be treated as a whole, in the same numerical procedure, because of the existence of such couplings.

Similarly, the fact that interstellar medium surrounds a star breaks the HE hypothesis quite quickly. For example, if $n_{\text{interst.}}=10^2$, if we assume the density low in the photosphere is following the HE law, one has:

$$n = n_0 e^{-Kh}, \tag{20}$$

where $n_0 = 10^{15}$ for $\tau \cong 1$, and $h=0$ (typical values).

Then, $n=10^2$ for $e^{-Kh}=10^{-13}$ or $Kh=30$. This is indeed not so far from the photosphere... And if we assume the Balmer continuum formed near $n=n_0$, the Lyman continuum may well be formed not far from the $n=100$ limit... a fact which changes strongly the transfer solution itself in the Lyman continuum. A typical example of this is the H II – H I transition region, where HE has completely ceased to be valid, and which corresponds roughly to the region where the Lyman photon escapes from.

That kind of effect can change considerably the classical models – as shown by Mihalas in preliminary computations dealing, truly enough, with extreme cases, but nevertheless rather demonstrative (Mihalas, 1972).

A soon as that kind of effects need to be taken into account, curvature has also to come in the picture; and a lot of usual approximation (the classical Eddington approximation $3K=J$ for example) are no more valid (we have to use $K=J$ instead): their use in the computing procedures might be very misleading.

We can therefore conceive of the computation procedure of a model in the following way:

(i) compute a $(T_{\text{eff}}, g, \text{classical composition})$ – classical photosphere;

(ii) use this first-approximation model to redetermine (out of LTE) new abundances, improved opacities..., and new T_e values;

(iii) compute the limitations of this model: where is the 'upper photosphere' starting, where is the 'chromosphere' starting... and, accordingly, treat properly, above the appropriate layers, the photon transport, and the mechanical transport.

(iv) at each of these successive points, introduce through the TCB, the correct NLTE opacities, the correct geometries, the correct energy input when feasible, the correct momentum balance (if feasible) – which implies some knowledge about the HD and MHD behaviour of the outer layers.

Of course, at this stage, the bad treatment of HD and MHD can force us to use the comparison between the observation and the best RE-model to infer some better knowledge of the HD – MHD physics of the model. And then, the limitation is basically strong: if we use models to improve the knowledge of their behaviour, how can we use them to calibrate observations?

3.4. CONCLUSIONS

Having expressed several times already in this report that rather pessimistic view, we should explain why the two-dimensional classical methods have been working so well – an actual fact that many authors are using to justify their reluctance to a change in approach to the 'generalized theory of atmospheres and environments' we have just been outlining.

The main reason of this success is that most of the abnormalities are located in the outer layers, non-photospheric, and therefore, it has been possible to describe with a certain set of T_{eff}, g_{eff}, $A_{i,eff}$ the photospheric regions. But we must be conscious that the parameters found from the best fit between models and observations might not necessarily be the good ones, the physical ones – as the deeper regions are counter-affected by the unobserved (or little observed) outer regions. They are just 'ad hoc' parameters.

The 'grobanalyse' of Unsöld, which essentially distinguishes between the T_{eff} of the deep photosphere and the T_0 characteristic of the regions where the lines are formed does not modify this conclusion, as the T_0 in question is some kind of excitation – ionization temperature, far from identical to an average electron temperature of these layers. Actually, one should describe an 'ad hoc' model with, at each layer, three kinds of temperature: $T_{exc-ion}$; $T_{radiation}$; T_e...

The adequacy of these parameters for a 'representation' of the observations is not in question. The success of the methods rely on the fact that the various layers are strongly coupled – which we certainly do not deny. After all, due to the kind of couplings that exist, the number of significant parameters is not large... What are they? How many?

Certainly, the $\mathscr{L}$, the $\mathscr{M}$, the R, the X, the Y, coupled by the *two* equations of internal structure (and together with a physical approach to the various physical phenomena linked with the escape of energy and momentum from the star's internal regions) are amongst what we need to describe the star, and in particular its atmosphere. These three parameters might be indeed replaced by T_{eff}, g, and the main chemical composition parameter, H/He.

But also the *age*, which may affect the outer layers very strongly and therefore, indirectly, the photosphere itself: envelopes are thicker for young stars, shells probably (as in the case of planetaries) for stars that have undergone unstabilities and matter ejection, and which are therefore rather evolved objects.

But also the *details* of the chemical composition (not only H/He, or H/(H + He + + others)), such as the ratios Fe/C, C/O, etc.... do play a part. Such details are a function of the age (through evolutionary processes involving nuclear reactions, diffusion processes, etc....) or of the stellar population (location in the Galaxy of the birth-place of the star, composition of protostellar matter...) – or a function of both.

This is, finally, more than two parameters. But the fact that most of the stars used in the calibrations are members, more or less, of the solar neighbourhood, explain essentially why age effects or stellar population effects have been of very little influence upon the apparent quality of the two-dimensional classifications (being admitted that the ratio H/He does affect little this calibration). This statement brings a 'serious warning': the calibrations valid near the Sun can be completely wrong for stars of the halo, or for stars in the Magellanic Clouds...

And our discussion brings now another 'serious warning': if, by using coherent grids of models, we found a value of T_{eff}, and g, for a given star, we still shall have to recalibrate these values, according everything else we may know about the star (envelopes and ages, populations and chemical composition, magnetic fields, etc....) into another set of values of T_{eff}, and g, that are really, physically, linked to the values of $\mathscr{L}$, $\mathscr{M}$, R, X, Y... characterizing really the star as a whole (as illustrated earlier by Figure 1).

The best conclusion is certainly that, as much as possible, we should treat stars as individuals, not judging a priori from any classification scheme their atmospheric parameters, – or, if forced to do so, doing it with a sufficient degree of self-defidence. By no means can we say that the diagnosis of effective temperatures and gravity is still in its quantitative stage; we can indeed obtain no more than acceptable approximations, and the model atmosphere theory is just at its start!...

Acknowledgements

The author is glad to express here his gratitude to the several colleagues with whom he has been fortunate enough to discuss about these problems, and especially to Dr Katherine Gebbie, Dr Françoise Praderie, Dr Dimitri Mihalas, and above all, to Dr Richard N. Thomas who has been a permanent source of inspiration, for several years.

References

Allen, D. and Swings, J. P.: 1971, *Astrophys. J.* **167**, L 41.
*Allen, C. W.: 1963, *Astrophysical Quantities*, 2nd edition, Athlone Press, London.
Allen, D. and Swings, J. P.: 1971, *Astrophys. J.* **167**, L 141.
Auer, L. H. and Mihalas, D.: 1972, *Astrophys. J. Suppl.* **24**, 193.
Avrett, E. H. and Krook, M.: 1963, *Astrophys. J.* **137**, 874.
*Barbier, D.: 1952, *Les atmosphères stellaires*, Flammarion, Paris.
Barbier, D.: 1960, *Ann. Astron.* **23**, 431.
Barbier, D. and Morguleff, N.: 1964, *Compt. Rend. Acad. Sci. Paris* **B258**, 4925.
Berger, J., Chalonge, D., Divan, L., and Fringant, A. M.: 1956, *Ann. Astrophys.* **19**, 267.
*Blaauw, A.: 1963, *Stars and Stellar Systems* 3, 383.
Bless, R. C.: 1960, *Astrophys. J.* **132**, 532.
Boyartchuk, A. A.: 1962, *Izv. Krymsk. Astrofiz. Obs.* **28**, 123.
Bradley, P. T. and Morton, D. C.: 1969, *Astrophys. J.* **156**, 687.
Briot, D.: 1972, private communication.
Brown, R. H. and Twiss, R. Q.: 1956, *Nature* **178**, 1046.
Brown, R. H. and Twiss, R. Q.: 1964, according anonymous note in *Sky Telesc.* **27**, 348.
Brown, R. H., Davis, J., and Allen, L. R.: 1967, *Monthly Notices Roy. Astron. Soc.* **137**, 375.
Burbidge, G. and Burbidge, M.: 1953, *Astrophys. J.* **117**, 407.
*Burnichon, M. L.: 1972, Thesis, Paris.

Carruthers, G. R.: 1968, *Astrophys. J.* **151**, 269.
Carruthers, G. R.: 1969, *Astrophys. Space Sci.* **5**, 387.
Cassinelli, J. P.: 1971, *Astrophys. J.* **165**, 265.
Cassinelli, J. P. and Hummer, D. G.: 1971, *Monthly Notices Roy. Astron. Soc.* **153**, 9.
Cayrel, R.: 1958, *Ann. Astrophys. Suppl.* No. 8.
Chalonge, D. and Divan, L.: 1972, submitted for publication.
Code, A. D.: 1954, in M. A. Wrubel (ed.), *NSF Conference on Stellar Atmospheres*, Univ. of Indiana Press.
Conti, P. S. and Althschuler, W. R.: 1972, *Astrophys. J.* **170**, 325.
*Curtiss, R. H.: 1932, *Handbuch der Astrophysik* **5**, 1.
Delplace, A. M.: 1970, *Astron. Astrophys.* **7**, 68, 459.
Delplace, A. M.: 1971, *Astron. Astrophys.* **10**, 246.
Divan, L.: 1963, *Compt. Rend. Acad. Sci. Paris* **257**, 615.
Doazan, V.: 1965, *Ann. Astrophys.* **28**, 1.
Dufton, P. L.: 1972, *Astron. Astrophys.* **16**, 301.
Dumont, S.: 1972, private communication to Burnichon, A. M.
Dyck, H. M. and Milkey, R. W.: 1972, Kitt Peak Nat. Obs., preprint.
Fawell, D. R.: 1970, Thesis, Univ. of Sussex.
Feautrier, P.: 1967, *Ann. Astrophys.* **30**, 125.
Feautrier, P.: 1968, *Ann. Astrophys.* **31**, 257.
*Fehrenbach, Ch.: 1958, *Handbuch der Physik* **50**, 1.
FitzGerald, M. P.: 1969, *Publ. Astron. Soc. Pacific* **81**, 71.
Gebbie, K. B. and Thomas, R. N.: 1971, *Astrophys. J.* **168**, 461.
Gehlich, U. K.: 1969, *Astron. Astrophys.* **3**, 169.
Gerbaldi, M.: 1972, *Compt. Rend. Acad. Sci. Paris.* **B274**, 669; **B275**, 295.
Greone, T. F.: 1969, *Astrophys. J.* **157**, 737.
Gros, M.: 1972, Thesis, 3rd cycle, Paris.
Hack, M.: 1963, *Mem. Soc. Astron. Ital.* **34**, 1.
Hardorp, J. and Scholz, M.: 1968, *Z. Astrophys.* **69**, 350.
*Harris, D. L. III: 1963, *Stars and Stellar Systems* **3**, 263.
Hayes, D. S.: 1970, *Astrophys. J.* **159**, 165.
Hearn, A. G.: 1970, *Monthly Notices Roy. Astron. Soc.* **150**, 227.
Hearn, A. G.: 1971, *Monthly Notices Roy. Astron. Soc.* **155**, p. 3.
Heintze, J. R. W.: 1968, *Bull. Astron. Inst. Neth.* **20**, 1.
Hickok, F. R. and Morton, D. C.: 1968, *Astrophys. J.* **152**, 203.
Hjellming, R. M.: 1968, *Astrophys. J.* **154**, 533.
Hummer, D. G. and Rybicki, G. B.: 1971, *Monthly Notices Roy. Astron. Soc.* **152**, 1.
Hunger, K.: 1955, *Z. Astrophys.* **36**, 42.
*Jager, C. de: 1955, *Principes fondamentaux de classification stellaire*, CNRS Colloquium No. 55, p. 141.
Jager, C. de and Neven, L.: 1967, *Solar Phys.* **1**, 27.
Jager, C. de and Neven, L.: 1968, *Solar Phys.* **3**, 159.
Jager, C. de and Neven, L.: 1972, private communication.
*Jefferies, J. T.: 1968, *Spectral Line Formation*, Blaisdell Publ. Co., Waltham, Mass., Toronto, London.
Jenkins, F. L.: 1952, *General Catalogue of Trigonometric Parallaxes*, Yale Univ. Obs.
Johnson, H. L.: 1966, *Ann. Rev. Astron. Astrophys.* **4**, 493.
Johnson, H. L. and Iriarte, B.: 1958, *Lowell Obs. Bull.* **4**, 47.
*Keenan, P. C. and Morgan, W. W.: 1951, in J. A. Hynek, (ed.), *Astrophysics*, McGraw-Hill, New York, p. 20.
*Keenan, P. C.: 1963, *Stars and Stellar Systems* **3**, 78.
Klinglesmith, D. A.: 1971, NASA Spec. Rep. 3065.
Kohl, K.: 1964, *Z. Astrophys.* **60**, 115.
Kopal, Z.: 1955, *Ann. Astrophys.* **18**, 379.
Kopylov, I. M.: 1955, *Izv. Krymsk. Astrofiz. Obs.* **15**, 153.
Kopylov, I. M.: 1958, *Izv. Krymsk. Astrofiz. Obs.* **20**, 56.
Krikorian, R.: 1972, Thesis, 3rd cycle, Paris.

Kuiper, G. P.: 1938, *Astrophys. J.* **88**, 429.
Kurusz,: 1969, in O. Gingerich (ed.), *Proc. 3rd Harvard Smithsonian Conf. on Stellar Atmospheres*, MIT Press, Cambridge, p. 375.
Labs, D.: 1951, *Z. Astrophys.* **29**, 199.
Latham, D. W.: 1971, Smithsonian Astrophys. Obs. Spec. Rep. 321.
Leibacher, J. W.: 1971, Thesis, Harvard Univ., Cambridge, Mass.
Low, F. J., Johnson, H. L., Kleinmann, D. E., Latham, A. S., and Geisel, S. L.: 1970, *Astrophys. J.* **160**, 531.
Lucy, L. B.: 1971, *Astrophys. J.* **163**, 95.
Maeder, A.: 1971, *Astron. Astrophys.* **13**, 444.
Maeder, A. and Peytremann, E.: 1970, *Astron. Astrophys.* **7**, 120.
Magnan, C.: 1972, submitted to *Astron. Astrophys.*
Melbourne, W. G.: 1960, *Astrophys. J.* **132**, 101.
Mihalas, D.: 1965, *Astrophys. J. Suppl.* **9**, 321.
Mihalas, D.: 1966, *Astrophys. J. Suppl.* **13**, 1.
*Mihalas, D.: 1970, *Stellar Atmospheres*, Freeman, San Francisco.
*Mihalas, D.: 1971, *Théorie des Atmosphères Stellaires* (1er cours avancé de la Société Suisse d'Astronomie et d'Astrophysique, Saas-Fee), p. 1.
Mihalas, D.: 1972, private communication.
Mihalas, D. and Auer, L. H.: 1970, *Astrophys. J.* **160**, 1161.
Morguleff, N. and Veron, M.: 1970, *Astron. Astrophys.* **4**, 391.
Morton, D. C.: 1969, *Astrophys. J.* **158**, 629.
Morton, D. C. and Adams, T. F.: 1968, *Astrophys. J.* **151**, 611.
Munch, G.: 1946, *Astrophys. J.* **104**, 87.
Mustel, E. R.: 1940, *Publ. Sternberg Inst.* **13**, 5.
Oke, J. B. and Schild, B. E.: 1970, *Astrophys. J.* **161**, 1014.
Osawa, K.: 1956, *Astrophys. J.* **123**, 517.
Osmer, P. S.: 1972, *Astrophys. J. Suppl.* **24**, No. 206, 247.
Pagel, B.: 1971, *Théorie des Atmosphères Stellaires* (1er cours avancé de la Société Suisse d'Astronomie et d'Astrophysique, Saas-Fee), p. 157.
Pease, F. G.: 1931, *Ergebn. d. exakt. Naturwiss.* **10**, 84.
Pecker, J. C.: 1950, *Ann. Astrophys.* **13**, 294, 319, 433.
Pecker, J. C.: 1951, *Ann. Astrophys.* **14**, 115, 152.
Pecker, J. C.: 1955, *Ann. Astrophys.* **18**, 145.
*Pecker, J. C.: 1955, *Principes fondamentaux de classification stellaire*, CNRS, Colloquium No. 55, p. 85.
*Pecker, J. C.: 1965, *Ann. Rev. Astron. Astrophys.* **3**, 135.
Pecker, J. C.: 1969, *IAU Colloq. No. 2*, NBS Spec. Publ. 332, p. 323.
Pecker, J. C.: 1971, (in French) in *17th Intern. Astrophys. Symp. Liège* (in press), (in English) JILA Rep. 109.
*Pecker, J. C. and Thomas, R. N.: 1961, in R. N. Thomas *et al.* (eds.), 'Aerodynamic Phenomena in stellar Atmospheres', *IAU Symp.* **12**, 1.
Pecker, J. C., Praderie, F., and Thomas, R. N.: 1973, in preparation.
Petterson, D. M. and Scholz, M.: 1971, *Astrophys. J.* **163**, 51.
Popper, D. M.: 1959, *Astrophys. J.* **129**, 647.
Rudkjøbing, M.: 1947, *Publ. og Mindre Meddelelser Københavns Obs.* No. 145.
Schild, R. E., Peterson, P. M., and Oke, J. B.: 1971, *Astrophys. J.* **166**, 95.
*Schmidt-Kaler, T.: 1965, *Landolt-Börnstein* 6, No. 1, 284.
Simonneau, E.: 1972, *Compt. Rend. Acad. Sci. Paris* **B274**, 85; **B275**, 169.
*Souffrin, P.: 1971, *Théorie des Atmosphères Stellaires* (1er cours avancé de la Société Suisse d'Astronomie et d'Astrophysique, Saas-Fee), p. 238.
Spite, F.: 1966, *Ann. Astrophys.* **29**, 601.
Stecher, T. P.: 1969, *Astron. J.* **74**, 98.
Stecher, T. P.: 1970, *Astrophys. J.* **159**, 543.
Strom, S. E., Gingerich, O. J., and Strom, K. M.: 1966, *Astrophys. J.* **146**, 880.
Strom, S. E. and Kurucz, R. L.: 1966, *J. Quantit. Spectrosc. Radiat. Transfer* **6**, 591.
*Strömgren, B.: 1963, *Stars and Stellar Systems* 3, 123.

*Thomas, R. N.: 1965, *Some Aspect of Non-Equilibrium Thermodynamics in the Presence of a Radiation Field*, Univ. Colorado Press, Boulder.
Thomas, R. N.: 1969, *IAU Colloq. No. 2,* NBS Spec. Publ. 332, p. 38 and 259.
Thomas, R. N.: 1970, JILA Rep. 106.
*Thomas, R. N. and Athay, R. G.: 1961, *Physics of the Solar Chromosphere*, Interscience, New York.
*Thomas, R. N. and Gebbie, K. B.: 1971, *The Menzel Symposium*, NBS spec. Publ. 353, p. 84.
Tsuji, T.: 1969, in Kumar, S. S. (ed.), *Low Luminosity Stars*, Gordon Breach No. 1, p. 457.
Underhill, A. B.: 1955, *Publ. Dominion Astrophys. Obs., Victoria* 10, 169.
Underhill, A. B.: 1957, *Publ. Dominion Astrophys. Obs., Victoria* 10, 357.
*Unsöld, A.: 1955, *Physik der Sternatmosphären*, 2nd edition, Springer-Verlag, Berlin.
Van Citters, G., Wayne, and Morton, D. C.: 1970, *Astrophys. J.* 161, 695.
*Van Regemorter, H.: 1959, *Ann. Astrophys.* 22, 249, 341, 363, 681.
*Vardya, M. S.: 1970, *Ann. Rev. Astron. Astrophys.* 8, 87.
Warner, B.: 1966, *Monthly Notices Roy. Astron. Soc.* 133, 389.
Weaver, H. and Ebert, A.: 1964, *Publ. Astron. Soc. Pacific* 76, 6.
Wijbenga, J. W. and Zwaan, C.: 1972, *Solar Phys.* 23, 265.
Wolff, B.: 1971, *Astron. Astrophys.* 10, 383.

Papers of historical interest, reference papers and general work on models are designated by the symbol *.

DISCUSSION

Thomas: In the case of early-type stars we know that those stars are loosing mass, which means that the atmospheres around them must be expanding. How good are the model atmospheres which take into account expansion velocities?

Pecker: The problem is a very complex one. For example, the transfer in some intense lines (a transfer which we know to affect the model through blanketing effects) is completely different, according to the velocity field, as the absorption coefficient is displaced with respect to the incident profile of the line, at any depth. Actually it has been possible to include that type of effects in relatively simple atmospheres, where other refinements have not been applied (for example, in grey atmospheres), A good example of what can be done is given by the work by Magnan (in the process of publication), who is using the Monte-Carlo method. I expect that this field will progress significantly in the coming years; the situation is not too good, but less 'discouraging' than in the case of convective layers for example. But the main problem will then be: 'what type of velocity field to introduce?'. And if one wants a physically consistent reply to this question, then, I must confess that I think we are still a very long way off.

Demarque: This is probably a discouraging question, but one of importance to stellar interior research. You have said very little about convection in stellar atmospheres.

Pecker: I fully agree with Dr Demarque. Indeed, the ionization of the main components of a star gives place to a convective zone, more or less deep in the photosphere, at least for spectral types later than A0. But at the present time not only do we not know how to compute well the T, p_e distribution in this zone (the mixing-length theory has well known drawbacks; improvements are, I believe, on the way through the work of Spiegel, Souffrin, Zahn and others, but it is still at a very preliminary stage), but we do also not know well how to compute the effects of the convective zone in the outer layers where mechanical energy transport starts to play a role. The important work by Leibacher gives certainly a better basis to this problem, but again we are only starting to understand the physics and still very far from being able to compute ready-to-be-used models! Indeed, the situation is 'discouraging'!!

McCarthy: Do you consider it to be possible that some kind of a scheme for classifying stellar models could be made as an aid to observational astronomy. Such a classification scheme could be analogous to that which has on several occasions proven so helpful in observational stellar spectroscopy.

Pecker: Why not? I think this would indeed be an excellent idea. Apart from a T_{eff}, g_{eff}, abundance classification, this classification would indicate in each case what kind of degeneracy has been introduced to improve the model compared to what I call 'classical' models in my paper. This would essentially make clear at what height (when proceeding from the inside) the model in consideration ceases to be valid.

Jaschek: That means essentially that one can explain stars which in the MK system are called normal, with perhaps the exception of the supergiants.

Pecker: You cannot, I think, say you 'explain' them. At least, you may 'represent' them by a set of parameters, such as T_{eff} or log g but the physical meaning of that set of parameters is still to be questioned, even for class V stars, in many areas of the HR diagram (early-type stars, late-type stars, Am stars...). With this restriction, I agree with you, and would say that for many stars of class V, no obvious contradiction can be detected between models derived in order to represent any given set of observed criteria.

Eggen: Is it possible that the difficulties you have been discussing are the cause of the observed spread in stellar metal abundance (omitting the extreme halo stars)?

Pecker: Probably partly, but only partly. In each specific case, a careful discussion should be done, to reply to the question: ''Is the spectral anomaly we measure in a given group of lines due to metal abundances differences, or is it due to abnormal NLTE effects, to shells, to effects of convection on temperature gradients and so forth...?''

Jaschek: I conclude from what you said that you would not trust abundance differences between stars smaller than a factor of 5. What worries me is that many photometries are detecting small metal abundances differences, which are calibrated on atmospheric analyses which apparently are somewhat shaky.

Pecker: I do certainly share your worry. Small differences in metal content may often be due to some undue simplification (differential methods?) of the determination of abundances in stars that may have different structures. I can refer to the already old discussion of WR stars: Underhill years ago, denied that the difference between WC and WN is an abundance difference; and we should not forget that when applying differential methods, one implies that many things are identical in the two stars in comparison – from NLTE effects to differences in gradients of the temperature (even in LTE), between stars having for example convective zone of different importance.

Hack: What do you think about the super-metal rich stars found by the spectrophotometric method of Spinrad? In a paper by Spinrad spectrograms and tracings are reported of a SMR star and of a so-called normal star, which look practically identical.

Pecker: I am puzzled, as you are, by the kind of facts you are stating. I would be tempted to say that the Spinrad classification scheme may be sensitive to metal abundance *and/or* other factors in the atmosphere. But I have not studied this scheme in detail and I do not consider my reply as satisfactory!

Newell: If I may add to the story of α Cen; while there are perplexing cyclic changes in the helium spectrum and in the metal line spectrum, there appears to be no corresponding change in the Balmer jump, Paschen slope, and hydrogen line spectrum. It is hard to admit significant changes in effective temperature and effective surface gravity during the cycle.

Golay: I should like to ask Pecker to give me some bibliographical information: I suggest that Pecker draws at the blackboard a HR diagram and gives the bibliographic references of the best models for each part of HR.

Pecker: I am not prepared to give list of models, so rich has been the production of 'rather good' models in the recent years. Only as examples of very good models, I could quote Auer-Mihalas models for early-type stars – Hummer-Mihalas models for nuclei of Planetary Nebulae... and on the other side of the diagram, Aumann or Tsuji models for cold stars. But good models produced by authors such as Peytremann, Carbon and associates, and many others, without being 'perfect' (none is perfect!) should obviously be quoted. I shall try to draw a 'conceptual' diagram of the sort you want, if possible for inclusion as an appendix to this discussion, in time for being published with it (Figure 21), I certainly do agree on the fact that such a diagram would be a very interesting guide to have under hand.

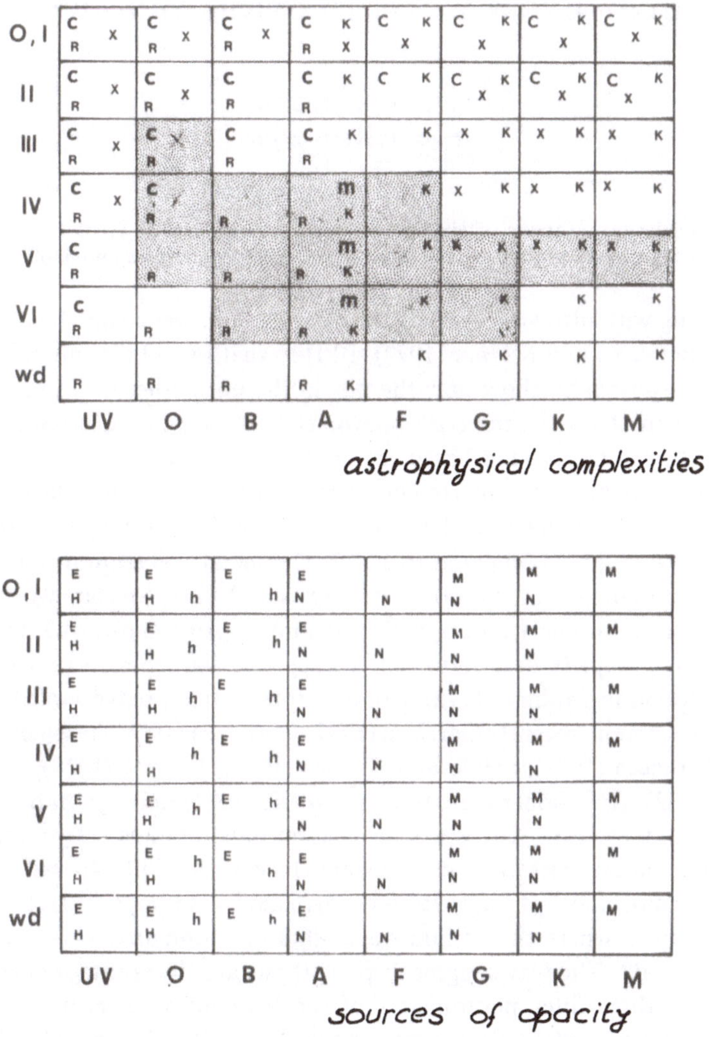

	UV	O	B	A	F	G	K	M
O,I	C R X	C R X	C R X	C R X K	C K X	C K X	C K X	C K X
II	C R X	C R X	C R	C R K	C K	C K X	C K X	C K X
III	C R X	C R X	C R	C R K	K	X K	X K	X K
IV	C R X	C R	R m K	R K	K X	X K	X K	X K
V	C R	R	R m K	R m K	K X	X K	X K	X K
VI	C R	R	R m K	R m K	K	K	K	K
wd	R	R	R	R K	K	K	K	K

astrophysical complexities

	UV	O	B	A	F	G	K	M
O,I	E H	E H h	E h	E h N	N	M N	M N	M
II	E H	E H h	E h	E h N	N	M N	M N	M
III	E H	E H h	E h	E N	N	M N	M N	M
IV	E H	E H h	E h	E N	N	M N	M N	M
V	E H	E H h	E h	E N	N	M N	M N	M
VI	E H	E H h	E h	E N	N	M N	M N	M
wd	E H	E H h	E h	E N	N	M N	M N	M

sources of opacity

Fig. 21. Astrophysical complexities; sources of opacity. *On abscissa,* the spectral type – including the very hot UV stars that can exist (for example in young galaxies); in ordinate, the luminosity class, including class 0, to which belong some stars of the Magellanic Cloud. *In the upper part,* the letters designate the complexities that have to be introduced in models: C is for curvature; R for non-grey radiative equilibrium, m for magnetism, K for convective layers, X for extended atmospheres. The shaded area correspond to those relatively well filled by honest-to-good models. *In the lower part,* E is for electron scattering, H for ionized helium, h for neutral helium, N for negative hydrogen ion, M for molecules; blanketing by impurities and atomic hydrogen have to be put in all cases. Clearly this is highly schematical; it has not been found suitable to put in the diagram any particular selection of models…

ULTRAVIOLET ENERGY DISTRIBUTION OF He-WEAK STARS

P. L. BERNACCA

Asiago Observatory, Italy

Abstract. Ultraviolet observations of Helium-weak stars carried out on OAO-2 indicate that temperature estimates from the ground based region of the spectrum may be unreliable.

A paper dealing with ultraviolet observations of Helium-weak stars has been recently prepared by Bernacca and Molnar (1972) (BM) so that I will here summarize only the results. The purpose is to show how the flux in the ultraviolet may in some cases be different than expected from the *UBV* photometry, placing an uncertainty on temperatures derived from the ground based region of the spectra.

The helium-weak stars can be roughly described by saying that their photometric spectral type S_Q is earlier by 2 subclasses or more than that given by the MK classification system, that is B3–5 compared to B7–9. The metal spectrum in these stars may also be stronger than normal but not as greatly as in the hot Ap stars which also show the color-spectrum anomaly and in which helium is also found to be weak for their colors. A fairly complete description of the spectral characteristics of helium-weak stars can be found in Molnar (1972). Contributors to the discovery of these peculiar objects have been more recently Garrison (1967), Bernacca (1968), Jaschek *et al.* (1969), Ciatti and Bernacca (1971), Schild and Chaffee (1971), Molnar (1972), and Bernacca and Ciatti (1972). Atmospheric analysis by Norris (1971) and by Molnar (1972) has given $0.25 \leqslant \theta_e \leqslant 0.35$, $3.3 \leqslant \log g \leqslant 4.5$ and helium underabundant by a factor of 2 to 15. Metal abundances exist only for one star (Hack 1969). I like to call the attention on the color-color plot of $m_\lambda(4250\text{\AA}) - m_\lambda(3320\text{\AA})$ vs $m_\lambda(1910\text{\AA}) - m_\lambda(3320\text{\AA})$ shown in Figure 1 where the helium-weak stars are compared with standard stars between B3 and B8. The wavelengths in parenthesis are the constant energy effective wavelengths of three filter photometers of the Wisconsin Experiment Package on board OAO-2 (Code *et al.* 1970). Figure 1 shows the results schematically. The detailed diagram can be found in the BM paper. The broken line represents what we may consider a temperature sequence. The scatter of the observations is shown by the shaded area. The spread observed for the standard stars has to be attributed to a combination of the uncertainty in the reddening correction and to a different degree of line blocking in the 1910Å filter for each star produced by the second and third spectrum of the metals. The correspondence between the spectral types of the standards and their location in the temperature band is shown schematically on the figure. Four of the helium-weak stars lie within the band (open circles) in good agreement with the spectral type predicted by the *UBV* photometry. They are: 3 Cen A (B3–5), HD 144334 (B3–5), α Scl (B5), and HD 144844 (B7). The stars 3 Sco (B3) and HD 21699 (B4) (open circles in the order) are apparently fainter by $0^{\text{m}}2$ at 1910Å with respect to the mean relation (broken line). However the effect can be qualitatively

B. Hauck and B.E. Westerlund (eds.), Problems of Calibration of Absolute Magnitudes and Temperature of Stars, 222–224.

explained in terms of additional line blocking in the 1910 Å filter due to Si II and Si III lines which are known to be stronger than normal in the visual spectrum. The possibility of having underestimated the reddening correction at 1910 should also be considered. It cannot be excluded that these two stars have ultraviolet fluxes which are normal for their *UBV* colours.

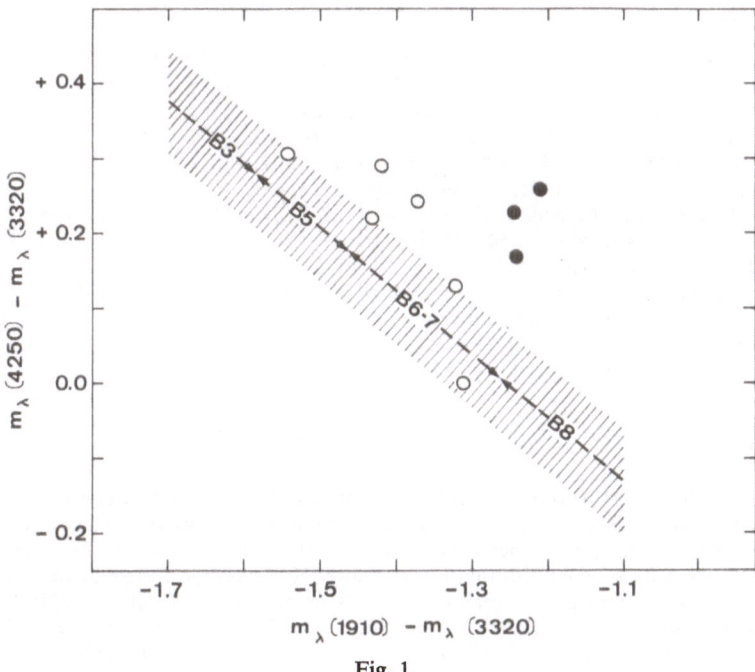

Fig. 1.

The filled circles on the figure represent the stars HR 8535 (B5, B8 IIIp), HR 8770 (B4, B8 III?) and HD 144661 (B5, B7 IIIp). The types given in parenthesis are S_Q and the MK spectral type respectively. The above three stars are fainter at 1910 Å by about $0^m.3$ than their ground based colors would predict. It does not seem possible to explain the anomaly in terms of line blocking or a wrong reddening correction (see BM). The flux at 1910 Å seems to be more appropriate for their MK type so that a possible conclusion is that the region covered by the *U* filter is anomalous showing an excess of radiation. If so, these stars can't longer be considered to have atmospheres deficient of helium since they should be given a lower temperature which would be more appropriate for their faint helium lines. It is interesting to note that according to Garrison (1972), HD 144661 is a Mn-star and that HR 8535 and HR 8770 are likely to be hot Ap stars of the Si-λ4200 type. The conclusion that these hot Ap stars are not helium-weak is suggestive but certainly not proved as of yet. Leckrone (1971) has analyzed a fairly large number of Si-λ4200 stars and He has found systematically that these objects have the far ultraviolet flux more in agreement with their MK types than with their S_Q.

Despite the implications that these ultraviolet observations might have concerning the helium abundance and the nature of the peculiar B type stars I restrict here the conclusion to emphasize how temperature determinations based on a limited portion of the continuous spectrum appear to be unreliable at least for these peculiar objects.

In particular it should be said that the fitting of the data to a model atmosphere in the near Balmer continuum does not give an unambiguous value of the effective temperature.

References

Bernacca, P. L.: 1968, *Contr. Oss. Astrof. Asiago*, No. 202.
Bernacca, P. L. and Ciatti, F.: 1972, *Astron. Astrophys.* **19**, 482.
Bernacca, P. L. and Molnar, M. R.: 1972, *Astrophys. J.* **178**, 1.
Ciatti, F. and Bernacca, P. L.: 1971, *Astron. Astrophys.* **11**, 425.
Code, A. D., Houck, T. E., McNall, J. F., Bless, R. C., and Lillie, C. F.: 1970, *Astrophys. J.* **161**, 377.
Garrison, R. F.: 1967, *Astrophys. J.* **147**, 1003.
Hack, M.: 1969, *Astrophys. Space Sci.* **5**, 403.
Jaschek, M., Jaschek, C., and Arnal, M.: 1969, *Publ. Astron. Soc. Pacific* **81**, 650.
Leckrone, D. S.: 1971, private communication.
Molnar, M. R.: 1972, *Astrophys. J.* **175**, 453.
Norris, J.: 1971, *Astrophys. J. Suppl.* **23**, 213.

DISCUSSION

Schild: Where is the reddening trajectory in your graph, and how have you corrected for reddening?

Bernacca: We have used $(B - V)_0$ given by the colour type and the mean reddening curve by Bless and Sovage (OAO Symposium, Amherst, Mass., August 1971). I have not here the slope of the reddening path on the colour-colour plot that I have shown. However, the correction at 1910Å is larger than that at 4250Å so that the reddening trajectory is nearly parallel to the m_λ (1910) $- m_\lambda$ (3320) axis. I like to mention that HR 8535, HR 8770 and HD 144661 are still fainter than normal at 1910 Å when Stecher's reddening curve is used, a curve which overestimates the reddening in the UV.

THE ABSOLUTE MAGNITUDE AND
STELLAR POPULATION OF A Ba II STAR

R. D. McCLURE*

Yale University Observatory, U.S.A.

Abstract. A star showing extreme features characteristic of the Ba II stars has been found in the old open cluster NGC 2420. A new colour-magnitude diagram has been constructed for the cluster, and a distance modulus of $11^\text{m}7$ is obtained by fitting the unevolved main sequence to the zero-age main sequence. The location of the Ba II star is consistent with its being a disc population star of absolute magnitude $M_v = -0^\text{m}3$, mass $1.4\,M_\odot$ and age 2.4×10^9 yr.

The Ba II stars were first recognized as a class by Bidelman and Keenan (1951) because of the enhancement, on low dispersion spectra of features such as CN, CH, and C_2 bands, and Sr II and Ba II lines. Normal luminosity indicators for G and K stars such as the $\lambda\,4215$ CN band, and the $\lambda\,4077$ line of Sr II can have strengths similar to those in supergiant spectra, but Bidelman and Keenan recognized that other features suggest luminosities closer to giant stars. Warner (1965) and Eggen (1972) summarize the available data on individual Ba II stars for which some estimate of absolute magnitude can be obtained from H and K emission widths, trigonometric parallaxes etc. These estimates place the Ba II stars, on the average, slightly more luminous than the normal G and K giants of luminosity class III. A recent determination of the absolute magnitude of the Ba II stars as a class has been made by MacConnell *et al.* (1972) who obtained a mean magnitude of $M_v = -0^\text{m}4$ from analysis of the proper motions of a sample of 120 stars, most of which have been discovered recently on Michigan objective prism plates. This determination again places the Ba II stars just slightly brighter than the normal K giants.

The origin of the peculiarities of the Ba II star spectra has been the subject of study of several authors who have carried out curve of growth analyses on high dispersion spectra of a few of the brighter examples. Burbidge and Burbidge (1957) and Warner (1965) conclude that carbon and the S process elements are overabundant. This indicates that heavy elements were synthesized in the interiors of the stars where the neutron capture process can occur, and in some way mixed to the surface in an advanced stage of evolution, perhaps during the carbon burning stage. Warner (1965) concludes that this process probably occurs in intermediate disc population stars of mass range $1-4\,M_\odot$. Eggen (1972) estimates from the space motions of the Ba II stars, that they belong to the old disc population and have masses of 1 to $1.5\,M_\odot$.

In order to determine an absolute magnitude more directly, and to pinpoint more accurately the type of stellar population to which a class of star belongs, it is always favourable to find a cluster that contains a star of the class. A Ba II star that, in fact, shows extreme enhancement of the features peculiar to this class has been found by

* Visiting Astronomer, Kitt Peak National Observatory which is operated by the Association of Universities for Research in Astronomy, Inc., under contract with the National Science Foundation.

B. Hauck and B. E. Westerlund (eds.), Problems of Calibration of Absolute Magnitudes and Temperature of Stars, 225–228.
All Rights Reserved. Copyright © 1973 by the IAU.

the author in the old open cluster NGC 2420. This is star 'X' in West's (1967) num-
bering system for the cluster. The spectrum of this star is similar to that of HD 24035
illustrated by MacConnell *et al.* (1972) in that the carbon features are stronger than
in most Ba II stars and the C_2 $\lambda 4737$ band is quite visible. The strength of the Ba II $\lambda 4554$
line would likely place the star in Warner's most extreme group (Ba 5).

Photographic photometry has been done on stars within a 5' radius of the cluster
center on 5 *B* and 4 *V* plates taken with the Yale 1-m telescope. A photoelectric
sequence of 40 stars already existed for the cluster due to photometry by Arp (see
West, 1967) and Sarma and Walker (1962). Twenty-four of these stars have been
reobserved photoelectrically and two stars have been added to the faint end of the
sequence from *UBV* photometry on the Kitt Peak 50-in. telescope.

The resulting colour magnitude diagram is shown in Figure 1. The proper motion
studies by Cannon and Lloyd (1970) and van Altena an Jones (1970) have been used

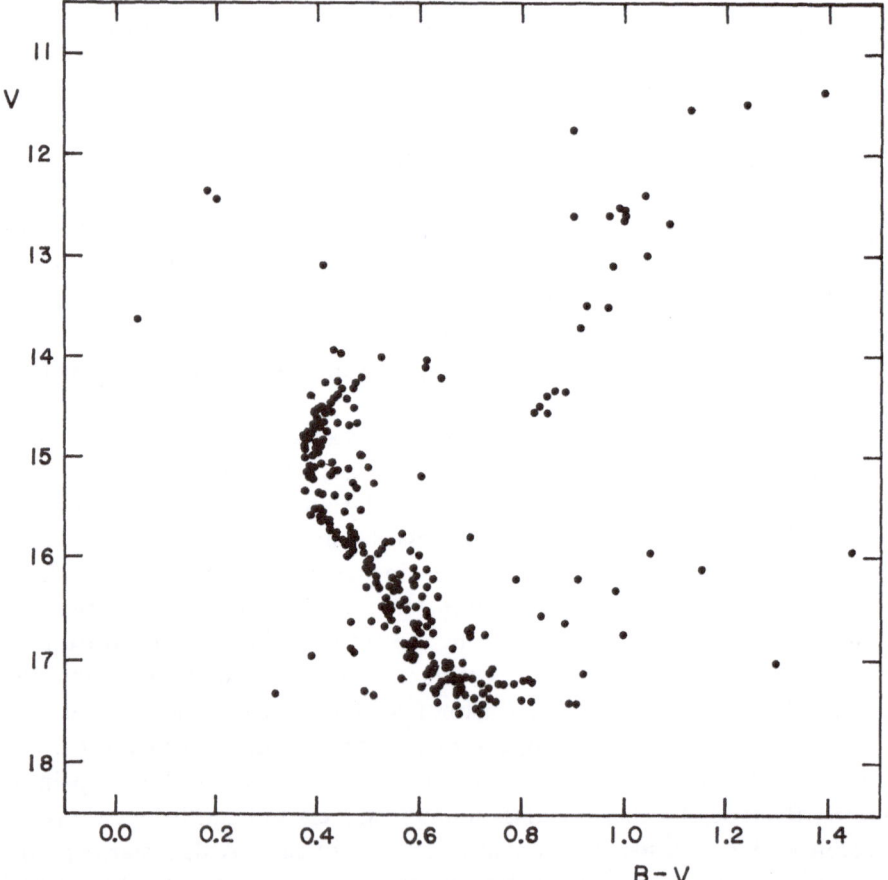

Fig. 1.   The colour-magnitude diagram of the open cluster NGC 2420 ($l^{II} = 198°$, $b^{II} = 20°$) from
photographic photometry of 9 Yale 1-m telescope plates. The Ba II star is the brightest
and reddest star on the diagram.

to eliminate non-members. Cannon and Lloyd's plates extended to $\sim 15.7 V$ magnitude, so between this limit and the faint limit of the photoelectric sequence, all stars for which iris photometry could be done are plotted in Figure 1. The Ba II star is the brightest and reddest star on the diagram. There are three other bright stars in the same region of the diagram, and the bluest of these also appears to have peculiar line strengths. It is possible that it may be a marginal Ba II star similar to those listed in Table II of MacConnell et al. (1972). The Ba II line at 4554 is not visible in its spectrum although the Sr II $\lambda 4077$ line is very strong.

Star X, the extreme Ba II star has a proper motion that is close to the cluster proper motion but not quite close enough to say that it is a member on this basis. To check further on membership, spectra at 78 Å mm^{-1} were taken on the Kitt Peak 2-m telescope of 3 of the 4 brightest stars in Figure 1. The Ba II star velocity from the best plate agrees well with the velocity of the other two bright stars, both of which lie very close to the cluster center and are difinite members on the basis of proper motion. The velocities of these stars fall within the range 86–96 km s^{-1}. With such a large velocity, this low-spread provides a high probability that the Ba II star is a member. A second poorer exposure of the Ba II spectrum gives a velocity of 134 km s^{-1}, but this higher value may be due to the quality of the plate. The mean error in velocity per plate, however, is only ~ 5 km s^{-1} so the possibility exists that there is a true variation.

Arp's photoelectric sequence revealed a rather large ultraviolet excess for an open cluster; $\delta (U-B) = +0\overset{m}{.}09$. Photoelectric photometry by the author confirms this large excess, and in addition narrow band photometry on the DDO photometric system (defined by McClure and van den Bergh, 1968) shows that the K stars in the clump of the giant branch have weak CN bands, such as one would expect for metal weak stars.

West (1967) has discussed the reddening of the cluster and concludes on the basis of the UBV photometry of the photoelectric sequence stars that the reddening is very small. He uses a value of $E(B-V) = 0\overset{m}{.}01$. The reddening can also be determined from DDO photometry using the method described by McClure and Racine (1969). Again, the reddening appears to be negligible, and no reddening corrections have been applied here.

In order to obtain a distance for the cluster, an attempt has been made to fit the main sequence to the zero-age main sequence (Table 1 of Eggen, 1965). Corrections have been made to the main sequence for line blanketing effects suggested by Wildey et al. (1962) for an ultraviolet excess of $\delta (U-B)_{0.6} = +0.09$. After these corrections the main sequence fitting results in a distance modulus of $11\overset{m}{.}7$ for the cluster. This places the cluster at a distance of 2200 pc, and because of its high galactic latitude of 20° it is located more than 800 pc above the galactic plane. It was this large distance from the plane that led van den Bergh (1958) to conclude that this was probably a very old cluster.

The high galactic latitude and remoteness of the cluster provides further evidence in favour of the membership of the Ba II star. MacConnell et al. (1972) estimate that about 1% of G and K giants in the solar neighborhood are Ba II stars. In order to investigate the probability of finding a red giant star of this brightness within the area of the cluster, iris photometry was done (by Mr T. Forrester) on a field centered $0°\!\!.5$ south of the cluster, and of area 1.5 times the area of the cluster photometry. No stars were found in this area with $B-V>0^{m}\!\!.70$ and $V<12^{m}\!\!.9$. Combining this fact with the estimate that 1% of G and K giants are Ba II stars leads to the conclusion that the probability of star X being a field star is negligible.

The derived distance modulus of the cluster places the Ba II star at an absolute magnitude of $M_v=-0^{m}\!\!.3$, in fortuitous agreement with the mean magnitude derived by MacConnell *et al.* (1972) for this class. The NGC 2420 colour-magnitude diagram can be converted to the $\log L$ vs $\log T_{\rm eff}$ plane and fitted to the theoretical isochrones computed recently by Demarque and Gisler (1972). This results in an age determination of 2.4×10^9 yr (assuming a chemical composition of $Y=0.25$ and $Z=0.01$) and a main sequence turn-off mass of $1.4 M_\odot$. This appears to confirm Eggen's (1972) conclusion that the Ba II stars are old disc stars of 1 to $1.5 M_\odot$.

Acknowledgements

The author wishes to thank Dr Pierre Demarque for fitting the cluster to the theoretical isochrones and for many inspiring discussions.

References

Altena, W. F. van and Jones, B. F.: 1970, *Astron. Astrophys.* **8**, 112.
Bergh, S. van den: 1958, *Z. Astrophys.* **46**, 176.
Bidelman, W. P. and Keenan, P. C.: 1951, *Astrophys. J.* **114**, 473.
Burbidge, E. M. and Burbidge, G. R.: 1957, *Astrophys. J.* **126**, 357.
Cannon, R. D. and Lloyd, C.: 1970, *Monthly Notices Roy. Astron. Soc.* **150**, 279.
Demarque, P. and Gisler, G.: 1972, *Bull. Am. Astron. Soc.* **4**, 326.
Eggen, O. J.: 1972, *Monthly Notices Roy. Astron. Soc.* **159**, 403.
MacConnell, D. J., Frye, R. L., and Upgren, A. R.: 1972, *Astron. J.* **77**, 384.
McClure, R. D. and Bergh, S. van den: 1968, *Astron. J.* **73**, 313.
McClure, R. D. and Racine, R.: 1969, *Astron. J.* **74**, 1000.
Sarma, M. K. S. and Walker, M. F.: 1962, *Astrophys. J.* **135**, 11.
Warner, B.: 1965, *Monthly Notices Roy. Astron. Soc.* **129**, 263.
West, F. R.: 1967, *Astrophys. J Suppl.* **14**, 384.

DISCUSSION

Keenan: The recognition of a barium star in a cluster and its luminosity classification are very exciting! Does it not appear now that there is a group of these barium-carbon stars?

McClure: Yes. Perhaps all old disc stars go through this stage in their evolution.

PART VII

STELLAR TEMPERATURE SCALE AND BOLOMETRIC CORRECTIONS

STELLAR TEMPERATURE SCALE AND
BOLOMETRIC CORRECTIONS

J. R. W. HEINTZE

Astronomical Institute Sonnenborgh Observatory, Utrecht, the Netherlands

Abstract. In chapter 1 basic methods are reviewed, and applications and suggestions for future work are presented. In chapter 2 a revision is given of the intrinsic-colour relation in the U, B, V system of hot main-sequence stars. Some temperature-colour relations are discussed in chapter 3, where also a correction formula is given for the effects of interstellar reddening on the effective temperatures of hot main-sequence stars. An empirical mass-luminosity relation is given in chapter 4.

1. Basis Methods, Future Work and Applications

1.1. INTRODUCTION

Pecker (these proceedings, page 173) has reviewed some problems concerning the determination of effective temperatures. From his paper it will be clear that no new scales of temperatures and of bolometric corrections can be given at this moment. Such scales would not be better than previous ones.

At the moment there is far more sense in trying to reduce some of the disagreements mentioned by Pecker. This can be done by improving the theory as well as the observational accuracy. The latter will be emphasized in this paper.

1.2. BASIC FORMULAE

The well known definition formula of the effective temperature, T_{eff}, is

$$\sigma T_{\text{eff}}^4 = \pi \int_0^\infty \mathscr{F}_\lambda \, d\lambda = \pi \int_0^\infty \mathscr{F}_\nu \, d\nu, \tag{1}$$

where σ denotes Stefan-Boltzmann's constant and $\mathscr{F}$ is the monochromatic emergent flux at wavelength λ (ergs cm^{-3} s^{-1}) or at frequency ν (ergs cm^{-2} s^{-1} Hz^{-1}).

The absolute bolometric magnitude, M_{bol}, can be written as

$$M_{\text{bol}} = -5 \log R/R_\odot - 10 \log T_{\text{eff}} + M_{\text{bol}\odot} + 10 \log T_{\text{eff}\odot}. \tag{2}$$

Combined with the definition of the bolometric correction, BC,

$$M_{\text{bol}} = M_v + BC, \tag{3}$$

one gets the classical expression:

$$BC = -M_v - 5 \log R/R_\odot - 10 \log T_{\text{eff}} + M_{v\odot} + BC_\odot + 10 \log T_{\text{eff}\odot}. \tag{4}$$

This formula gives the possibility to determine the BC in an empirical way (see

Section 1.8). It may already be stated here, that $T_{\text{eff}\odot}(=5800\,\text{K}$, see Code's paper, these proceedings page 131), $BC_\odot$ (see Section 1.8) and $M_{v\odot}(=4.87$ mag., Gallouet, 1964) can be assumed to be rather well known at present; that sometimes T_{eff} can be determined from spectral scans and spectra (Sections 1.6 and 1.7); that the radius $R/R_\odot$ sometimes can be found if the star is a component of a well studied eclipsing binary, which has well determined radial velocity curves, or sometimes can be calculated from the parallax and the angular diameter as could be obtained from interferometric measurements or lunar occultations; that M_v sometimes can be found from narrow band photometry, spectroscopic parallaxes or from astrometric parallaxes and V_0, the latter being given by

$$V_0 = V - R \cdot E(B - V). \tag{5}$$

Unfortunately, the colour excess, $E(B-V)$, (see chapter 2) as well as the ratio of the visual interstellar extinction to the colour excess, R, is often rather uncertain.

Other well known expressions for the BC are

$$BC = 2.5 \log \frac{\int_0^\infty S_\lambda \mathscr{F}_\lambda \, d\lambda}{\int_0^\infty \mathscr{F}_\lambda \, d\lambda} + c \quad \text{or} \tag{6a}$$

$$= 2.5 \log \frac{\int_0^\infty S_\lambda f_\lambda \, d\lambda}{\int_0^\infty f_\lambda \, d\lambda} + c \quad \text{or} \tag{6b}$$

$$= 2.5 \log \frac{\int_0^\infty S_\lambda F_\lambda \, d\lambda}{\int_0^\infty F_\lambda \, d\lambda} + c, \tag{6c}$$

where S_λ denotes a normalized photovisual sensitivity function as given by Matthews and Sandage (1963) for example (see the last but one paragraph of Section 2.7);

f_λ the monochromatic stellar flux at the Earth in a relative scale,

F_λ the absolute monochromatic stellar flux measured at the Earth, and

c a constant. For the Sun F_λ is measured over a sufficiently large wavelength region and $BC_\odot$ is known from Equation (3), $M_{v\odot}$ and the adopted zero point of the bolometric scale. So c is known. However, here a hitherto unsolved difficulty arises, which can be shown most easily by combining the Equations (6a) and (1) into:

$$BC = 2.5 \log \int_0^\infty S_\lambda \mathscr{F}_\lambda \, d\lambda - 10 \log T_{\text{eff}} - 2.5 \log \int_0^\infty S_\lambda \mathscr{F}_\lambda (\odot) \, d\lambda +$$
$$+ 10 \log T_{\text{eff}} (\odot) + BC (\odot). \tag{7}$$

Now $\mathscr{F}_\lambda(\odot)$, as obtained from direct measurements or from an *empirical* solar model, is not equal to $\mathscr{F}_\lambda(\odot)$ as obtained from a *stellar* model with $T_{\text{eff}} = 5800\,\text{K}$ and $\log g = 4.44$ (cgs units), calculated in hydrostatic and radiative equilibrium and with constancy of the integrated flux throughout those parts of the model that are important for producing the emergent spectrum $\mathscr{F}_\lambda$. Reference may be made to Weidemann *et al.* (1967) and to the discussion remarks of Weidemann and Popper on page 276 of Gingerich (1969). It may be suggested that in order to determine an usable c the adopted $BC_\odot$ should be used is formula (6a) and that the $\mathscr{F}_\lambda(\odot)$ should be obtained from a stellar model with $T_{\text{eff}} = 5800\,\text{K}$ and $\log g = 4.44$.

1.3. DIRECT EMPIRICAL DETERMINATION OF THE BC AND FUNDAMENTAL DETERMINATION OF T_{eff}

For some unreddened stars F_λ is already known over a sufficiently large wavelength region. In such cases BC can be determined empirically with Equation (6c). In the near future much more complete flux envelopes will be known of stars for which the BC then can be found with Equation (6c). A main problem will be the accurate correction of the observed flux envelope for interstellar and/or circumstellar reddening.

Once a BC is obtained in this way, the T_{eff} of that star can be found from Equation (4) rewritten as:

$$10 \log T_{\text{eff}} = - M_v - BC - 5 \log R/R_\odot + M_{v\odot} + BC_\odot + 10 \log T_{\text{eff}\,\odot}. \tag{8}$$

1.4. THEORETICAL DETERMINATION OF THE BC

In most cases the BC's have still to be derived from model atmospheres with (6a), where $\mathscr{F}_\lambda$ is the calculated flux. A consistent grid of models should be used and this grid should include a model with $T_{\text{eff}} = 5800\,\text{K}$ and $\log g = 4.44$. It is necessary that the influence of variation of the line strengths is properly taken into account; the calculations have to be performed for a set of $V \sin i$ values. The constant c in Equation (6a) could be determined as suggested at the end of Section 1.2. A possibility of checking these BC's will be indicated in Section 1.8.

1.5. DETERMINATION OF T_{eff} FROM THEORETICAL BOLOMETRIC CORRECTIONS

The theoretical BC's according to Equation (6a) are a function the line strengths, $V \sin i$, $\log g$ and T_{eff}. These BC's can be plotted against T_{eff} together with (BC, T_{eff}) relations according to Equation (4) for stars for which $R/R_\odot$ and M_v are known. The intersection of the lines representing the Equations (6a) and (4) yields an estimate of the effective temperature. Note that T_{eff} can be estimated even if the star is reddened by interstellar extinction if $E(B-V)$ and R (see Equation (5)) are rather well known. Uncertainties in $BC_\odot$ do not matter too much if the $BC_\odot$ adopted in formula (4) is also used in Equation (6a) to calculate c.

An example of this method is given in Figure 14 in Heintze (1968) from which the effective temperatures of Vega and Sirius are estimated to be 9650 ± 550 K and 10800 ± 150 K respectively. In this Figure Balmer- and metal-line blanketed models of Strom (1968) were used for which $BC_\odot = -0.11$. Models with a 10 times higher metal abundance than the Sun give 10600 ± 150 K for Sirius, whereas Hanbury Brown et al. (1967) found 10350 ± 180 K from their interferometric measurements.

This Figure also shows that while for $T_{eff} < 10000$ K the BC practically does not depend on gravity, it does for $T_{eff} > 10000$ K. So in applying this method the gravity has to be known for the A and later type stars. Sometimes unexpected differences in gravity can occur. For example from

$$\log g = \log g_\odot + \log M/M_\odot - 2 \log R/R_\odot \qquad (9)$$

and the radii found by Hanbury Brown et al. (1967) for Sirius and Vega it follows that $\log g_{Sirius} - \log g_{Vega} \approx 0.45$ assuming equal masses for both stars. Yet both stars were classified as main sequence stars for a long time (see discussion remark of Miss Divan in these proceedings, page 267).

1.6. The determination of T_{eff} from observed energy distributions in the visual part of the spectrum

1.6.1. The Paschen Continuum

The observed absolute energy distribution of Vega according to Hayes (1967) and to Oke and Schild (1970) agree very well with each other in the Paschen continuum (see Code's paper in these proceedings, p. 131). With Vega as a standard, the energy distribution in the visual part of the spectrum of all kinds of stars can be determined fairly accurately with photoelectric scanner methods (for a general review article see Oke, 1965). Interstellar reddening changes the slope of the Paschen continuum and corrections have to be made for this. Comparison between the corrected and the theoretical energy distributions yields T_{eff}.

The slope of the Paschen continuum calculated from model atmospheres depends on the line strengths (whatever the cause of differences in the line strengths may be: abundance or NLTE effects, microturbulence influences), $V \sin i$, $\log g$ and T_{eff}. The blocking of the continuous energy flow of a star by lines causes backwarming, which changes the temperature distribution in some parts of the stellar atmosphere and therefore the energy distribution.

Bernacca (these proceedings, page 222) reported UV flux variations of α Scl. It would be of great interest to investigate whether linestrengths variations can be found in α Scl and whether they are correlated with the variations in the continuum flux.

The early grids of models did not include any blanketing effect. Nowadays grids exist of models with Balmer-line blanketing; with H- and He-line blanketing; with

H-, He- and metal-line blanketing and as reported by Schild (communication during the symposium) an extensive grid of models including all known absorbing mechanisms and all possible kinds of blanketing is in preparation at the Harvard-Smithsonian Astrophysical Observatory at Cambridge, Mass. As soon as this grid of models will be available an attempt can be made to revise the temperature scale and the bolometric corrections.

In general the computed slope of the Paschen continuum becomes the steeper the more line blanketing is put into the models. This causes a lowering of the effective temperature of a star of which the observed energy distribution is matched with theoretical energy distributions. At the same time the agreement between observed and calculated UV fluxes is being improved. Figure 8 in Underhill (1972) is an example: the energy distribution of a fully line-blanketed model with $T_{eff} = 10000$ K and $\log g = 4$ from Maran *et al.* (1968) gives much better agreement with the measured UV fluxes of Sirius between 2000 and 3000 Å than a H-line blanketed model of $T_{eff} = 9750$ K and $\log g = 4$ from Klinglesmith' (1971) grid.

1.6.2. *The Balmer Jump*

For the hot stars the Balmer jump, BJ, is a rather good temperature indicator. However, apart from the still existing differences in the observed BJ of Vega (see Code's paper, these proceedings page 131) there are some difficulties in the determination of the BJ from the observations and from the models, which should be carried out in the same way. The Balmer and Paschen continua have to be extra-polated to $\lambda 3700$. A rather good procedure is to plot monochromatic magnitudes (fluxes in units of erg cm^{-2} s^{-1} Hz^{-1}; $\mathscr{F}_\nu$) as a function of $1/\lambda$ and to extrapolate linearly from $1/\lambda = 2.2 - 2.4\ \mu^{-1}$ and $1/\lambda > 2.7\ \mu^{-1}$ to $1/\lambda = 2.7\ \mu^{-1}$ and to take the magnitude difference at $1/\lambda = 2.7\ \mu^{-1}$ (see also the answer to Schmidt-Kaler's comment on page 269 of these proceedings and the discussion remark of Miss Divan on page 267). BJ's determined in this way are independent of interstellar reddening, because the extinction (in mag., and flux-decrease in erg cm^{-2} s^{-1} Hz^{-1}) is a linear function of the frequency for $2.2 < 1/\lambda(\mu^{-1}) < 3$.

1.7. THE SIMULTANEOUS DETERMINATION OF T_{eff} AND $\log g$

In the $(T_{eff}, \log g)$ plane the line representing the possible combinations of T_{eff} and $\log g$ for which the model-Paschen continuum matches the observed one can be drawn as a function of abundance, line strengths etc. The observed H- and He-lines for example can provide other $(T_{eff}, \log g)$ relations in the $(T_{eff}, \log g)$ plane so that (an) intersection point(s) may be found (see Section 1 of Pecker's paper, in these proceedings page 173). Regarding the H-lines: the position of the line representing the possible combinations of T_{eff} and $\log g$ that can explain the observed hydrogen line profiles (see Figures 1 and 2 of Heintze, 1968) depends on whether the Edmonds *et al.* (1967) broadening function or the Griem (1967) broadening theory is used. Up till now no conclusive observational evidence has been obtained for deciding which broadening function has to be preferred. Moreover, the profiles of the published (observed) H profiles of a star depend on the chosen height of the continuum above the H line and vary greatly from one observer to the other. As soon as these problems are solved one possible cause of discrepancies in determining T_{eff} with $(T_{eff}, \log g)$ relations can be removed.

A nice example of the use of the $(T_{eff}, \log g)$ plane is given in Figure 2 in Klinglesmith (1972). $(T_{eff}, \log g)$ relations as derived from the Paschen continuum, the H-lines and the He I-lines of 29 Psc (rather B6 V than B8 III) are given as functions of the He content using the model grid of Klinglesmith (1971). The results are $T_{eff} = = 15150 \pm 700$ K, $\log g = 4.08 \pm 0.15$ and $n(He)/n(H) = 0.027$ assuming the star to be unreddened. In Section 3.2 it will be shown that 29 Psc is slightly reddened. Correction for it results in $T_{eff} = 15400$ K and $\log g = 4.14$.

1.8. INDIRECT EMPIRICAL DETERMINATION OF THE *BC*

Kuiper's (1938) scale of *BC*'s was scaled to radiometric magnitudes of F2V and later type stars. Popper (1959) has improved this scale using the same radiometric magnitudes of Pettit and Nicholson (1928). Morton and Adams (1968) adopted Popper's (1959) scale for the stars cooler than F2, in which $BC_\odot = -0.07$ mag. According to Aller (1963, p. 293) $BC_\odot = -0.11$ mag. In this paper $BC_\odot = -0.09$ is adopted as a reasonable compromise.

For the hotter stars Morton and Adams scaled their published (theoretical) *BC*'s to Popper's (1959) value of the *BC* at F0 V. They stress, that if one uses formula (7) with $\mathscr{F}_\lambda(\odot)$ as given by Allen (1963) and $BC_\odot = -0.07$ mag., their published *BC*'s have to be corrected by -0.29. According to Hyland (1969) this has indeed to be done (see Section 1.10).

In the future, *BC*'s calculated from models can possibly be scaled and checked with observed *BC*'s as indicated in Section 1.3. There the old radiometric measurements are replaced by complete flux envelopes F_λ. Another possibility is that as soon as T_{eff} and $\log g$ are known fairly well for a star the models and the method to obtain c (see last paragraph of Section 1.2) can be tested by equating the result of formula (6a) with to that of formula (4). Whenever the radius of a star is not known, Equation (4) can be combined with Equation (9) giving:

$$BC = - 2.5 \log \imath/\imath_\odot - (M_v + 10 \log T_{eff} - 2.5 \log g) + \\ + (M_{v\odot} + BC_\odot + 10 \log T_{eff\odot} - 2.5 \log g_\odot). \tag{10}$$

From this formula it is clear that components of binaries of which the distance and the individual masses are known and for which T_{eff} and $\log g$ can be determined from scans and spectra could be suitable objects for this purpose. In this way a rather reliable empirical mass-luminosity relation (see chapter 4) might be constructed.

Formula (10) can be used in an attempt to find in an empirical way the influence of the line strength on the *BC*. For, it can be used for stars with strong as well as weak metallic lines of which the $R - I$ colors are the same and the H-line profiles are identical. Possibly in this way the *BC* can be found empirically as a function of T_{eff}, $\log g$, $\delta(U - B)$ and $V \sin i$.

Strom (1968) tried to calculate BC's for models with different line strengths due to changes in the metal abundance. In Figure 1 some preliminary results are shown. Note that the absolute value of the BC decreases with increasing line strengths.

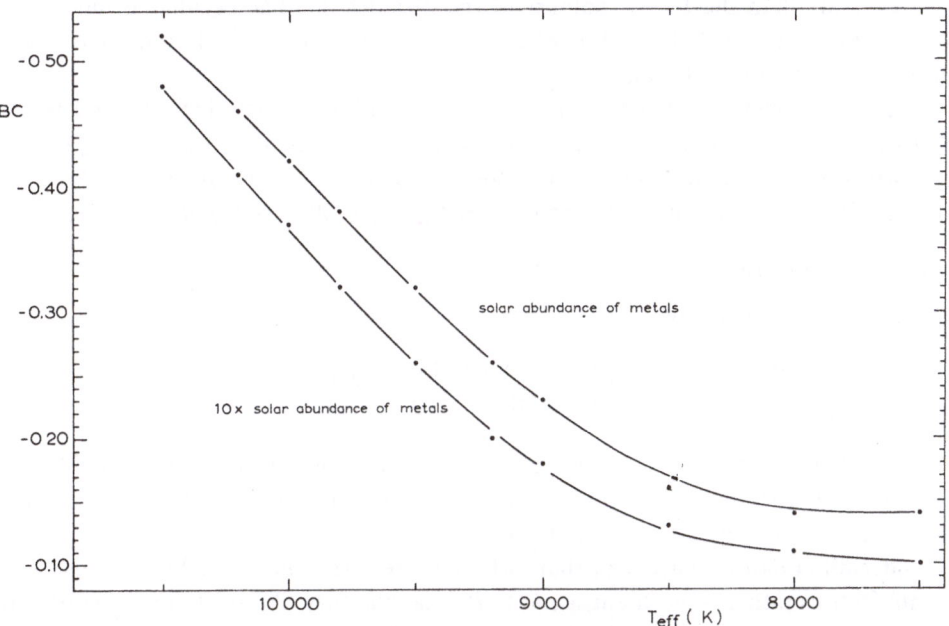

Fig. 1.   Bolometric correction as a function of the effective temperature according to
Strom's (1968) models.

1.9. THE HYADES' PROBLEM

A decrease of the absolute value of the BC could help to solve the Hyades' problem. The Hyades are known to be metal rich (see Parker *et al.* (1961) and Wallerstein *et al.* (1965) for example). It is quite well possible that when the influence of the metal-line strengths, $V \sin i$ and gravity on the BC will be known better, that the resulting absolute bolometric magnitude of the Hyades will decrease in such a way that they will conform to the mass-luminosity relation of the field stars without changing their masses and thus their distance (Hodge and Wallerstein (1966) and Wallerstein and Hodge (1967)).

This idea is supported by Eggen's (1969) measurements of M_I and $R-I$ for 24 M type main sequence members of the Hyades cluster and for 25 field stars of large parallax and in the same colour range as the cluster stars. The $(M_I, R-I)$ relation of both groups of stars are the same to within less than 0.1 mag. if the distances of the Hyades stars as obtained by Wayman *et al.* (1965) with the convergent-point method are used. Theoretically it is conceivable that the absolute I magnitude is much less, if any, affected by blanketing. With respect to the field stars the absolute visual magnitude of the Hyades is too bright. If the absolute values of the BC's of the Hyades indeed have to be smaller (see end of Section 1.8) the difference in M_{bol} between field stars and Hyades becomes smaller than the difference in M_v.

It is interesting to note that according to Figure 2 of Schild's paper (these proceedings, page 31) the $U_2 - V$ colours of the Hyades behave much more similar to those of the field stars than the $U_2 - V$ colours of the Pleiades do. The peculiar behaviour of the Pleiades can probably be explained by the presence of circumstellar shells (Jones, 1972b; Hobbs, 1972).

It may be remembered that at any case the mild subdwarfs have moved to the Hyades-man sequence in the (M_v, $B - V$ plane as well as in the (M_{bol}, $\log T_{eff}$) plane by correcting the magnitudes and the colours of the subdwarfs for the weakness of the metal lines in their spectra (Sandage and Eggen, 1959; Wildey et al., 1962).

1.10. DETERMINATION OF STELLAR RADII

Formula (4) can be written as:

$$\log R/R_\odot = -0.2\,(M_v + BC + 10 \log T_{eff}) + \\ + 0.2\,(M_{v\odot} + BC_\odot + 10 \log T_{eff\odot}). \tag{11}$$

Hyland (1969) has shown that at least for the hot stars the BC's published by Morton and Adams (1968) give erroneous results. Application of a correction of -0.29 mag. (see Section 1.8) gives much better results.

Underhill (1966) has stressed, that although the introduction of blanketing in the models causes rather large changes in the BC the quantity $(BC + 10 \log T_{eff})$ does not change much. (The T_{eff} of a blanketed model is less than the T_{eff} of an unblanketed model of which the Paschen continua match an observed energy distribution, see last paragraph of section 1.6.1.) Therefore according to Equation (11) $R/R_\odot$ does not depend much on the grid of models used. Moreover the exact zero point of the BC scale does not have to be known if $BC_\odot$ is taken from a suitable model of the grid. Underhill did not investigate the dependence of $(BC + 10 \log T_{eff})$ on $\log g$. If there is any, the gravity of the star to be studied should be determined first.

1.11. MASS DETERMINATION

The gravity and the radius provide the mass of a star. Formula (10) can be rewritten as

$$\log \imath/\imath_\odot = -0.4\,(M_v + BC + 10 \log T_{eff} - 2.5 \log g) + \\ + 0.4\,(M_{v\odot} + BC_\odot + 10 \log T_{eff\odot} - 2.5 \log g_\odot). \tag{12}$$

The same remarks can be made as in Section 1.10.

For the blue horizontal branch stars in M67 Sargent (1968) found in this way $\mathcal{M} = (0.7 \pm 0.2)\,\mathcal{M}_\odot$. Sargent did not mention whether the BC's were derived from the models used or were taken from a list.

Possibly Equation (12) can be used to determine masses of Cepheids. There is still a discrepancy between the masses of Cepheids as derived from stellar pulsation theory and those derived from stellar evolution computations, which are 1.5–2 times larger

(Fricke *et al.*, 1971). M_v can be found from the empirical period-colour-luminosity (Sandage and Tamman, 1969); T_{eff} and $\log g$ can be determined in the same way as done by Oke (1961); the models used provide the BC. As soon as observed radii of Cepheids become available from interferometric measurements for example, this method can be tested. For then the observed radii can be compared with calculated ones according to Equation (11). At the same time, however, the search for and the astrometric study of Cepheids in binary systems should not be neglected.

2. Intrinsic Colours

2.1. INTRODUCTION

Quite a number of astronomers still use the Johnson and Morgan (1953) intrinsic-colour relation. However, in 1963 at least three authors (Crawford, Serkowski and Westerlund) suggested that the 1953 intrinsic-colour relation is not blue enough for the late B-type stars. Moreover, α Vir, the hottest apparently unreddened star Johnson and Morgan could use in 1953 had at that time a published $(B-V) = -0.26$, whereas it now is -0.23 (being the mean of the values given in the catalogue of Blanco *et al.* (1972), hereafter called BDDF catalogue). For this reason I decided to reinvestigate the $[(U-B)_0, (B-V)_0]$ relation of the hot main sequence stars, unaware of the fact that Schmidt-Kaler (1965) had already revised it (see his comment on page 269 of these proceedings).

In this paper not only stars listed by Johnson and Morgan, but also stars listed by Serkowski (1963) and Westerlund (1963) are used. For the colours mean values determined from the BDDF catalogue are taken. Moreover all the O5–B1 stars of the BDDF catalogue and the stars of three young clusters (see Section 2.3) are used.

2.2. REDDENING LINES

2.2.1. *Basic Formulae*

The interstellar reddening for stars of the same spectral type can be represented (see Fitzgerald (1970) for example) by

$$\frac{E(U-B)}{E(B-V)} = a_1 + a_2 \cdot E(B-V). \tag{13a}$$

Assuming $(B-V)_0$ to be the intrinsic $(B-V)$ colour of stars of a certain spectral type, $(U-B)_0$ can be found by first adopting an $(U-B)_0'$ and then solving the equation

$$\begin{aligned}(U-B) - (U-B)_0' = a_0 &+ a_1 \left[(B-V) - (B-V)_0\right] + \\ &+ a_2 \left[(B-V) - (B-V)_0\right]^2\end{aligned} \tag{13b}$$

for the constants a_0, a_1 and a_2 by the method of least-squares. Then

$$(U-B)_0 = (U-B)_0' + a_0. \tag{14}$$

In this way a set of possible combinations of $(B-V)_0$ and $(U-B)_0$ can be found for each spectral type. This was carried out for the O5–B1 main sequence stars of the BDDF catalogue. Table I shows some results. It turns out that a_2 does not change if $(B-V)_0$ changes from -0.24 to -0.34. However, a_1 does depend on $(B-V)_0$. Sometimes a_2 is put equal to zero. This is necessary if only a few stars are available (see Figure 2). From Equation (13a) it follows that possible combinations of $(B-V)_0$ and $(U-B)_0$ can be found in this case from

$$U - B = Q + a_1 (B - V), \qquad (15a)$$

where $Q=(U-B)_0-a_1(B-V)_0$. This quantity can be determined directly from the observations without knowing $(B-V)_0$ and $(U-B)_0$.

Once $(B-V)_0$ and $(U-B)_0$ are known the quadratic form (13b) can be written as

$$U - B = Q + A_1 (B - V) + a_2 (B - V)^2, \qquad (15b)$$

where

$$\left. \begin{array}{l} Q = (U - B)_0 - a_1 (B - V)_0 + a_2 (B - V)_0^2 \\ \text{and} \\ A_1 = a_1 - 2a_2 (B - V)_0. \end{array} \right\} \qquad (16)$$

a_1 or A_1 is the slope of the reddening line. The final choice will be represented by S (see Figure 8). a_2 is the curvature of the reddening line and the term $a_2(B-V)^2$ can be omitted if $|(B-V)| > 0.25$.

In Section 2.7 the final values of $(B-V)_0$ and $(U-B)_0$ will be determined.

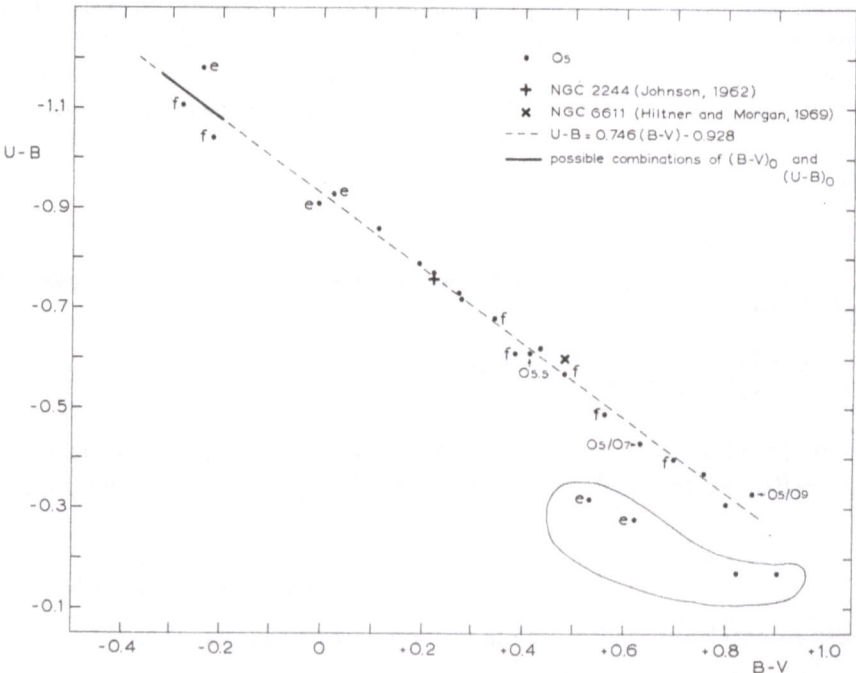

Fig. 2.   Two colour diagram of the O5 stars of the BDDGF catalogue. Encircled stars are omitted for the determination of the reddening line.

2.2.2. *Application to the O5–B1 V Stars*

2.2.2.1. *The O5 stars.* In Figure 2 the $[(U-B), (B-V)]$ diagram of the O5 stars of the BDDF catalogue is given. The 4 encircled fairly reddened stars in the lower part of Figure 2 are not included for the determination of Q and a_1 in Equation (15a).

2.2.2.2. *The O6, O7 and O8 V stars.* The two-colour diagram of these stars of the BDDF catalogue is given in Figure 3. The linear expression (15a) and the quadratic

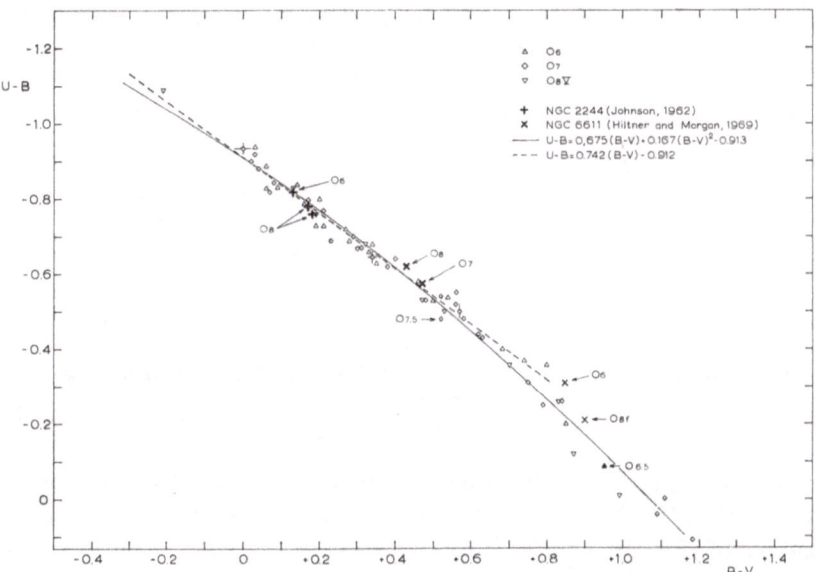

Fig. 3.   Two colour diagram of the O6, O7 and O8 V stars of the BDDF catalogue.

expression (13b) of the reddening line give quite different sets of possible combinations of $(B-V)_0$ and $(U-B)_0$. This is not the case for the O8 V stars alone as can be seen from Table I.

2.2.2.3. *The O9/O9.5 IV/V stars.* An interesting result obtained for these stars is that the possible combinations of $(B-V)_0$ and $(U-B)_0$ are nearly identical whether or not the Cygnus stars are left out, included or partly included (see Table I). So mean values could be used to represent the line of possible combinations of $(B-V)_0$ and $(U-B)_0$ of the O9/9.5 IV/V stars in the two-colour diagram (heavy short line in the left top part of Figure 4).

Note that v Ori, classified as B0 V, apparently is an O9 V or an O9.5 V star according to the Figures 4 and 5.

2.2.2.4. *The B0/0.5 IV/V stars.* Because of their large deviations from the mean the

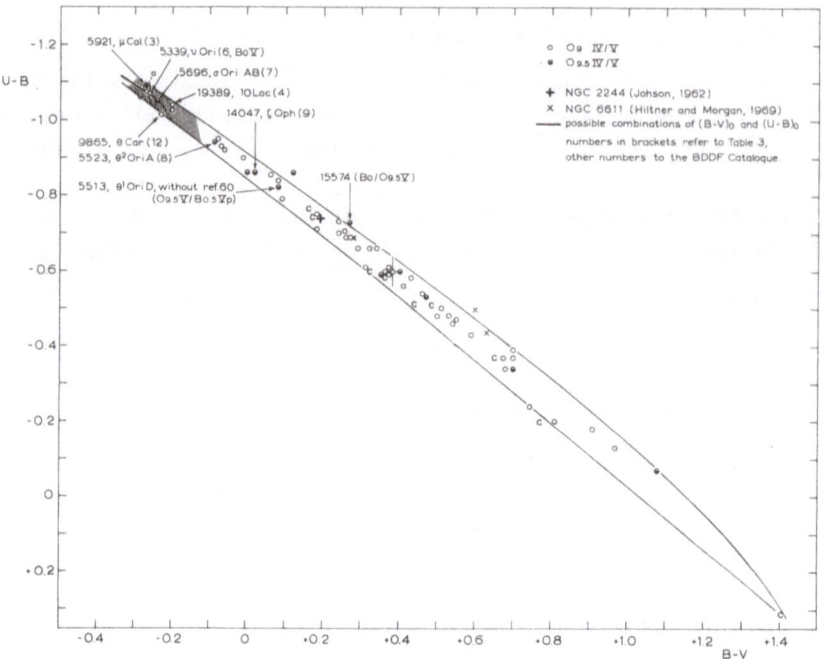

Fig. 4. Two colour diagram of the O9/9.5 IV/V stars of the BDDF catalogue. Cygnus stars are indicated by C.

encircled stars of Figure 5 are not used for the determination of the linear or quadratic relation between $(U-B)$ and $(B-V)$, which give nearly the same set of possible combinations of $(B-V)_0$ and $(U-B)_0$ (see Table I).

μ Col is rather an O9.5 V star than a B0 V star. If the classification of γ Cas and o Pup is correct then these stars are either peculiar or doubles.

2.2.2.5. *The B1 IV/V stars.* The two-colour diagram of these stars is shown in Figure 6. Encircled stars are omitted for the determination of the $[(U-B), (B-V)]$ relations.

2.3. INTRINSIC COLOURS FROM CLUSTERS

2.3.1. *A Preliminary $(B-V)_0$-Spectral Type Relation*

Three young clusters [NGC 2244 (O5–B1 V) (Johnson, 1962); NGC 6611 (O5–B2 V) (Hiltner and Morgan, 1969) and NGC 2232 (B2 V–A0) (Claria, 1972)] were chosen in order to determine a preliminary $[(U-B)_0, (B-V)_0]$ relation. Figure 2 of Johnson's paper shows that NGC 2244 must be very young, so does Figure 1 of Hiltner and Morgan's paper for NGC 6611. The $(B-V)_0$ spectral type relation of the first two clusters was hand-drawn, that for NGC 2232 is represented by a straight line obtained by means of least-squares. The three groups were shifted together by eye to one graph, going through λ Lep (B0.5 V) which is assumed to be unreddened (see Figure 5). The

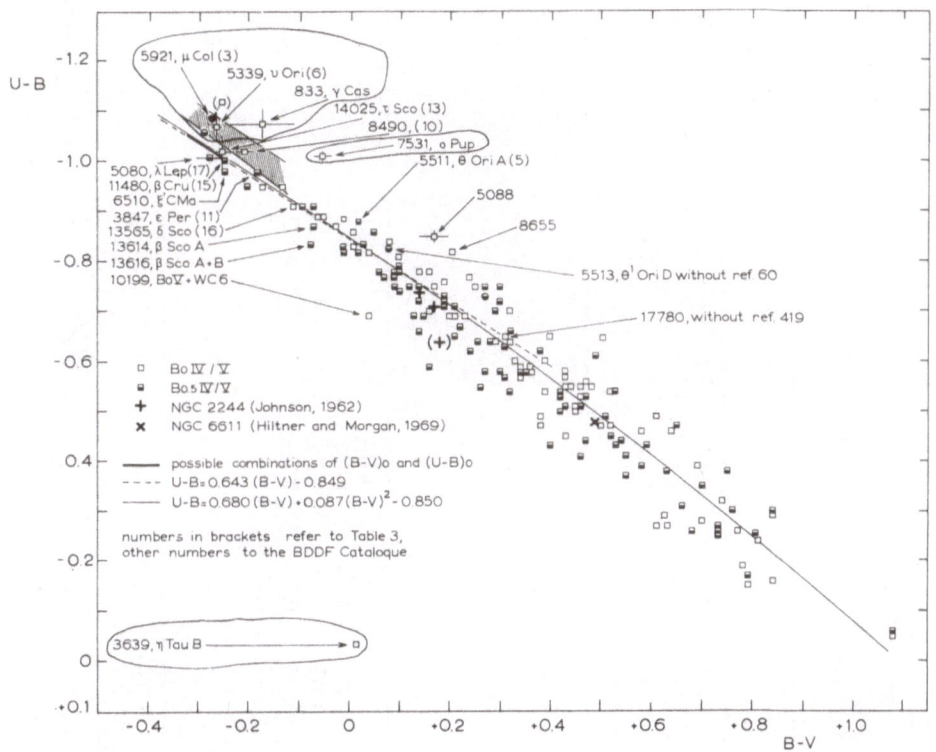

Fig. 5.   Two colour diagram of the B0/0.5 IV/V stars of the BDDF catalogue. Encircled stars are omitted for the determination of the reddening line.

result is shown in Figure 7a. The thin crosses in this figure have not been used as these stars probably are not members of NGC 6611. $E(B-V)$ as determined from Figure 7a is 0.45 mag. for NGC 2244 (0.46 mag. according to Johnson), 0.74 mag. for NGC 6611 (0.80 mag. acc. to Hiltner and Morgan) and 0.05 mag. for NGC 2232 (0.01 mag. acc. to Claria).

2.3.2. A Preliminary $(U-B)_0$-Spectral Type Relation

Preliminary values of $E(U-B)$ were obtained with Equation (13a) by taking a_1 values from Table I for which determination the just derived $(B-V)_0$ values were used. For NGC 2232 a_2 is taken equal to zero and $a_1 = 0.645$ (see Section 2.4). The results are shown in Figure 7b.

In Figure 7 the position of the hot subdwarf HD 49798 (Jaschek and Jaschek, 1963) is indicated. Figure 7 suggests that HD 49798 could quite well be regarded as a normal O6 star.

2.3.3. A Preliminary Intrinsic-Colour Relation

From Figures 7a and 7b an $[(U-B)_0, (B-V)_0]$ relation can be determined and

TABLE I

Some data of the main-sequence stars used to determine reddening functions in the two-colour diagram and some results (see Section 2.2.1)

	B1 IV/V		B0/0.5 IV/V		O9/9.5 IV/V			O8 V		O8 V/O7/O6		O5/O5e/O5f
					All stars	Without most red (Cygnus) stars	Without Cygnus stars					
Number of stars available	118	86	164	83	67	66	59	11	6	65	53	23
range of B−V (min)	-0.25_5	-0.25_5	-0.26	-0.26	-0.27_5	-0.27_5	-0.27_5	-0.21	-0.21	-0.21	-0.21	-0.24
range of B−V (max)	$+0.88$	$+0.48$	$+1.08$	$+0.36$	$+1.40_5$	$+1.08$	$+1.08$	$+1.22_5$	$+0.53$	$+1.22_5$	$+0.80$	$+0.80$
a_2	0.036	0	0.087	0	0.078	0.046	0.048	0.101	0	0.167	0	0

$(U-B)_0$ — a_1

$(B-V)_0 =$	B1 (118)	B1 (86)	B0 (164)	B0 (83)	All (67)	W.m.r. (66)	W.Cyg (59)	O8V (11)	O8V (6)	(65)	(53)	(23)
-0.34		-0.97	-1.05	-1.05_5	-1.10_5	-1.11_5	-1.11	-1.18_5	-1.20	-1.12	-1.16_5	-1.18
-0.32			-1.04	-1.04	-1.09	-1.10	-1.10	-1.16_5	-1.18	-1.11	-1.15	-1.16_5
-0.30	-0.96		-1.03_5	-1.03	-1.08	-1.08_5	-1.08_5	-1.16	-1.16_5	-1.10	-1.13_5	-1.15
-0.28	-0.95_5		-1.01	-1.01_5	-1.06_5	-1.07	-1.07	-1.14_5	-1.15	-1.09	-1.10_5	-1.13_5
-0.26	-0.95		-1.00	-1.00								
-0.24	-0.94_5											

A_1

$(B-V)_0 =$	B1 (118)	B1 (86)	B0 (164)	B0 (83)	All (67)	W.m.r. (66)	W.Cyg (59)	O8V (11)	O8V (6)	(65)	(53)	(23)
-0.34			0.623		0.672	0.721	0.704	0.714		0.561		
-0.32			0.627		0.675	0.713	0.706	0.718		0.568		
-0.30	0.688		0.630		0.678	0.715	0.708	0.722		0.575		
-0.28	0.689		0.633		0.681	0.717	0.710	0.726		0.582		
-0.26	0.690		0.637									
-0.24	0.692											
(adopted A_1)	0.710	0.674	0.680	0.643	0.721	0.740	0.740	0.782	0.808	0.675	0.742	0.746

Q

	B1 (118)	B1 (86)	B0 (164)	B0 (83)	All (67)	W.m.r. (66)	W.Cyg (59)	O8V (11)	O8V (6)	(65)	(53)	(23)
adopted Q-values	-0.763	-0.768	-0.850	-0.849	-0.890	-0.880	-0.885	-0.934	-0.924	-0.913	-0.912	-0.928
(adopted, braced)	-0.765		-0.850			-0.885		-0.930				

adopted values of

	B1 IV/V	B0/0.5 IV/V	O9/9.5 IV/V	O8 V	O5/O5e/O5f
$(B-V)_0$	-0.26	-0.27_5	-0.29_5	-0.34	-0.28
$(U-B)_0$	-0.94_5	-1.03	-1.09_5	-1.18_5	-1.14_5

} results do not change whether the first or the second set is used

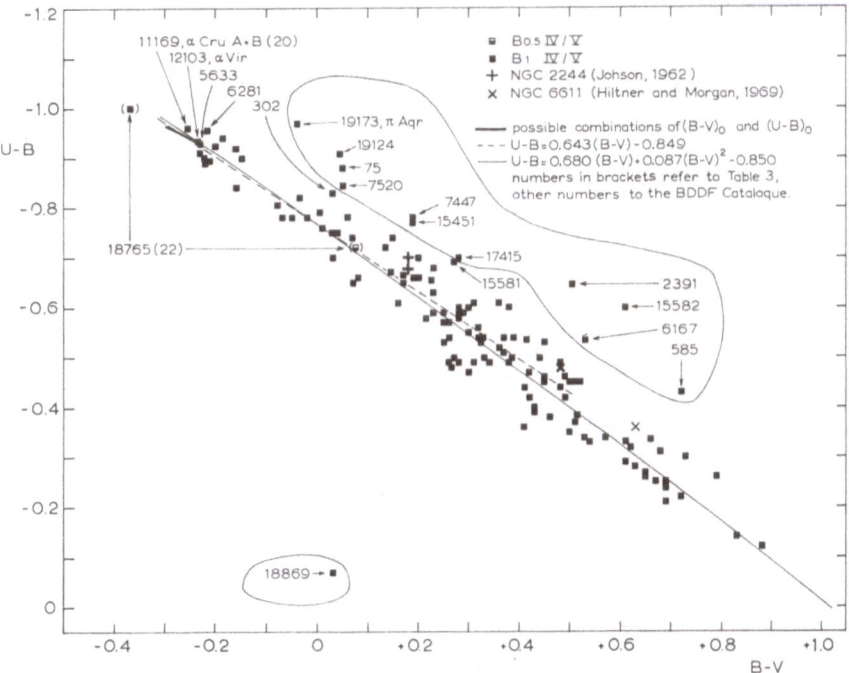

Fig. 6.   Two colour diagram of the B1 IV/V stars of the BDDF catalogue. Encircled stars are omitted for the determination of the reddening line.

drawn in a two-colour diagram in which also the lines of possible combinations of $(B-V)_0$ and $(U-B)_0$ for each spectral type between O5–B1 stars (see Section 2.2) can be drawn (see the four short heavy lines in the top part of Figure 10). So preliminary values of $(B-V)_0$ and $(U-B)_0$ could be obtained for the stars of the Figures 2, 3, 4, 5 and 6 and by substituting into Equation (16) also Q and A_1 values of formula (15b)

2.4. THE SLOPE OF THE REDDENING LINE

Values of the slope of the reddening line, S, as derived by Johnson (1958), Serkowski (1963), and Fitzgerald (1970) for hot main sequence stars are plotted in Figure 8. In the same figure values of A_1 and $a_1(a_2=0)$ as obtained for the stars of the Figures 2–6 (see Section 2.2.1) are plotted. For the O8 V, B0/0.5 IV/V and the B1 IV/V stars the A_1 and a_1 values are connected with each other by vertical lines. The vertical line at O9/9.5 represents the spread in the A_1 values of the different groups of O9/9.5 IV/V stars (see Table I).

The heavy line in Figure 8 can be regarded as a reasonable compromise and is a proposal for S values to be used in the future. Numerical values are given in Table II.

This proposal includes that the Q value of not too much reddened stars of spectral type B1 V–A1 V should be calculated with

$$Q = (U - B) - 0.645\,(B - V) \qquad (17)$$

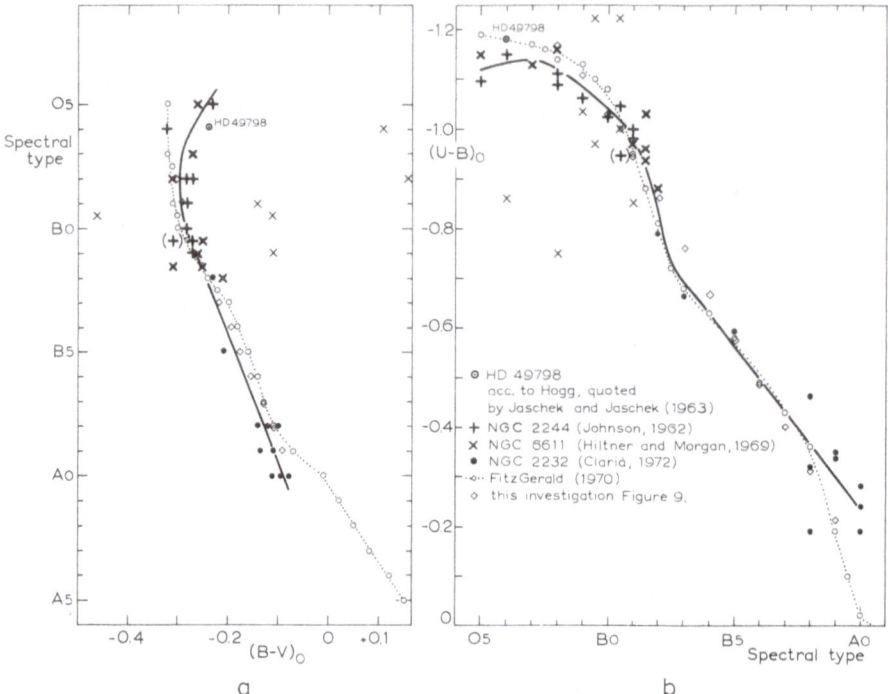

Fig. 7. $(B-V)_0$ (Figure a) and $(U-B)_0$ (Figure b) as a function of spectral type for the stars of the clusters NGC 2244, NGC 6611 and NGC 2232. The adopted $E(B-V)$ values are 0.45 mag., 0.74 mag. and 0.05 mag. respectively. See Section 2.31.

and not with Johnson and Morgan's (1953) formula:

$$Q = (U - B) - 0.72\,(B - V). \tag{18}$$

For some 80 B2V–A1V stars Q values were calculated with Equation (17). These values are plotted against spectral type in Figure 9a. The stars are listed in Table III. Among these stars are the apparently unreddened ones used by Johnson and Morgan (1953), Morgan *et al.* (1953) and the bluest stars listed by Serkowski (1963) and Westerlund (1963). The Q values of the O5, the O5/O7/O8 V, the O8 V, the O9/9.5 IV/V, the B0/0.5 IV/V and the B1 IV/V stars were taken from Table I. By least-squares straight line was fitted through these points, represented by

$$Q = -1.24 + 0.08\,n \quad \text{for} \quad 5 \leqslant n < 15, \tag{19}$$

where the spectral type indicator n is equal to 5 for B0/V and increases to $n = 15$ for A0 V.

2.5. The $[Q, (B-V)_0]$ and $[Q, (U-B)_0]$ relations

In Figure 9b the Q values of the B2 V–A1 V stars of Table III according to Equation (17) are plotted against $(B-V)$. No $(B-V)_0$ values are assigned to the Q values of the

O8 V and O9/9.5 IV/V stars of Table I whereas for the B0/0.5 IV/V and B1 IV/V stars the Q values are plotted against the $(B-V)_0$ values of section 2.3.3 from which the Q values were calculated with Equation (16) (filled squares in Figure 9b). By least-squares a straight line was fitted through these two filled squares and the bluest stars,

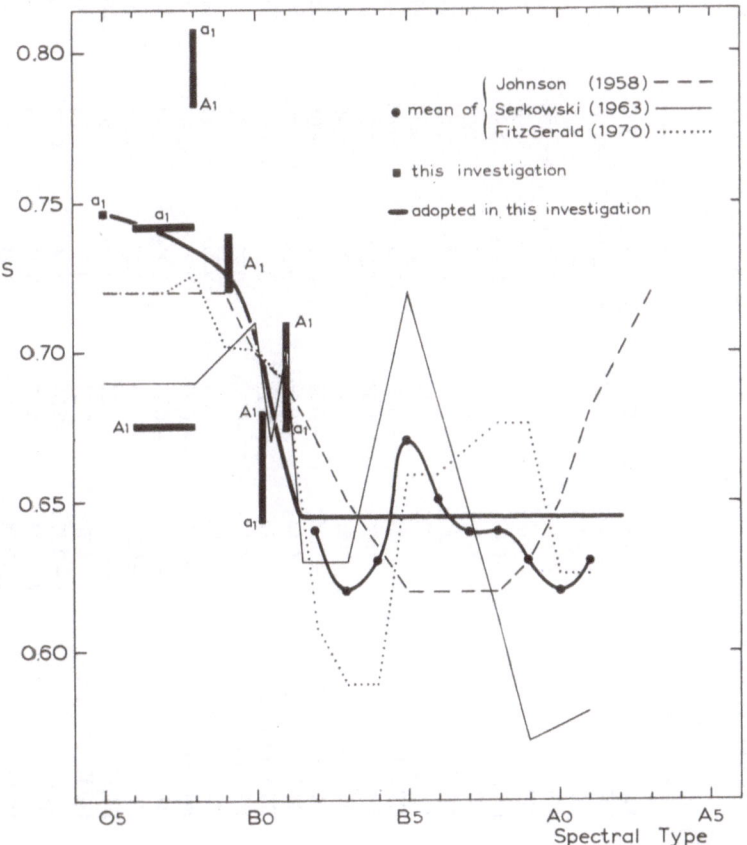

Fig. 8. The slope of the reddening line as a function of spectral type.

TABLE II

Proposed S values for hot
main-sequence stars

Spectral type	S
O5/O6/O7	0.74
O8 V	0.73_5
O9 V	0.72_5
O9.5 V	0.72
B0 V	0.70_5
B0.5 V	0.68
B1 V–A0 V	0.64_5

TABLE III

Hot main sequence stars used to determine the final intrinsic-colour relation

The $(B-V)$, $(U-B)$ and V values are mean values from the BDDF catalogue. Q values are calculated from these values with formula (17). Underlined stars (large symbols in Figure 9) are in first approximation assumed to be unreddened

Name	No. This paper	BDDFcat	Spectral type	Q	$B-V$	$(B-V)_0$	$U-B$	$(U-B)_0$	V
λ Cep	1	18994	O6 f	-0.92	$+0.25$	-0.30	-0.74	-1.15	5.04
—	2	2451	O8/O8 V	-0.93_5	-0.21	-0.30_5	-1.09	-1.16	7.46
μ Col	3	5921	{ B0 V / O9.5 V	-0.94_5 / -0.85_5	-0.25 / -0.28_5	-0.30_5 / -0.28	-1.12 / -1.06	-1.16 / -1.05_5 }	5.17
10 Lac	4	19389	O9 V	-0.89_5	-0.20	-0.29_5	-1.04	-1.10_5	4.88
θ^1 Ori A	5	5511	B0.5 V	-0.89_5	$+0.02$	-0.29_5	-0.88	-1.09_5	6.72
υ Ori	6	5339	B0 V	-0.88_5	-0.26	-0.29	-1.07	-1.09	4.62
σ Ori AB	7	5696	O9.5 V	-0.87_5	-0.24	-0.28_5	-1.05	-1.08_5	3.73
θ^2 Ori A	8	5523	O9.5 V	-0.87_5	-0.09	-0.28_5	-0.94	-1.08	5.08
ζ Oph	9	14047	O9.5 V	-0.87_5	$+0.02$	-0.28_5	-0.86	-1.08	2.57
—	10	8490	B0 Vn	-0.87_5	-0.20_5	-0.28_5	-1.02	-1.08	5.15
ε Per	11	3847	B0.5 IV	-0.86	-0.18	-0.28	-0.98	-1.05	2.89
θ Cas	12	9865	O9.5 V	-0.85	-0.23	-0.28	-1.01_5	-1.05	2.76
τ Sco	13	14025	B0 V	-0.84_5	-0.25	-0.28	-1.02	-1.04	2.83
—	14	5488	B0 V(p)/B0.5 V	-0.84	-0.25	-0.28	-1.01	-1.03	4.78
β Cru	15	11480	B0 III/B0.5 V	-0.83_5	-0.24_5	-0.27_5	-1.00_5	-1.02_5	1.26
δ Sco	16	13563	B0 IV/B0 V	-0.83	-0.11	-0.27_5	-0.91	-1.02_5	2.30
λ Lep	17	5080	B0.5 IV	-0.82	-0.27_5	-0.27	-1.01	-1.01	4.29
—	18	5366	B1 V	-0.82	-0.18_5	-0.27_5	-0.94	-0.99_5	5.34
—	19	6281	B1 V	-0.81_5	-0.21_5	-0.27	-0.95_5	-0.99	5.91
α Cru A+B	20	11169	B1/B1 V	-0.79	-0.25_5	-0.26_5	-0.96	-0.96_5	0.79
—	21	5633	B1 V	-0.78	-0.23	-0.26	-0.93	-0.95	6.03
—	22	18765	B1 V	{ -0.77 / -0.75_5	$+0.07_5$ / -0.37	-0.26 / -0.25_5	-0.72 / -1.00	{ -0.94 / -0.92_5 }	6.66
κ Sco	23	15027	B2 IV	-0.74_5	-0.21_5	-0.25	-0.88_5	-0.91	2.41
—	24	5132	B1 V/B5 V/B2 V	-0.74_5	-0.21_5	-0.25	-0.88	-0.90	5.69
α Lup	25	12788	B2/B1 V	-0.74	-0.21	-0.25	-0.88	-0.90_5	2.30
δ Cet	26	2626	B2 V/B2 IV	-0.72	-0.21_5	-0.24_5	-0.86	-0.88	4.05
v Cen	27	12315	B2 IV/B2 V	-0.71	-0.23	-0.24	-0.86	-0.87	3.41
γ Peg	28	128	B2/B2 V/B2 IV	-0.71	-0.22	-0.24	-0.85	-0.86_5	2.86

Table III (Continued)

Name	No. This paper	BDDFcat	Spectral type	Q	B−V	(B−V)$_0$	U−B	(U−B)$_0$	V
μ² Sco	29	14254	B2 IV	−0.69	−0.21$_5$	−0.23$_5$	−0.83$_5$	−0.85	3.56
γ Lup	30	13314	B2n/B2 Vn	−0.68$_5$	−0.20$_5$	−0.23$_5$	−0.81$_5$	−0.83$_5$	2.77
χ Cen	31	13000	B3/B2 V	−0.64$_5$	−0.21$_5$	−0.22$_5$	−0.78$_5$	−0.79	3.13
—	32	5673	B3 V	−0.64$_5$	−0.15$_5$	−0.22$_5$	−0.74$_5$	−0.79	6.84
η Hya	33	8422	B3/B3 V	−0.61$_5$	−0.19$_5$	−0.21$_5$	−0.74	−0.75$_5$	4.29
σ Sgr	34	16048	B3/B3 V/B2 V	−0.61	−0.21	−0.21$_5$	−0.74$_5$	−0.74$_5$	2.05
α Pav	35	17659	B3 IV	−0.58$_5$	−0.20	−0.20$_5$	−0.71$_5$	−0.72	1.93
ι Her	36	15035	B3/B3 V	−0.57$_5$	−0.18	−0.20$_5$	−0.69	−0.70$_5$	3.80
τ Lib	37	13346	B2 V/B2.5 V	−0.56	−0.17$_5$	−0.20	−0.67$_5$	−0.69	3.65
η UMa	38	12316	B3 V	−0.55$_5$	−0.18$_5$	−0.20	−0.67$_5$	−0.68$_5$	1.88
η Aur	39	4878	B3 V	−0.55$_5$	−0.18	−0.20	−0.67	−0.68	3.18
—	40	8467	B4 IV	−0.55$_5$	−0.17$_5$	−0.20	−0.66$_5$	−0.68	4.86
—	41	287	B3/B5 IV	−0.51$_5$	−0.12	−0.19	−0.59$_5$	−0.64	5.58
—	42	11156	B6/B7 V	−0.51	−0.14$_5$	−0.18$_5$	−0.60	−0.62$_5$	10.65
ν And	43	672	B5 V	−0.48	−0.15	−0.18	−0.57$_5$	−0.59$_5$	4.52
κ Hya	44	9190	B3/B5 V	−0.47	−0.15$_5$	−0.17$_5$	−0.57	−0.58	5.05
τ Her	45	13871	B5 V	−0.46	−0.15$_5$	−0.17	−0.56	−0.57	3.89
α Scl	46	907	B8 III(p)/B5	−0.44$_5$	−0.16$_5$	−0.17	−0.55	−0.55	4.30
α Gru	47	18893	B2 V/B5 V	−0.44$_5$	−0.15	−0.17	−0.54	−0.55	1.74
ψ² Aqr	48	19937	B3 V/B5 V	−0.44$_5$	−0.15	−0.17	−0.54	−0.55	4.40
β Sex	49	9684	B6 V	−0.43	−0.13$_5$	−0.16$_5$	−0.52	−0.54	5.07
29 Psc	49–50	20638	B8 III/B6 V	−0.41$_5$	−0.13$_5$	−0.16	−0.50	−0.52	5.10
19 Tau	50	3570	B6 V	−0.39$_5$	−0.10$_5$	−0.15$_5$	−0.46	−0.49	4.30
23 Tau	51	3607	B6e/B6 IVnn	−0.38	−0.06	−0.15	−0.42	−0.48	4.18
π Cet	52	2717	B5 V/B7 V	−0.35	−0.14	−0.14	−0.44	−0.44	4.24
18 Tau	53	3565	B8 V	−0.31	−0.07$_5$	−0.13	−0.36	−0.39$_5$	5.65
β Lib	54	13171	B8 V	−0.30	−0.11	−0.13	−0.37	−0.38	2.61
—	55	11355	A0/B7 V	−0.29$_5$	−0.13$_5$	−0.13	−0.38$_5$	−0.38	8.76
α Leo	56	9452	B8 V/B7 V	−0.29	−0.11$_5$	−0.12$_5$	−0.36$_5$	−0.37	1.35
φ Eri	57	2360	B8 V	−0.28	−0.12$_5$	−0.12$_5$	−0.36	−0.36	3.55
β CMi	58	7201	B7 V/B8 V	−0.23	−0.10	−0.11	−0.29$_5$	−0.30	2.90
53 Tau	59	4169	B9 V	−0.21$_5$	−0.10	−0.10$_5$	−0.28	−0.28$_5$	5.35

Table III (Continued)

Name	No. This paper	BDDFcat	Spectral type	Q	$B-V$	$(B-V)_0$	$U-B$	$(U-B)_0$	V
ζ Peg	60	19416	B8.5 V/B8 V	-0.21	-0.08_5	-0.10_5	-0.26_5	-0.27_5	3.30
21 Tau	61	3586	B8 V	-0.20_5	-0.04	-0.10	-0.23	-0.27	5.76
η Aqr	62	19306	B8 V	-0.20	-0.09	-0.10	-0.26	-0.26_5	4.02
α Del	63	17924	B9 V	-0.17	-0.05_5	-0.09	-0.20_5	-0.23	3.77
–	64	11825	B9 V	-0.15	-0.08_5	-0.08_5	-0.20_5	-0.20_5	5.20
22 Tau	65	3595	B9.5 V	-0.13_5	-0.02	-0.08_5	-0.15	-0.19	6.43
134 Tau	66	5940	B9 V	-0.13	-0.07	-0.08	-0.17_5	-0.18	4.89
θ Crt	67	10459	B9 V	-0.12	-0.08	-0.08	-0.17	-0.17	4.75
ε Sgr	68	15626	A0 V/B9 IV	-0.09_5	-0.02_5	-0.07	-0.11	-0.14	1.84
ω² Aqr	69	20387	B9 V/A0/B9.5 V	-0.09	-0.04	-0.07	-0.11_5	-0.13_5	4.47
μ Ser	70	13452	A0 V	-0.08_5	-0.04	-0.07	-0.11	-0.13	3.54
58 Aql	71	16962	A0 n	-0.08_5	$+0.10$	-0.07	-0.02	-0.13	5.61
θ Hya	72	8842	B9.5 V/A0 p	-0.07_5	-0.07	-0.06_5	-0.12	-0.12	3.88
–	73	11442	A0 V	-0.06	-0.04	-0.06	-0.08_5	-0.10	8.29
α CMa	74	6709	A0 V/A1 V	-0.04_5	0.00	-0.06	-0.04_5	-0.08	-1.46
–	75	11319	A1 V	-0.04_5	-0.04	-0.06	-0.07	-0.08	8.88
–	76	11314	A1 V	-0.03	-0.05_5	-0.05_5	-0.06_5	-0.06_5	1.31
α Peg	77	19775	B9 V	-0.02_5	-0.04	-0.05	-0.05	-0.06	2.49
109 Vir	78	12888	A0 V	-0.02_5	-0.01	-0.05	-0.03	-0.06	3.72
o Peg	79	19425	A1 V	-0.02	$+0.00_5$	-0.05	-0.01_5	-0.05	4.81
ρ Peg	80	19647	A1 V	-0.01_5	$+0.00_5$	-0.05	-0.01	-0.04_5	4.90
β UMa	81	10075	A1 V	-0.00	-0.01	-0.04_5	-0.01_5	-0.04	2.38
ε Aqr	82	17998	A1 V	$+0.02$	$+0.00_5$	-0.04_5	$+0.02_5$	-0.00_5	3.77
θ Vir	83	11889	A1 V	$+0.00_5$	-0.01	-0.04_5	$+0.00$	-0.02	4.37
α Lyr	84	15842	A1 V	$+0.00_5$	-0.00_5	-0.04_5	-0.00_5	-0.02	0.03

which are indicated by large filled symbols in Figure 9b. The thus obtained $[Q, (B-V)_0]$ relation is

$$(B - V)_0 = 0.277 \, Q - 0.045 \tag{20}$$

and might replace Johnson's (1958) formula

$$(B - V)_0 = 0.332 \, Q. \tag{21}$$

For each of the stars of Table III, $(B-V)_0$ was calculated with Equation (20) and $E(B-V)$ was determined. $E(U-B)$ was found from

$$E(U - B) = S \cdot E(B - V), \tag{22a}$$

where S is taken from Table II. For fairly reddened stars $E(U-B)$ was calculated with

$$E(U - B) = [S + 0.06 \, E(B - V)] \, E(B - V), \tag{22b}$$

where the value 0.06 is the mean of the a_2 values of Table I for the O9–B1 stars. In this way $(U-B)_0 = (U-B) - E(U-B)$ is known. Q values of stars of Table III were plotted against these $(U-B)_0$ values in Figure 9c. For the B0/0.5 IV and the B1 IV/V stars the Q values of Table I and the $(U-B)_0$ values of Section 2.3.3 were used. For the O8 V and O9/9.5 IV/V stars no $(U-B)_0$ values are plotted, because the uncertainties in them were regarded too large. By least-squares the following relation was obtained:

$$(U - B)_0 = 1.242 \, Q + 0.013 \quad \text{for} \quad Q < -0.6 \tag{23a}$$
$$(U - B)_0 = 1.175 \, Q - 0.030 \quad \text{for} \quad Q > -0.6. \tag{23b}$$

A less accurate expression for the entire spectral range under consideration is given by

$$(U - B)_0 = 1.200 \, Q - 0.021. \tag{23c}$$

The differences in $(U-B)_0$ with (23a) and (23b) are of the order of 0.005 mag. and never exceed 0.009 mag.

2.6. α VIRGINIS

It is worthwhile to note that the BDDF value of $(B-V)$ of α Vir (indicated in Figure 9 by 1972) does not correspond to the above mentioned $[Q, (B-V)_0]$ relation. The 1953 value of $(B-V)$, used by Johnson and Morgan (1953), lies nearly on the $[Q, (B-V)_0]$ relation of Figure 9b. However, the most recent value of $(B-V)$ has to be regarded as the better one. The colour excess $E(B-V)=0.03$ is most probably due to the binary nature of α Vir.

2.7. THE FINAL INTRINSIC-COLOUR RELATION

A plot similar to Figure 10 was made of $(U-B)_0$ and $(B-V)_0$ values of the stars of Table III (not shown here). For the O8 V to A0 V stars the intrinsic-colour relation

J.R.W. HEINTZE

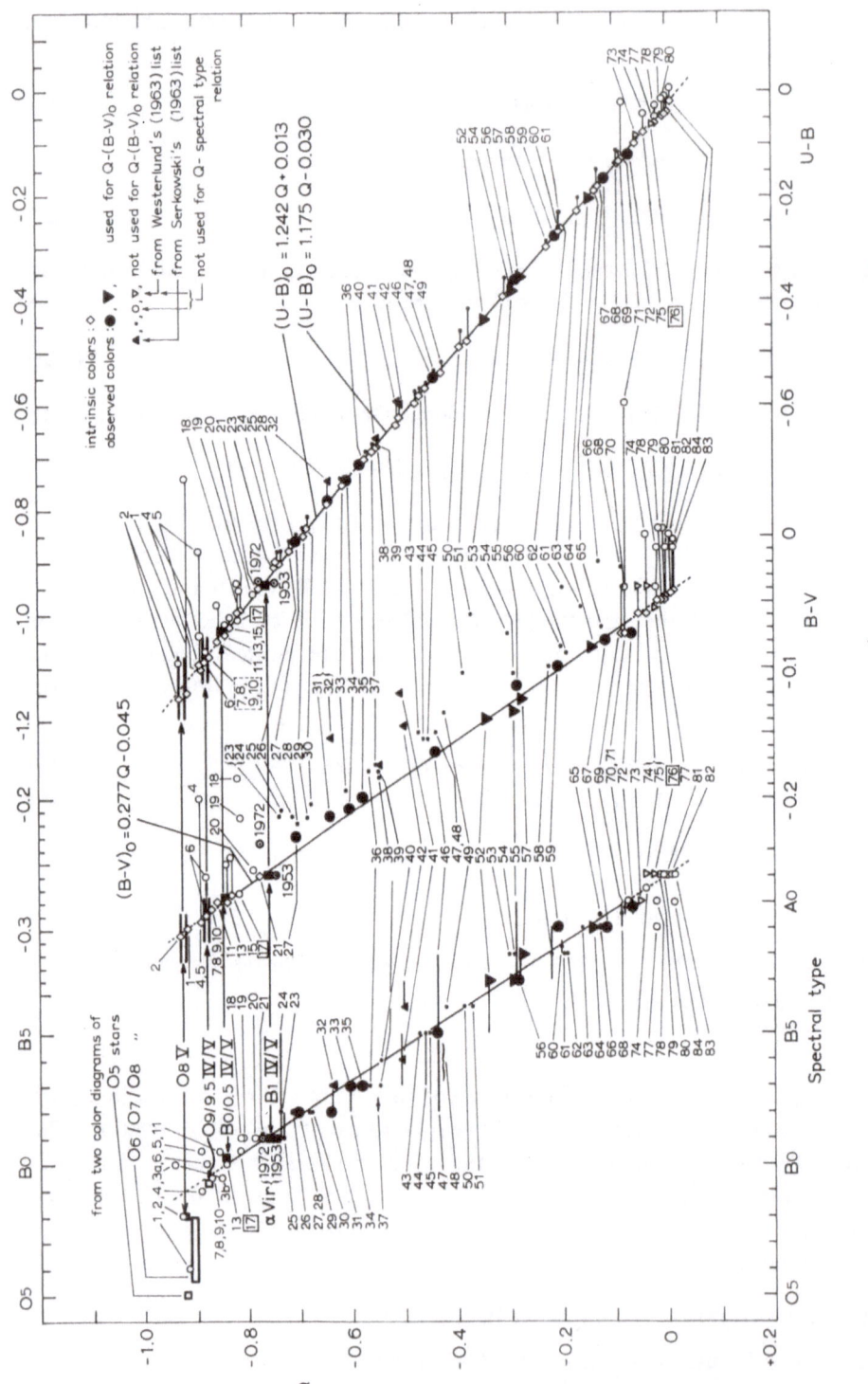

Fig. 9. Q as a function of (a) spectral type, (b) $(B-V)$ and (c) $(U-B)_0$.

as found by least-squares can be expressed by

$$(U - B)_0 = 4.636 (B - V)_0 + 0.252 \quad \text{for} \quad -0.32 < (B - V)_0 < -0.18 \qquad (24a)$$
$$(U - B)_0 = 3.971 (B - V)_0 + 0.130 \quad \text{for} \quad -0.18 < (B - V)_0 < -0.05 \qquad (24b)$$

or less accurately for the entire spectral range under consideration by

$$(U - B)_0 = 4.246 (B - V)_0 + 0.161. \qquad (24c)$$

The latter formula differs at maximum by 0.02 mag. in $(U-B)_0$ from Equations (24a) and (24b).

In Figure 10 the intrinsic-colour relation according to the Equations (24a) and (24b) is shown together with the results from the three young clusters (open squares in Figure 10, see Section 2.3). It turns out that the observed colours of λ Lep, star No. 17 of Table III and star No. 76 of Table III, which both can be assumed to be unreddened according to Figure 9, are nearly exact solutions of formula (24c) (differences in $U-B$ less than 0.005 mag.). This is not the case with the Equations (24a) and (24b). Perhaps formula (24c) has to be regarded as a better representation of the intrinsic-colour relation for the B type main-sequence stars than formulae (24a) and (24b). In that case the S values of Table II used to obtain $(U-B)_0$ (according to Equation (22a)) need some correction. Here still some work has to be done. However, Johnson's (1966) relation (dashed line in Figure 10) provides at any case too red $(U-B)_0$ values for given $(B-V)_0$ values in the region $-1.10 < (U-B)_0 < -0.90$. The open squares in Figure 10 give an indication that for $(U-B)_0 < -1.10$ the intrinsic-colour relation bends back again. It is conceivable that this part of the relation is going through HD 49798 (see Section 2.3 and Figure 7).

In Table IV the proposed intrinsic-colour relation for the B type stars is given. Although no unique relation with spectral type exists, tentative spectral types are given in the last column of Table IV. These were obtained from Figure 9 and should be used only as a rough estimate.

It should be remarked that for $-0.40 < (B-V)_0 < -0.05$ the proposed intrinsic-colour relation agrees very well with curve B of Figure 1 of the paper by Rufener and Maeder (these proceedings page 156). For $-0.05 \leqslant (B-V)_0 < +0.05$ curve B of Rufener and Maeder lies slightly higher than the proposed $[(U-B)_0, (B-V)_0]$ relation.

For $-0.05 < (B-V)_0 < +0.3$ the intrinsic-color relation of Figure 10 was drawn by eye through the lowest points such that it merges better into the relation for the Hyades main-sequence (Eggen, 1962) than Johnson's (1966) relation.

In Figure 10 the observed colours of the faint blue stars HZ 22 (Kowal, quoted by Young et al., 1972) and of HZ 29 (Ostriker and Hesser, 1968) are also plotted. The observed colours of HZ 22 are a solution of Equation (24c). Assuming HZ 22 indeed to be unreddened, this supports the idea that the atmosphere of HZ 22 could be quite similar to that of normal B-type stars. From the hydrogen-line profiles Greenstein (1969) found the surface gravity to be $\log g = 3.9$. Young et al. however found for

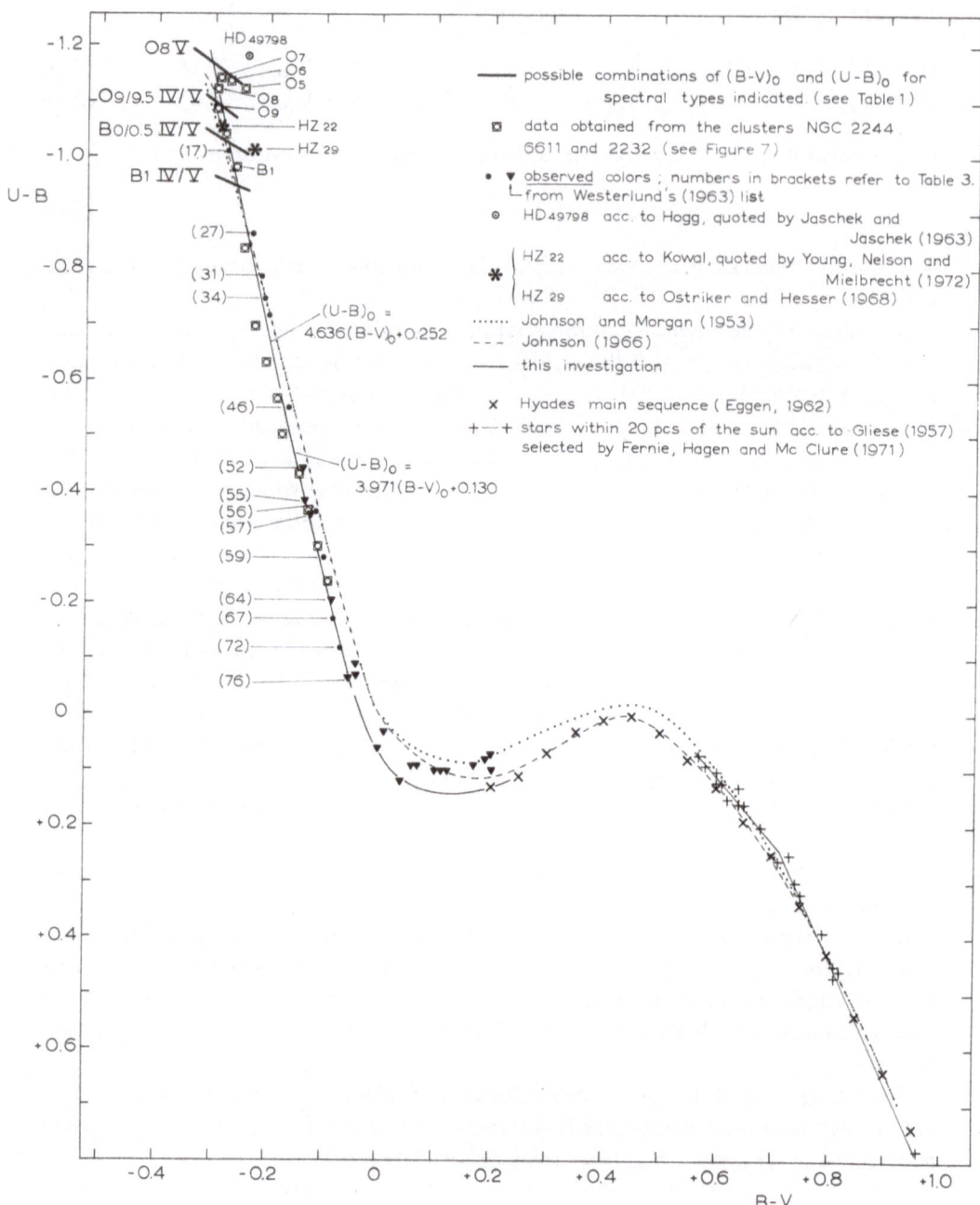

Fig. 10.   Proposed intrinsic colour relation (full-drawn line) compared with some
other intrinsic-colour relations.

the mass of the only visible primary component of this binary system $(P=0\overset{d}{.}573703)=$
$=0.2<\mathcal{M}/\mathcal{M}_\odot<0.7$.
The nature of HZ 29 $(P=1051.118\pm0.015$ s) is not yet known. According to Ostriker
and Hesser it could be a hot subdwarf undergoing radial pulsations, a magnetic variable or a typical member of a new class of variable stars. If not reddened its position
in the two-colour diagram is of importance for a better understanding of this class of
stars.

TABLE IV

The intrinsic-colour relation according to formula (24c) for hot main-
sequence stars

$(B-V)_0$	$(U-B)_0$	Q	Spectral type
-0.29_5	-1.09	-0.90	O9
-0.29	-1.07	-0.87_5	O9.5
-0.28	-1.03	-0.85	B0
-0.27	-0.98_5	-0.81	B0.5
-0.26	-0.94_5	-0.77_5	B1
-0.24	-0.86	-0.70	B2
-0.22	-0.76_5	-0.62	B3
-0.19_5	-0.67	-0.54_5	B4
-0.17_5	-0.58	-0.46_5	B5
-0.15_5	-0.49_5	-0.39	B6
-0.13	-0.39_5	-0.31_5	B7
-0.11	-0.30_5	-0.23_5	B8
-0.09	-0.22	-0.16	B9
-0.08	-0.18	-0.12	B9.5

Q values are obtained from Equation (20) and spectral types from Figure 9a.

2.8. COMPARISON WITH CRAWFORD'S (1963) WORK

Crawford (1963) has measured $(B-V)$ and $(U-B)$ colours of 501 B8 and B9 stars
brighter than 6.5 mag. and has compared the results with the intrinsic-colour relation
of Johnson and Morgan (1953). From his table it follows that 12% of the stars have
an $E(U-B)\approx0$ and that 20% are definitely too blue in $(B-V)$ with respect to Johnson
and Morgan's relation. (The $E(U-B)$ values of these stars are: -0.01 mag. for 11%;
-0.02 mag. for 5%; -0.03 mag. for 1.8%; -0.04 mag. for 0.8%; -0.05, -0.06,
-0.08 and -0.12 mag. each for 0.2%.)
With the proposed intrinsic-colour relation it is found that

$$
\begin{array}{ll}
\text{for 13 stars (2.6\%)} & |E(B-V)| \leqslant 0.005 \\
\left.\begin{array}{l}\text{for 3 stars}\\ \\ \text{for 5 stars}\end{array}\right\}(1.6\%) & \begin{array}{l} -0.01 < E(B-V) < -0.005 \\ \left\{\begin{array}{l}-0.03 < E(B-V) < -0.01\\ -0.017 \leqslant E(U-B) \leqslant -0.008\end{array}\right\}\end{array}
\end{array}
$$

Of the latter 5 stars (BS 1638, 2033, 4082, 3059, 7911) the first three are peculiar
stars (11 Ori, A0 Si; 137 Tau, Ap; 25 Sex, Ap).

3. Effective Temperatures as a Function of $(U-B)$ for B Type Stars

3.1. THE OBSERVATIONS

It has been known for a long time that $[T_{eff}, (U-B)]$ relations of B type stars are very smooth if T_{eff} is determined from a comparison of observed energy distributions with theoretical ones of a grid of models (see Figure 3 in Heintze (1969) for example). This is true even if no corrections for reddening are applied. Clearly the apparent lower effective temperature determined from the reddened energy distribution is with the reddened $(U-B)$ colour still a point of the $[T_{eff}, (U-B)]$ relation for unreddened stars. The same seems to be true for the influence of rotation.

Hyland (1969) found a rather linear $[\theta_{eff}=5040/T_{eff}, Q]$ relation for B type stars.

Also Schild *et al.* (1971) found a fairly linear relation between θ_{eff} and Q for $\theta_{eff} < <0.46 (Q < -0.1)$.

It turns out that different published temperature relations can nicely be expressed as a linear relation between θ_{eff} and $(U-B)$. The following expressions could be derived by means of least-squares, for Hyland's (1969) results:

$$\theta_{eff} = 0.314_5 (U - B) + 0.513 \tag{25a}$$

for Heintze's (1969) results:

$$\theta_{eff} = 0.306 (U - B) + 0.507 \tag{25b}$$

for the results of Schild *et al.* (1971)

$$\theta_{eff} = 0.304_5 (U - B) + 0.502. \tag{25c}$$

3.2. TEMPERATURE CORRECTIONS FOR INTERSTELLAR REDDENING AND ROTATION

From Equations (25a, b, c) it follows that

$$\Delta\theta_{eff} = - 0.31 \times E (U - B) \tag{26a}$$

or

$$\Delta\theta_{eff} = - 0.31 \times S \times E (B - V). \tag{26b}$$

According to Johnson and Morgan's (1953) intrinsic-colour relation, 29 Psc (B6 V rather than B8 III, see last paragraph of Section 1.7) is unreddened. However, according to the proposed intrinsic-colour relation $E(B-V)=0.03$, so $\Delta\theta_{eff}=-0.006$. This means that $T_{eff}=15400$ K rather than 15150 K and $\log g=4.14$ rather than 4.08.

Theoretical work has shown the dependence of Mv, M_{bol}, $(U-B)$, $(B-V)$ etc. on $v_R \sin i$ or v_R, where v_R denotes the equatorial rotational velocity (see Collins (1970) for a recent review article). The effective temperature of a rotating star is lower than the effective temperature of the same star in case it would not be rotating. Here just as in the case of reddening the observed effective temperature obtained by fitting the observed energy distribution of the rotating star to a theoretical energy distribution of a non rotating star provides together with the observed $(U-B)$ value a point on the $[T_{eff}, (U-B)]$ relation of unreddened stars (see Figure 3 in Heintze, 1969).

From Equation (26a) it follows that

$$\varDelta\theta_{\text{eff}} = -0.31 \times \varDelta(U-B),$$

where in this case $\varDelta(U-B)$ denotes the changes in $(U-B)$ due to the rotational velocity. From Figure 15 in Collins and Harrington (1966), Figure 10 of this paper and formula (24c) it is possible to derive for the hot main-sequence stars the following relation between the changes in $(U-B)$ and $(B-V)$ due to rotation

$$\varDelta(U-B) = 4.246\,\varDelta(B-V).$$

Model calculations of Maeder and Peytremann (1970) show that

$$\varDelta(B-V) = k_\theta(v_R\sin i)^2,$$

where k_θ is a constant.

Combination of these three equations yields:

$$\varDelta\theta_{\text{eff}} = -1.32\,k_\theta(v_R\sin i)^2. \tag{27}$$

For comparison, Roxburgh and Strittmatter (1966) found in a theoretical approach as a first approximation

$$\varDelta\log T_{\text{eff}} = k_T v_R^2$$

and

$$\varDelta M_{\text{bol}} = k_{Mb}v_R^2,$$

where k_T and k_{Mb} depend on the type of the star. Some dependence on $\sin i$ is present but very little. Strittmatter (1966) found from a study of the Praesepe cluster that

$$\varDelta M_v = b_{Mv}(v_R\sin i)^2$$

at a given colour.

3.3. TEMPERATURE RELATIONS FROM MODEL PREDICTIONS

In Figure 11 the $[\theta_{\text{eff}}, (U-B)]$ relations (25a, b and c) are shown together with some results of model-calculations. For Mihalas' (1965) unblanketed models one finds

$$\theta_{\text{eff}} = 0.303(U-B) + 0.513 \quad \text{for} \quad 0.18 \leqslant \theta_{\text{eff}} \leqslant 0.28. \tag{28a}$$

For $\theta_{\text{eff}} \geqslant 0.28$ the Mihalas' (1965) results are nearly identical to those of Klinglesmith (1971), which are hydrogen-line blanketed models. However, the results of Klinglesmith do not agree with the results of the Balmer-line blanketed models of Mihalas (1966). For $\theta_{\text{eff}} \geqslant 0.40$ the results of the latter models practically coincide with the temperature relation of Schild *et al.* (1971) (see Figure 11). Moreover if the $(U-B)$ values of Mihalas (1965) are corrected by $+0.03$ mag the extension of the linear $[\theta_{\text{eff}}, (U-B)_{\text{corr}}]$ relation between $0.18 \leqslant \theta_{\text{eff}} \leqslant 0.28$ to greater values of θ_{eff} fits nicely into Mihalas' (1966) results. Heintze (1968) found that Mihalas' 1965 $(U-B)$ values have to be corrected by $+0.02$ mag. A correction of $(U-B)$ by $+0.025$ mag. causes a correction of -0.08 in the constant c of formula (28a), giving

$$\theta_{\text{eff}} = 0.303\,(U - B) + 0.505 \quad \text{for} \quad 0.18 \leqslant \theta_{\text{eff}} \leqslant 0.40. \tag{28b}$$

This formula is in very good agreement with the Equations (25b) and (25c).

3.4. Uncertainties in the $[\theta_{\text{eff}}, (U-B)]$ relation

3.4.1. The Main-Sequence Stars Used by Schild et al. (1971)

In Figure 11 also θ_{eff} values as found by Schild *et al.* (1971) for the eight main-sequence B type stars studied by them are plotted against $(U-B)$ values as derived from the BDDF catalogue for these stars. In this case the relation between θ_{eff} and $(U-B)$ turns out to be

$$\theta_{\text{eff}} = 0.326\,(U - B) + 0.509. \tag{29a}$$

It has to be investigated whether or not the $(U-B)$ values used by Schild *et al.* are the correct ones because these stars were also used for the determination of the intrinsic-colour relation proposed here (see Table III and Section 2.5).

3.4.2. Effective Temperatures Derived by Auer and Mihalas (1972)

In Figure 11 also the effective temperatures of three hot stars (10 Lac, υ Ori and τ Sco) as determined by Auer and Mihalas (1972) are plotted against their $(U-B)_0$ values given in Table 3. These points seem to support formula (29a) as adding of these three stars to the eight main sequence stars of Schild *et al.* results in a least-squares solution

$$\theta_{\text{eff}} = 0.329\,(U - B) + 0.511, \tag{29b}$$

which is nearly identical to Equation (29a).

Auer and Mihalas included in their models a complete NLTE treatment of the H, He I and He II lines (departures from LTE for the first five levels of a 15-level hydrogen atom, allowing explicitly for six line transitions, and for departures from LTE in the groundstate continua of He I and He II; $N(\text{He})/N(\text{H})$ is assumed to be 0.10). They did not compare theoretical with observed energy distributions. Therefore their results are independent of interstellar reddening.

Auer and Mihalas did not give a final T_{eff} of υ Ori as they did not have reliable observations of the He II line profiles of this star and just these lines are very sensitive to T_{eff} in this temperature range. However, they mention that a model of $T_{\text{eff}} = 30\,000$ K and $\log g = 4.0$ (assuming υ Ori to be a B0 V star!, see section 2.2.4) gives a better fit for the He I line profiles (especially $\lambda\lambda 5015$, 5048 and 6678) than in the case of τ Sco ('another' B0 V star) for which they publish $T_{\text{eff}} = 31\,500$ K and $\log g = 4.15$. If $30\,000$ K is indeed a fairly representative effective temperature for υ Ori, which according to Section 2.2.4 is rather an O9.5 V than a B0 V star, then υ Ori fits nicely to the temperature-relations of Hyland (Equation (25a): $29\,600$ K); Heintze (Equation (25b): $30\,400$ K) and Schild *et al.* (Equation (25c): $29\,600$ K). So τ Sco, which indeed is a B0 V star according to Figure 5, should have a lower effective temperature than

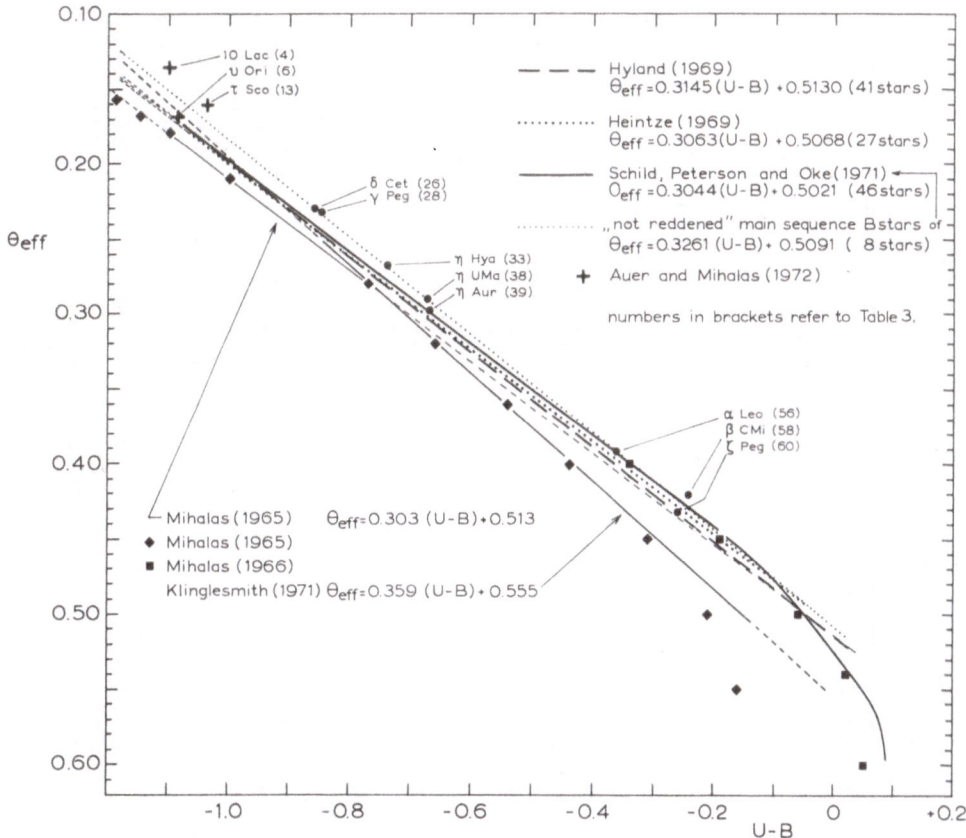

Fig. 11. θ_{eff} as a function of $U-B$ according to some grids of models (Section 3.3) and according to some observational results (Section 3.1 and 3.4)

30000 K in the order of 26900 K, being the mean of the values obtained from the Equations (25a, b and c).

A final remark: Auer and Mihalas' (1972) models do not include any line blanketing. It is possible that UV-line blanketing once included into these models, will slightly lower the effective temperatures of the stars studied by them.

4. The Mass-Luminosity Relation

In Table V the most modern determinations of masses and radii of 70 components of eclipsing binaries, of 48 components of reliable astrometric binaries and available data of α Leo, α PsA, α Aql and the Sun are compiled.

Effective temperatures were assigned to these stars according to the published spectral types. For the B type stars formula (19) is combined with formula (23c) and with the formulas (25c) or (29b) giving

$$\theta_{eff} = 0.028\, n + 0.043 \qquad (30a)$$

TABLE V

Most recent masses and/or radii, M_v and M_{bol} (if published) of 70 components of eclipsing binaries, 48 components of astrometric binaries and 5 single stars

In column 10: M_{bol} calculated from $R/R_\odot$ and an adopted temperature scale (see Section 4) are given and in column 11: M_{bol} from a mass-luminosity relation derived from the data of column 5 and 10

No.	Name	Main ref.	Spectral type	$\mathcal{M}/\mathcal{M}_\odot$	$R/R_\odot$	log g	M_v published	M_{bol}	M_{bol} acc. to(2)	M_{bol} acc. to mass-lum.rel. (31b)	(31a)
1	V444 Cyg B	[6]	O6	29.6	9.8	3.93				−7.85	−7.25
2	UW CMa A	[6]	O7f	19.0	18.6	3.18				−6.75	−6.28
3	UW CMa B	[6]	O7f	23.0	14.8	3.46				−7.25	−6.72
4	AO Cas A	[6]	O9 III	19.0	13.9	3.43				−6.75	−6.28
5	AO Cas B	[6]	O9 III	23.0	8.9	3.90				−7.25	−6.72
6	γ Cyg A	[9]	O9.5 V	17.4±0.8	5.9	4.13±0.03			−6.7 − 6.2	−6.50±0.13	−6.07±0.11
7	γ Cyg B	[9]	O9.5 V	17.2±0.8	5.9	4.13±0.03			−6.7 − 6.2	−6.47±0.13	−6.05±0.11
8	AH Cep A	[6]	B0 Vn	16.5	6.1	4.09			−6.4 − 6.0	−6.36	−5.94
9	AH Cep B	[6]	B0 Vn	14.2	6.1	4.02			−6.4 − 6.0	−5.93	−5.57
10	μ₁ Sco A	[5]	B1.5 V	14.1	5.25	4.15		−4.9 [5]	−5.0 − 4.7	−5.91	−5.55
11	μ₂ Sco B	[5]	(B3)	9.3	5.75	3.89		−4.2 [5]	−4.3 − 4.2	−4.63	−4.39
12	32 Cyg B	[6]	B3 V	8.1	3.9	4.17		3.4 [5]	−3.45 − 3.3	−4.18	−3.97
13	σ Aql A	[6]	B3 V	6.9	4.3	4.02			−3.6 − 3.5	−3.63	−3.46
14	σ Aql B	[6]	B3 V	5.5	3.4	4.12			−3.2 − 3.0	−2.81	−2.68
15	V539 Ara A	[24]	B3 [1]	6.8	4.45	3.97			−3.8 − 3.6	−3.57	−3.40
16	V539 Ara B	[24]	B5 V [1]	5.8	3.73	4.06			−3.4 − 3.2	−3.00	−2.87
17	31 Cyg B	–	B4 V [12]	6.6±0.9 [9]	4.7 [6]	3.91±0.06		{−4.1 [12], −3.8 [5]}	−3.4 − 3.3	−3.4±0.5	−3.3±0.5
18	Z Vul A	[9]	B4 V [4]	5.4±0.3	4.7	3.82±0.03			−3.4 − 3.3	−2.75±0.2	−2.6±0.2
19	U Oph A	[13]–[19]	B4 V [12]	5.30±0.36	3.4	4.10±0.03		−2.6 [12]	−2.7 − 2.6	−2.65±0.25	−2.5±0.25
20	U Oph B	[13]–[19]	B5 V [12]	4.65±0.34	3.1	4.12±0.03		−2.4 [12]	−2.1 − 2.0	−2.2±0.25	−2.1±0.25
21	ζ Phe A	[26]	B6 V	3.85±0.2	2.96±0.05	4.08±0.04		−1.3 [12]	{−1.55 −1.53, ±0.05 ±0.03}	−1.45±0.2	−1.44±0.2
22	ζ Aur B	–	B7 V [9]	5.6±0.6 [9]	7.0 [6]	3.49±0.05			−3.09	−2.85±0.4	−2.7±0.35
23	α Leo	[7]	B7 V		3.8±1.0				−1.68±0.58		
24	ζ Phe B	[26]	B8	2.5±0.1	1.96±0.06	4.28±0.03		+0.5 [12]	+0.02±0.07	+0.31±0.17	+0.31±0.17
25	AR Aur A	[13]	B9 V [12]	2.55±0.19	1.9	4.28±0.03		+0.6 [12]	+0.38	+0.24±0.32	+0.26±0.31
26	AR Aur B	[13]	B9 V [12]	2.30±0.19	1.7	4.34±0.04		+0.9 [12]	+0.61	+0.74±0.41	+0.76±0.35
27	TW Cas A	[21]	B9 V	1.84/2.4	2.40/259	4.06/3.88			−0.295/−0.13	+1.65/+0.49	+1.65/+0.50
28	RX Her A	[9]	B9.5 V [12]	2.75±0.06	2.4	4.12±0.01		+0.6 [12]	+0.03	−0.09±0.09	−0.07±0.09
29	RX Her B	[9]	B9.5 V [12]	2.33±0.03	2.0	420±0.005		+1.0 [12]	+0.42	+0.61±0.06	+0.62±0.05
30	V451 Oph A	[?]	A0	2.78±0.06	2.6±0.1	4.10			+0.35±0.08	−0.14±0.09	−0.11±0.08

No.	Star	Ref.	Sp. type							
31	α Lyr	[7]	A0 V	–	3.03±0.19	<3.9 [8]	0.45		−0.14±0.02	–
32	XZ Pup A	[23]	A0	2.6	3.5	3.76			−0.30	+0.15
33	AS Eri A	[13]	A0 V [12]	1.6	1.6	4.25			+1.42	+2.22
34a	Sirius A		A1 V	{2.2 [19]	1.76±0.04 [7]	{4.29±0.02	1.4		+1.05	+0.87
34b				1.6 [8]}		4.15±0.02}			+1.05	+2.91
35	β Aur A	[25]	A2 IV [11]	2.3 [13]	2.49±0.12	4.00±0.04		+0.4 [5]	+0.69±0.11	+0.67
36	β Aur B	[25]	A2 IV [11]	2.3 [13]	2.50±0.12	4.00±0.04		+0.6 [5]	+0.68±0.10	+0.67
37	CM Lac A	[9]	A2 V [13]	1.88±0.09	1.6	4.30±0.02		+1.7 [12]	+1.65	+1.56±0.21
38	Z Vul B	[9]	A2–3 III [4]	2.3±0.1	2.0	4.20±0.02			+1.24	+0.68±0.20
39	V477 Cyg A	[13]	A3 V [12]	1.78	1.5	4.33		+2.3 [12]	+1.93	+1.80
40	α PsA	[7]	A3 V	–	1.56±0.15	–			+1.86±0.21	–
41	V451 Oph B	[22]	A5? [12]	2.36±0.05	2.1±0.1	4.17±0.05		+1.7 [12]	+1.48±0.10	+0.56±0.09
42	α Aql	[7]	A7 IV/V		1.65±0.10	–			+2.24±0–13	
43	WW Aur A	[13]	A7 V [12]	1.8	1.95	4.12		+1.8 [12]	+1.87	+1.70
44	WW Aur B	[13]	A7 V [12]	1.7	1.90	4.12		+1.8 [12]	+1.93	+1.91
45	TX Her A	[13]	A7 V [12]	1.6	1.5	4.28		+2.6 [12]	+2.44	+2.33
46	TX Her B	[13]	A7 V [12]	1.4	1.4	4.31		+3.2 [12]	+2.59	+2.86
47	RU UMi A	[18]	F0 V	1.7±0.2	1.7±0.1	4.2±0.1			+2.54±0.13	+2.04±0.54
48	EI Cep A	[22]	F0	1.68±0.03	2.3±0.2	3.94±0.09			+1.88±0.19	+2.06±0.09
49	EI Cep B	[22]	F0	1.78±0.03	3.0±0.3	3.74±0.09			+1.31±0.22	+1.80±0.08
50	τ Cyg A	[5]	F0 IV	1.23		–	2.29	+2.25	–	+3.53
51	γ Vir A	[5]	F0 V	1.18	–	–	3.50	+3.46	–	+3.75
52	γ Vir B	[5]	F0 V	1.12	–		3.52	+3.48		+3.98
53	ZZ Boo A	[16]	F2 IV–V	1.75	1.75	4.19			+2.21	+1.88
54	ZZ Boo B	[16]	F2 IV–V	1.68	1.70	4.20			+2.78	+2.06
55	TV Cet A	[9]	F2	1.5						+2.58
56	CM Lac B	[13]–[9]	F2?	1.47±0.04	1.4	4.31±0.01		+3.0	+3.20	+2.68±0.13
57	ψ Vel A	[5]	F2 IV	1.29			+3.1	+3.1		+3.30
58	RS CVn A	[13]	F4 (12)	1.35	1.7	4.09		+2.8 [12]	+3.00	+3.08
59	Z Her A	[13]–[9]	F4 (12)	1.22±0.06	1.6	4.12±0.02		+2.9 [12]	+3.13	+3.58±0.24
60	Procyon A	–	F5 IV–V	1.78 [19]	2.17±0.15 [7]	4.02±0.06		+3.8 [12]	+2.61±0.15	+1.80
61	α Com A	[5]	F5 V	1.4	–	–	+3.70	{+3.66, +3.70 [10]}		+2.86
62	α Com B	[5]	F5 V	1.45	–	–	+3.73	+3.69		+2.76
63	CD Tau A	[13]–[9]	F5 V	1.33±0.05	–	–				+3.15
64	CD Tau A	[13]–[9]	F5 V	1.40±0.05	–	–				+2.91
65	V 477 Cyg B	[13]	F5(?) (12)	1.35	1.2	4.41		+3.18 [12]	+8.89	+3.09
66	HR 7484 A	[13]	F5 V (12)	1.3	1.3	4.32		+3.4 [12]	+3.71	+3.20
67	HR 7484 B	[13]	F5 V (12)	1.3	1.3	4.32		+3.7 [12]	+3.71	+3.20
68	VZ Hya A	[13]–[9]	F5 V	1.23±0.03	1.25	4.33±0.01		+3.6 [12]	+3.80	+3.53
69	VZ Hya B	[13]–[9]	F5 V	1.12±0.03	1.05	4.34±0.02		+4.1 [12]	+4.18	+3.99±0.13
70	42° 1956 A	[5]	F5 V	0.65	–	–	+3.51			+6.80
71	{δ Equ A	[17]			{1.26±0.1					
72	δ Equ B}		F7 V	1.23±0.08	(estim.)}	4.32±0.04	+4.0±0.1	+3.47	+3.92±0.02	+3.54±0.31

Table V (Continued)

No.	Name	Main ref.	Spectral type	$\mathcal{M}/\mathcal{M}_\odot$	$R/R_\odot$	log g	M_v (published)	M_{bol}	M_{bol} acc. to (2)	M_{bol} acc. to mass-lum.rel. (31b)	(31a)
73	99 Her A	[5]	F7 V	1.18	–	–	+4.04	{+3.99, +4.02 [10]}	–	+3.75	
74	WZ Oph A	[13]–[9]	F8 V	1.13±0.04	1.33	4.24±0.02		+3.8	+3.87	+3.95±0.17	
75	WZ Oph B	[13]–[9]	F8 V	1.11±0.04	1.36	4.21±0.02		+3.7	+3.82	+4.03	
76	ξ Cnc A	[5]	F8 V	1.00	–	–	+4.51	+4.46	–	+4.54	
77	β 648 A	[5]	G0 V	1.95	–	–	+3.97		+3.64	+1.40	
78	ζ Her A	[20]	G0 IV	1.22	–	–				+3.53	
79	η Cas A	[5]	G0 V	0.87	–	–	+4.67	+4.61 [10]		+5.24	
80	9 Pup A	[5]	G0 V	0.56	–	–	+4.8			+7.68	
81	26 Dra A	[5]	G1 V	1.1	–	–	+4.49	+4.42 [10]		+4.20	
82	α Cen A	[19]	G2 V	1.06	–	–	{+4.5, +4.36 [10]}	+4.29 [10]	–	+4.26	
83	UV Leo A	[9]	G2 V	1.02±0.04	1.09	4.36±0.02	+4.87		+4.57	+4.45±0.2	
84	Sun		G2 V	1.00	1.00	4.44			+4.79	+4.54	
85	UV Leo B	[9]	G2	0.95±0.04	1.05	4.37±0.02			+4.65	+4.81±0.2	
86	η CrB A	[5]	G2 V	0.66	–	–	+4.69	+4.62 [10]	–	+6.68	
87	85 Peg A	[20]	G3 V	0.87	–	–				+5.24	
88	Σ 3062 A	[5]	G4	1.3	–	–	+4.69	+4.60	–	+3.20	
89	TW Cas B	[21]	G5 IV	0.96/1.1	1.85/2.00	3.94/3.82	+5.47		+3.39/3.56	+4.75/+4.08	
90	Σ 3062 B	[5]	G8	1.9	–	–		+5.33	–	+1.50	
91	MM Her A	[24]	G8 V	1.22	2.8	3.63			+2.82	+3.57	
92	MM Her B	[24]	G8 V	1.19	1.5	4.16			+4.18	+3.70	
93	Σ 2173 A	[15]	G8 IV–V	1.14	–	–	+4.55 [5]	+4.48 [5]		+3.90	
94	Σ 2173 B	[15]	G8 IV–V	1.08	–	–	+4.66 [5]	+4.58 [5]		+4.17	
95	AR Lac B	[9]	K0	1.31±0.07	3.0	3.60±0.02			+2.84	+3.23±0.25	
96	Z Her B	[9]	K0	1.10±0.03	2.6	3.65±0.01			+3.14	+4.07±0.13	
97	85 Peg B	[20]	–	0.69	–	–				+6.4	
98	ζ Her B	[20]	d K0	0.66	–	–				+6.5	
99	70 Oph A	[19]	K1	0.95	–	–	+5.70	+5.54 [5]		+4.8	
100	ξ Boo B	[19]	K4 V	0.72	–	–	+7.64	+6.89		+6.2	
101	Σ 3121 A	[5]	d K4	0.62	–	–	+6.7	+6.3		+7.0	
102	−34. 11626 A	[5]	d K5	0.78	–	–	+7.0	+6.5		+5.8	
103	RU UMi B	[18]	K5 V	0.6±0.1	0.9±0.1	4.31±0.18			+6.24	+7.25±0.9	
104	61 Cyg A	[19]	K5	0.58	–	–	+7.5			+7.4	
105	α Cen B	[19]	K6	0.87	–	–	+5.9			+5.2	
106	70 Oph B	[19]	K6	0.69	–	–	+7.5			+6.4	
107	61 Cyg B	[19]	K7	0.57	–	–	+8.3			+7.5	
108	η Cas B	[5]	d M0	0.54	–	–	+8.42	+7.35		+7.8	
							+7.9	+7.1		+7.8	

No.	Name	Ref	Sp								
110	Hu 575 A	[5]	–	0.62	–	–	+7.9	+7.1	–	+7.0	+ 8.6
111	YY Gem A	[9]	M1	0.58±0.02	0.60	4.64±0.02		+7.7 [12]	+7.92	7.4±0.2	+ 9.4
112	YY Gem B	[9]	M1	0.58±0.02	0.60	4.64±0.02		+7.7 [12]	+7.92	7.4±0.2	+11.0
113	Hu 575 B	[5]	–	0.47	–	–	+ 8.2	+7.2	–	+ 8.5	+11.9
114	Σ 2398 A	[19]	M4	0.41	–	–	+11.2		–	+ 9.3	+12.3
115	Fu 46 A	[2]	M4	0.31	–	–	+10.96	+8.72	–	+10.9	+13.8
116	Krü 60 A	[19]	M4	0.27	–	–	{+11.7 / +11.84 [2]}	+9.60 [2]	–	+11.7	
117	Fu 46 B	[2]	M4	0.25	–	–	+11.34	+9.10		+12.2	
118	40 Eri C	[19]	M4e	{0.20 / 0.21 [2]}	–	–	{+12.8 / +12.62 [2]}	+10.10 [2]	–	+13.6	
119	Σ 2398 B	[19]	M5	0.41	–	–	+12.0		–	+ 9.3	+ 9.4
120	Ross 614 A	[19]	M5e	0.12 [27]	–	–	+13.3 [5]	+11.1 [5]	–	+16.8	+17.2
121	Krü 60 B	[19]	M6	0.16	–	–	{+13.2 / +13.39 [2]}	+10.58 [2]	–	+15.0	+15.2
122	L 726-8AB	[19]	M6e	0.15	–	–	{A +15.3 / B +15.8}	+12.68 [2]	–	+15.4	+15.7
123	Ross 614 B	[19]	–	0.06 [27]	–	–	+15.35 [2] / +16.8 [5]	+13.0 [5]	–	+21.6	+22.2

References to Table V

[1] Sahade, J. and Dessey, J. L.: 1952, *Astrophys. J.* **115**, 53.
[2] Van de Kamp, P.: 1959, *Astron. J.* **64**, 236.
[3] Popper, D. M.: 1959, *Astrophys. J.* **129**, 647.
[4] Harris III, D. L.: 1963, *Stars and Stellar Systems* 3, 267.
[5] Harris III, D. L., Strand, K. Aa. and Worley, C. E.: 1963, *Stars and Stellar Systems* 3, 273.
[6] Wood, F. B.: 1963, *Stars and Stellar Systems* 3, 377.
[7] Hanbury Brown, R. and Davis, J., Allen, L. R.: 1967, *Monthly Notices Roy. Astron. Soc.* **137**, 393.
[8] Heintze, J. R. W.: 1968, *Bull. Astron. Inst. Neth.* **20**, 1.
[9] Popper, D. M.: 1967, *Ann. Rev. Astron. Astrophys.* **5**, 85.
[10] Morton, D. C. and Adams, T. F.: 1968, *Astrophys. J.* **151**, 611.
[11] Toy, L. G. S.: 1969, *Astrophys. J.* **158**, 1099.
[12] Koch, R. H.: 1970, in K. Gyldenkerne and R. M. West (eds.), *IAU Colloquium No. 6*, Copenhagen University Publication Fund, p. 65.
[13] Popper, D. M.: 1970, in K. Gyldenkerne and R. M. West (eds.), *IAU Colloquium No. 6*, Copenhagen University Publication Fund, p. 1.
[14] Popper, D. M.: 1970, *Astrophys. J.* **162**, 925.
[15] Batten, A. H., Fletcher, J. M., and West, R. M.: 1971, *Publ. Astron. Soc. Pacific* **83**, 149.
[16] McNamara, D. H., Hansen, L., and Wilcken, S. K.: 1971, *Publ. Astron. Soc. Pacific* **83**, 192.
[17] Dworetsky, M. M., Popper, D. M., and Dearborn, D. S.: 1971, *Publ. Astron. Soc. Pacific* **83**, 207.
[18] Wood, F. B.: 1971, *Publ. Astron. Soc. Pacific* **83**, 286.
[19] Van de Kamp, P.: 1971, *Ann. Rev. Astron. Astrophys.* **9**, 103.
[20] Feierman, B. H.: 1971, *Astron. J.* **76**, 73.
[21] Cook, G. P.: 1971, *Astron. J.* **76**, 449.
[22] Popper, D. M.: 1971, *Astrophys. J.* **166**, 361.
[23] Johansen, K. T., Jørgensen, H. E., and Behr, V.: 1971, *Astron. Astrophys.* **11**, 20.
[24] Imbert, M.: 1971, *Astron. Astrophys.* **12**, 155.
[25] Johansen, K. T.: 1971, *Astron. Astrophys.* **12**, 165.
[26] Dachs, J.: 1971, *Astron. Astrophys.* **12**, 286.
[27] Kuiper, G. F. G.: 1971, *Astron. Astrophys.* **14**, 70.
[28] Lippincott, S. L.: 1972, *Astron. J.* **11**, 679.

or

$$\theta_{\text{eff}} = 0.030\,n + 0.015 \quad \text{respectively}, \tag{30b}$$

where the spectral-type indicator n is equal to 5 for B0 V and increases to $n = 15$ for A0 V.

Formula (30a) gives the temperatures according to Schild *et al.* (1971) and formula (30b) according to formula (29b) in which Auer en Mihalas' (1972) results are included. For the A0 V and later type stars mean effective temperatures were determined from the data published by Johnson (1962a), Keenan (1963), Morton and Adams (1968), and Schmidt (1972).

With these temperatures M_{bol} is calculated according to formula (2) for those stars of which the radii are known except for V 444 Cyg B, UW CMa A and B, A0 Cas A and B. For the latter stars no reliable temperatures are thought to be available. The calculated M_{bol} values are given in the 10th column of Table V.

For the 70 components of eclipsing binaries the mass-luminosity relation could be represented as follows if Equation (30a) is used:

$$M_{\text{bol}} = 4.54 - 11.54 \log \mathcal{M}/\mathcal{M}_{\odot} + 2.41 \log^2 \mathcal{M}/\mathcal{M}_{\odot}, \tag{31a}$$

if Equation (30b) is used:

$$M_{\text{bol}} = 4.54 - 11.45 \log \mathcal{M}/\mathcal{M}_{\odot} + 2.05 \log^2 \mathcal{M}/\mathcal{M}_{\odot}. \tag{31b}$$

With these relations absolute bolometric magnitudes were calculated for the 48 components of astrometric binaries and for the first five stars given in column 11 of Table V. At present no detailed conclusions can be drawn from the differences between the published and calculated M_{bol} values. Much more reliable observations of M_v and $(U-B)_0$ values are necessary and at the same time the determination of such a mass-luminosity relation has to be done in a much more refined way. The uncertainties in the hot part of this empirical mass-luminosity relation cause rather large differences in the very cool part of this relation. However in the cool part of the relation up to star No. 112 of Table V this empirical mass-luminosity relation gives quite reasonable results.

From the extrapolated M_{bol} the effective temperatures of V 444 Cyg B (O6) and A0 Cas B (O9 III) are 34 000 K and 31 000 K respectively according to equation (31b) and 30 000 K and 28 000 K respectively according to Equation (31a). For comparison: Auer and Mihalas find for HD 54662 (06.5): 41 000 K, for S Mon (O7 Vf): 39 000 K and for λ Ori (O8 IIIf): 37 500 K.

Acknowledgements

I wish to thank Dr E. P. J. van den Heuvel for a critical reading of the manuscript.

References

Auer, L. H. and Mihalas, D.: 1972, *Astrophys. J. Suppl.* **24**, 193.
Allen, C. W.: 1963, *Astrophysical Quantities*, 2nd ed. The Athlone Press, London.
Aller, L. H.: 1963, *The Atmospheres of the Sun and the Stars*, 2nd ed., The Ronaical Press Company, New York.
Blanco, V. M., Demers, S., Douglas, G. G., and FitzGerald, M. P.: 1968, *Publ. U. S. Naval Obs.* **21**, 2nd series.
Clariá, J. J.: 1972, *Astron. Astrophys.* **19**. 303.
Collins, G. W.: 1970, in A. Slettebak (ed.), *Stellar Rotation*, Reidel Publishing Co., Dordrecht, p. 85.
Collins, G. W. and Harrington, J. P.: 1966, *Astrophys. J.* **146**, 152.
Crawford, D. L.: 1963, *Astrophys. J.* **137**, 530.
Edmonds, F. N., Schlüter, H., and Wells, D. C.: 1967, *Mem. Roy. Astron. Soc.* **71**, 271.
Eggen, O. J.: 1962, *Quart. J. Roy Astron. Soc.* **3**, 259.
Eggen, O. J.: 1969, *Astrophys. J.* **158**, 1109.
Fernie, J. D., Hagen Jr., J. P., Hagen, G. L., and McClure, L.: 1971, *Publ. Astron. Soc. Pacific* **83**, 79.
FitzGerald, M. P.: 1970, *Astron. Astrophys.* **4**, 234.
Fricke, K., Stobie, R. S., and Strittmatter, P. A.: 1971, *Astrophys. J.* **171**, 593.
Gallouet, L.: 1964, *Ann. Astrophys.* **27**, 423.
Gingerich, O.: 1969, *Theory and Observation of Normal Stellar Atmospheres*, M.I.T. Press, Cambridge.

Gliese, W.: 1957, *Mitt. Astr. Rechen-Inst. Heidelberg Ser. A.*, No.8.

Gray, D. F.: 1967, *Astrophys. J.* **149**, 317.

Greenstein, J. L.: 1969, *Stellar Astronomy,* Vol. 1, Gordon Breach New York, pp. 89, 90.

Griem, H. R.: 1967, *Astrophys. J.* **147**, 1092.

Hanbury Brown, R., Davis, J., Allen, L. R., and Rome, J.: 1967, *Monthly Notices Roy. Astron. Soc.* **137**, 393.

Hayes, D.: 1967, Thesis UCLA; see also 1970, *Astrophys. J.* **159**, 165.

Heintze, J. R. W.: 1968, *Bull. Astron. Inst. Neth.* **20**, 1.

Heintze, J. R. W.: 1969, *Bull. Astron. Inst. Neth.* **20**, 154.

Hiltner, W. A. and Morgan, W. W.: 1969, *Astron. J.* **74**, 1152.

Hobbs, L. M.: 1972, *Astrophys. J. Letters,* **175**, 145.

Hodge, P. W. and Wallerstein, G.: 1966, *Publ. Astron. Soc. Pacific* **78**, 411.

Hyland, A. R.: 1969, in O. Gingerich (ed.), *Theory and Observation of Normal Stellar Atmospheres,* M.I.T. Press, Cambridge, p. 271.

Jaschek, M. and Jaschek, C.: 1963, *Publ. Astron. Soc. Pacific* **75**, 365.

Johnson, H. L. and Morgan, W. W.: 1953, *Astrophys. J.* **117**, 313.

Johnson, H. L.: 1958, *Lowell Observ. Bull.* **4**, 37.

Johnson, H. L.: 1962, *Astrophys. J.* **135**, 69.

Johnson, H. L.: 1962, *Astrophys. J.* **136**, 1135.

Johnson, H. L.: 1966, *Ann. Rev. Astron. Astrophys.* **4**, 193.

Jones, B. F.: 1972, *Astrophys. J. Letters* **171**, 57.

Keenan, P. C.: 1963, *Stars and Stellar Systems* 3, 91.

Klinglesmith, D. A.: 1971, *Hydrogen Line Blanketed Model Stellar Atmospheres,* NASA SP-3065.

Klinglesmith, D. A.: 1972, *Astrophys. J,* **171**, 79.

Kuiper, G. P.: 1938, *Astrophys. J.* **88**, 429.

Maeder, A. and Peytremann, E.: 1970, *Astron. Astrophys.* **17**, 120.

Maran, S. P., Kurucz, R. L., Strom, K. M., and Strom, S. E.: 1968, *Astrophys. J.* **153**, 147.

Matthews, T. A. and Sandage, A. R.: 1963, *Astrophys. J.* **138**, 30.

Mihalas, D.: 1965, *Astrophys. J. Suppl.* **9**, 321.

Mihalas, D.: 1966, *Astrophys. J. Suppl.* **13**, 1.

Morgan, W. W., Harris, D. L., and Johnson, H. L.: 1953, *Astrophys. J.* **118**, 92.

Morton, D. C. and Adams, T. F.: 1968, *Astrophys. J.* **151**, 611.

Oke, J. B.: 1961, *Astrophys. J.* **133**, 90; **134**, 214.

Oke, J. B.: 1965, *Ann. Rev. Astron. Astrophys.* **3**, 23.

Oke, J. B. and Schild, R.: 1970, *Astrophys. J.* **161**, 1015.

Ostriker, J. P. and Hesser, J. E.: 1968, *Astrophys. J. Letters,* **153**, 151.

Parker, R., Greenstein, J. L., Helfer, H. L. and Wallerstein, G.: 1961, *Astrophys. J.* **133**, 101.

Pettit, E. and Nicholson, S. B.: 1928, *Astrophys. J.* **68**, 279.

Popper, D. M.: 1959, *Astrophys. J.* **129**, 647.

Roxburgh, I. W. and Strittmatter, P. A.: 1966, *Monthly Notices Roy. Astron. Soc.* **133**, 345.

Sandage, A. R. and Eggen, O. J.: 1959, *Monthly Notices Roy. Astron. Soc.* **119**, 278.

Sandage, A. R. and Tamman, G. A.: 1969, *Astrophys. J.* **157**, 683.

Sargent, W. L. W.: 1968, *Astrophys. J.* **152**, 885.

Schild, R., Peterson, D. M., and Oke, J. B.: 1971, *Astrophys. J.* **166**, 95.

Schmidt, E. G.: 1972, *Astrophys. J.* **174**, 605.

Serkowski, K.: 1963, *Astrophys. J.* **138**, 1035.

Strittmatter, P. A.: 1966, *Astrophys. J.* **144**, 430.

Strom, S. E.: 1968, unpublished.

Underhill, A. B.: 1966, *Vistas Astron.* **8**, 41.

Underhill, A. B.: 1972, NASA GSFC Preprint X-670-72-102.

Wallerstein, G., Herbig, G. H., and Conti, P. S.: 1965, *Astrophys. J.* **141**, 610.

Wallerstein, G. and Hodge, P. W.: 1967, *Astrophys. J.* **150**, 951.

Wayman, P. A., Symms, L. S. T., and Blackwell, K. C.: 1965, *Roy. Observ. Bull.* No. 98.

Weidemann, V. and Bues. I.: 1967, *Z. Astrophys.* **67**, 416.

Westerlund, B. E.: 1963, *Monhtly Notices Roy. Astron. Soc.* **127**, 83.

Wildey, R. L., Burbidge, E. M., Sandage, A. R., and Burbidge, G. R.: 1962, *Astrophys. J.* **135**, 94.

Young, A., Nelson, B., and Mielbrecht, R.: 1972, *Astrophys. J.* **174**, 27.

DISCUSSION

Schild: The $Q-\theta_{\text{eff}}$ relation published in the contribution by Drs Peterson and Oke and myself shows a linear relation reminiscent of your linear $(U-B)_0-\theta_{\text{eff}}$ relation.

Our $Q-\theta_{\text{eff}}$ relation makes use of the most reliable available data for these stars rather than an average based upon all published values. Most of the stars for which we measured energy distributions are in the list of secondary standards of the *UBV* system, as published by Harris and Johnson (1961 or 1962). I therefore believe that our *UBV* photometry is free of systematic effects and should have negligible random errors.

I should like to make a final comment with which I expect concurrence from both Dr Heintze and Prof. Pecker. In my experience of deriving an effective temperature scale based upon detailed energy distribution compared to blanketed stellar models, and in my comparison of the effective temperature scale with the works of others based upon different data and models, I find that the difference between effective temperature scales appear to be less than the probable systematic errors of these scales. As we have seen best from the results presented by Prof. Code, large discrepancies still exist between observed and model-predicted far UV fluxes, and these differences will cause errors in our bolometric corrections far larger than our effective temperature scales would lead us to expect.

I propose that we concentrate our efforts upon deriving bolometric corrections and then effective temperatures, from direct measurements of far UV fluxes such as those presented by Prof. Code. The spacecraft in orbit, and others being prepared as well as an already large amount of available UV data, make it possible for most astronomers to use far UV data without becoming specialists in spacecraft instrumentation. Furthermore, as noted by Prof. Code, use of UV data is the *only* way to determine temperatures of the hottest stars, for which the concept of effective temperature breaks down.

Divan: **Remarques sur la définition des discontinuités de Balmer de la spectrophotométrie BCD.**

Théoriquement, si λ_0 est la longueur d'onde à laquelle se produit la discontinuité de Balmer, $D = \log(I_{\lambda_{0+}}/I_{\lambda_{0-}}, I_{\lambda_{0+}}$ et $I_{\lambda_{0-}}$ étant les intensités du rayonnement continu à la longueur d'onde λ_0 du côté rouge et du côté bleu de la discontinuité. Dans la pratique, pour les rayonnements stellaires, la discontinuité de Balmer ne se produit pas à une longueur d'onde bien définie λ_0; elle apparaît progressivement sur un intervalle de longueurs d'onde plus ou moins large et se termine du côté ultraviolet à une longueur d'onde λ'_0, variable d'une étoile à l'autre, mais toujours plus grande que la limite théorique $\lambda_0 = 3647$ Å. On est donc conduit à donner de D une définition arbitraire:

$$D = \log(I_{3700+}/I_{3700-}).$$

La valeur $\lambda = 3700$ Å a été choisie car à cette longueur d'onde la discontinuité de Balmer de la plupart des étoiles est déjà terminée; le continu ultraviolet est alors observable et l'on peut mesurer sans difficulté I_{3700-}, à condition que le pouvoir de résolution du détecteur utilisé ne soit pas trop faible.

La difficulté réside dans la détermination de I_{3700+}, car le continu visible cesse généralement d'être observable entre les raies spectrales dès $\lambda = 4000$ Å même pour les étoiles B et A, et il faut l'extrapoler de 4000 à 3700 Å. La valeur précise trouvée pour D dépend de la manière dont cette extrapolation est faite, et si l'on veut comparer les valeurs de D données par une grille de modèles aux valeurs observées on doit dans les deux cas faire l'extrapolation de la même manière; c'est pourquoi nous donnons ici quelques précisions sur la procédure utilisée dans la spectrophotométrie BCD.

Les opérations se passent en deux temps:

(a) pour chaque étoile on détermine la *différence* ΔD entre sa discontinuité de Balmer et celle d'une étoile étalon;

(b) le *zéro de l'échelle* ainsi obtenue est déterminé à l'aide d'une seule étoile, bien choisie, dont on mesure directement la discontinuité de Balmer sur l'enregistrement microphotométrique d'un spectre à grande dispersion, calibré en intensité.

L'opération (a) telle que nous la décrivons ci-dessous, est justifiée par le fait d'observation suivant: si l'on porte $\log(I_2/I_1)$ en fonction de $1/\lambda$ pour deux étoiles O ou B de températures différentes mais non rougies, dans le domaine 6200–4000 Å où le continu est observable on obtient des points alignés sur une droite. L'extrapolation de cette droite de 4000 à 3700 Å donne la différence ΔD entre les discontinuités de Balmer des deux étoiles. Pour les étoiles A et F on opère de la même façon mais le domaine spectral doit être réduit à l'intervalle 4600–4000 Å (au détriment de la précision sur ΔD) car les points situés au-delà de 4600 Å ne se placent plus sur la droite. Nous insistons sur le fait que l'extrapolation ne se fait pas sur la fonction $\log I = f(1/\lambda)$ qui est très loin d'être une droite et ne pourrait donner qu'une définition complètement arbitraire des discontinuités, ni même sur la fonction $\log [I/B\lambda(T)]$ que l'on ne peut déterminer que par des comparaisons d'une étoile à un corps noir terrestre, comparaisons si difficiles à réaliser que les discontinuités ainsi obtenues ont encore une marge d'incertitude relativement grande. Tout ce qui précède ne concerne que les étoiles sans rougissement interstellaire. Dans le cas général (étoiles rougies) la relation $\log(I_2/I_1) = f(1/\lambda)$ n'est pas linéaire et la simple extrapolation précédente ne peut plus être faite ; on peut cependant encore calculer le ΔD que l'on observerait en l'absence de matière interstellaire si l'on connaît la forme de la loi d'absorption pour les deux étoiles et une valeur au moins approximative de leur rougissement. Pour les discontinuités de la classification BCD, ce calcul est toujours fait car c'est à cette condition seulement que le paramètre D est réellement indépendant du rougissement interstellaire ; il est conduit de manière à utiliser l'ensemble de tous les points du domaine 6200–4000 Å pour éviter la perte de précision sur ΔD qu'entraînerait la réduction du domaine spectral au petit intervalle dans lequel la loi d'absorption interstellaire est sensiblement linéaire en $1/\lambda$.

L'opération (b) a été réalisée sur des spectres (dispersion 40 Å mm^{-1}; domaine spectral 6200–3000 Å) de ε Ori. Cette supergéante B0 a été choisie à la fois pour la finesse de ses raies (sur les spectres utilisés le continu visible est observable jusque vers 3850 Å) et pour la faible valeur de sa discontinuité de Balmer. Cette faible valeur de D réduit la marge d'erreur possible en facilitant l'extrapolation sur les enregistrements microphotométriques du continu visible: celui-ci doit en effet présenter à 3700 Å une température de couleur (donc, sur les enregistrements, une pente) très voisine de celle du continu ultraviolet, ce qui ne serait pas du tout le cas pour une étoile à grande discontinuité. L'incertitude sur la valeur de D pour ε Ori semble ainsi être seulement environ ± 0.005.

Une incertitude supplémentaire de ± 0.015 sur le zéro de l'échelle des discontinuités BCD publiées jusqu'ici vient d'un rattachement insuffisant de ε Ori à l'ensemble des étalons habituels ; ce zéro n'est donc défini qu'à ± 0.02 près alors que les ΔD sont connus en général à ± 0.01 près. Un meilleur rattachement de ε Ori permettra de faire disparaître en grande partie cette incertitude supplémentaire ; le travail est en cours.

Il résulte de tout ceci que si l'on veut comparer les discontinuités de la classification BCD à une grille de modèles il y a intérêt, à déterminer une échelle relative des discontinuités de la grille par la même méthode que dans les observations, c'est-à-dire en comparant chaque modèle de la grille à l'un d'entre eux et en extrapolant les $\log(I_2/I_1)$ obtenus de 4000 à 3700 Å. On peut ensuite comparer les discontinuités individuelles des modèles aux discontinuités observées, mais en se rappelant les incertitudes inhérentes à la définition du zéro des échelles de discontinuités.

Remarque. Les méthodes décrites pour l'extrapolation du continu sous les dernières raies de la série de Balmer peuvent servir également à déterminer le continu sous une raie ou une bande large; par exemple, on peut déterminer le continu d'une étoile rouge sous la bande 4430 en prenant sur la même plaque le spectre d'une étoile non rougie de type voisin; à partir des valeurs de $\log(I_2/I_1)$ en dehors de la bande et du continu de la deuxième étoile on peut reconstituer celui de la première.

Van den Bergh: Could Dr Fitzgerald perhaps comment on the proposed revisions of the intrinsic $U-B$ vs $B-V$ relation for main sequence stars. This point is important because the proposed revisions are particularly large in the region $-0.5 < U-B < 0.0$ which is used to obtain the reddening of globular clusters via observations of cluster horizontal branch stars. With the proposed new intrinsic colour-colour relation halo clusters that are now regarded as unreddened would have $E_{B-V} \approx 0.03$.

Fitzgerald: The colours presented by Heintze for $(U-B)_0 < 0.00$ are consistent with my colours. At most they differ systematically by 0.02 to the ultraviolet in $(B-V)_0$. This amount is within as yet undetermined intrinsic scatter of the two-colour relation for class V stars. This scatter and the real values of $(B-V)_0$ should be determined from cluster studies. For the colours $(U-B)_0 > 0$ Heintze's sequence is from Westerlund's 1963 list of 'bluest' stars. Again because of intrinsic scatter, 'bluest' stars should *not* be used. Plots of $(B-V)$ vs m_v are able to give good $(B-V)_0$ if a few stars of given spectral class are unreddened. This was the procedure I used for A0 V stars and later for stars later than A0 V and I see no reason to change my colours, (given in *Astron. Astrophys.* 4).

The $(U-B)_0$ colours of O-stars show an apparent turn toward smaller negative values. This turn appears to rely perhaps too much on the very young clusters. If the $(B-V)_0$ colours based on 7 visual binaries are correct, then the turn off referred to above is in approximate agreement with the two-colour reddening lines.

Note added after meeting. For the intrinsic colours of stars later than about B8 the colours indicated by Heintze represent an envelope around the observed colours. The intrinsic colours as a function of spectral type have a cosmic scatter in them as demonstrated in an earlier paper (FitzGerald, 1970, *Astron Astrophys*. **4**, 234,), so envelopes around the observed colours should not be used for mean intrinsic colours of class V stars. At present I see no reason to change the colours for stars later than B8 from those given in the paper cited above.

Hauck: I agree with your comment concerning the inhomogeneity of the photometric data; it is not only the case for the UBV system! It would be urgent to ask observers for more homogeneous data. Perhaps a by-product of the Blanco *et al.* catalogue is the astronomer's variability more than star's variability.

Garrison: I would like to make two general comments concerning the use of observational data for transformation to theoretical parameters. First, we assume that by taking means of many observers, we improve the accuracy. I would like to suggest that by submitting to this 'tyranny of the mean' we usually decrease the accuracy. This is because of difficulties of transformation to a common system, as well as to observational errors. If I were interested in such a calibration, I would choose one careful observer who has observed a large sample and calibrate *his* system and filters, because the internal accuracy would be higher.

The second comment is that, too often, it is assumed that there is a unique relation between spectral type and colour and that one or the other is 'wrong' if they don't agree. In such a relation, the spread is greater than the acceptable errors in either. An extreme example is the Bp class of stars. The spectra are slightly peculiar and the difference between the spectrum and the colour gives an additional piece of information. But, there are other examples in which the spectra are normal, but significantly different from the colours. When colours and spectra are different, most spectroscopists know better than to say that the colours are wrong, and some photometrists have learned not to conclude that the classification is wrong. They are just different.

Jaschek: I would like to write down a few figures to pin down what Garrison said before. There does not exist such a thing as *the* intrinsic colour of a given spectral type. In early B-type stars, classified by a single astronomer, usually the dispersion is of the order of 0^m07 in $(U-B)_0$. This is not simply a question coming in because of the use of spectral type; if we analyse only the colours, we find that for a given $(B-V)$ colour we get (for A type dwarfs) a dispersion of about 0^m04. Therefore one should not think of the main sequence as being a curve, but rather a band. The Q method which is often used, assumes implicitly, that there exists such a curve and should therefore be used very cautiously. (The results quoted here are from a paper by the Jaschek's in *IAU Symp*. **50**.)

Schild: Your estimated intrinsic plus observational scatter in $(U-B)_0$ for early type stars is really an upper limit to the true value, since your derivation assumes all stars of given spectral type to have the same $(B-V)_0$; the well known scatter of several hundredths in intrinsic $(B-V)$ at given spectral types immediately implies an even greater scatter in intrinsic $U-B$. The procedure of de-reddening all stars of given spectral type to a single value of $(B-V)_0$ guarantees an intrinsic scatter of several hundredths in $(U-B)_0$ even if there are no errors in photometry or spectral classification.

Schmidt-Kaler: I have a remark and two questions. First the remark: I did almost exactly the same 11 yr ago (cf. *Astron. Nachr.* **286**, 113 and Landolt Börnstein, 1965). The resulting intrinsic colours have been given in the Landolt-Börnstein tables. The difference in procedure may have been, e.g. to retain the quadratic term, to use only reliable series of *UBV* measurements like, of course, Hiltner's, to check on residual reddening of apparently blue unreddened stars by means of measurements of interstellar polarization etc. I have just made a quick check of the $(U-B)_0$ vs $(B-V)_0$ relation and I find full agreement with your results within at most ± 0.01 for the B5–A0 stars; for the earlier types my $(U-B)_0$'s are ca. 0^m06 redder than yours for a given $(B-V)_0$ but still bluer than Johnson's for a given spectral type. This difference is probably due to the different treatment of the interstellar reddening quadratic term. Regarding the scatter of the $U-B$'s I might say that we observe $U-B$ to, say, $\pm 0^m005$, but we do not know whether we are exactly on Johnson's system. This is due to the fact that Johnson had the ultraviolet cut-off by the atmosphere, and since the height of the observatories above sea level is different, in general, there remains some basic ambiguity in the $U-B$ definition.

I have now two questions to you as a theoretician: Firstly – since I have dealt with quite a few spectra: Do you have a receipe to draw the continuum?

Secondly: The temperatures of B-stars you gave are 'cooling off', so does also the total energy output? This would be very important for stellar evolutionary considerations.

Heintze: Not feeling like a theoretician I should say in general no. For the hot B type stars a very good first approximation is to dɪaw by eye or by computer a mean lines through the graininess of the microphotometer trace. Only the very weak lines will be missed in this way.

These weak lines could be found by superposing several spectra of the same star on each other by computer. The traces have to be made with a microphotometer-comparator with high positional accuracy which gives the intensity of the spectrum (in a relative scale) as a function of plate position. Such a machine, built by Faul-Coradi, Scotland, is just installed at the Utrecht Observatory; (the minimum step size is one micron). I think each spectrum has to be splitted up in parts with at any case at both ends a spectral line of which the centre can be determined rather accurately from the profile. The spectrum intensities of the different spectra at the positions where the intensities will be added have in general to be found by interpolation between two successive measured points laying at both sides of the 'adding'-position. A linear interpolation is possible if the distance between two successive measurements is small enough (on low dispersion spectra with high resolution a few micron, on high dispersion spectra 10–50 μ).

It seems to me that this procedure can be used also for spectra of later type stars. It is quite well possible that windows (parts in the spectrum without lines) will show up much more clearly in this way.

I do not know how many spectra will be necessary; perhaps 5–10. Whenever long exposure times are possible much can already be done by widening the spectra to 0.5 or even 1 mm.

A check whether the continuum above symmetric lines with extended wings is drawn correctly is possible by superposing on each other the long- and short wavelength part of the line (plotted on transparant semi-logarithmic paper, one part being mirrored against wavelength). The red and violet relative intensities at each distance $\Delta\lambda$ from the centre have to be the same. If not a correction can easily be found now.

In this way also the point where the H-lines start to confluence can be found. Going one by one to higher members of the Balmer series by example it is possible in this way to find the continuum above the confluencing Balmer lines. The point $1/\lambda = 2.4\ \mu^{-1}(\lambda = 4167\ \text{Å}$; see Section 1.6.2 of my paper) is chosen such that no difficulties caused by the blending of the Balmer lines arise.

Only the effective temperatures of the O9 and the early B type stars are now generally believed to be less than according to the Morton and Adams' (1968) scale. If no other forms of energy transport than radiation are working then indeed the total energy output of these stars has diminished at the same time.

Around A0 V however the effective temperatures aɪe now believed to be a little bit higher: ≈ 10000 K instead of 9500 K.

Keenan: So much has been said about the relevance of spectral types that it should be remembered that although the photometric indices often have much smaller errors, frequently they do not distinguish between the physical variables as well as spectral types can – provided that the spectroscopic observer is careful to point out all the peculiarities that he can see on the spectrograms.

Jaschek: I agree completely with Dr. Keenan and would only like to add that the values I gave refer to dwarfs, which are recognized as such by spectroscopy. Therefore we are not using all the spectroscopic information available. The advantage consists of course largely in the possibility of sorting out everything which is not what you want (i.e. main sequence stars).

Maeder: I have a comment about the scatter, about which you are quite right, i.e. the scatter between the different UBV measurements of the same star by different authors. At least a part of this scatter comes from neglecting the colour-terms in the reduction for atmospheric extinction, as has been described by Rufener (1964, *Publ. Obs. Genève* 66).

Crawford: I feel the dispersion that Jaschek finds is small rather than large, in light of the difference in $(U-B)_0$ between spectral types. Remember that spectral type is a quantized parameter; $(U-B)_0$ is not. Therefore a unique $(U-B)_0$ does not exist for any given spectral type for that reason alone.

The luminosity effects on the Balmer jump (or on $(U-B)_0$) that Miss Divan mentioned earlier, and with which I agree, also fuzz up the relation, if the stars you are averaging or plotting, are not of the same age. In a cluster, this spread is small, and the $(U-B)/(B-V)$ relations (from homogeneous data, i.e. not averaged) have very little scatter.

Fitzgerald: I thought it might be relevant to discuss briefly the internal consistency of the 'polyglot' of the observations in the Blanco-Demers-Douglass-Fitzgerald Photoelectric Catalogue (BDDF). In a paper to be published in Astronomy and Astrophysics Supplement Series I have analysed the internal consistency of observations of stars observed two or more times. Of the 20705 stars in BDDF about 7000 have two or more observations. Of these 532 have discordant results in one or more of V, $B-V$, $U-B$, or between colour and spectral-luminosity class. These discrepancies have been excluded from the following analysis. (They are listed in the *Astronomy and Astrophysics Supplement Series* article). The discrepancies are attributable to transcription errors (10%), contamination in visual binary stars (16%), variables (22%), reported photometry (53%), and classification or photometry (15%)

The resulting standard deviations from the mean magnitudes and colours were formed as a function of reference number and spectral class. These are shown in Figures 1 and 2 respectively of the *Astronomy and Astrophysics Supplement Series* article. In figure one we plot the number of references having a certain standard deviation from the mean (σ) versus σ for references with 15 or more stars in common with other references.

Obviously some references give poor photometry. The standard deviations for each reference will be given in the *Astronomy and Astrophysics Supplement Series* article by reference number (in order to preserve anonymity). The standard deviations as a junction of spectral class clearly show the greater tendency of M and Carbon stars to be variable. Possibly O stars are also variable but here the number of O stars with observations in common is only 20. (Perhaps other observers should reobserve some of the O stars to improve these statistics).

On average the consistency between magnitudes and colours is considerably better than I expected, but more than might be desired. The average standard deviations from the mean parallax observatons (excluding the discrepancies) is ± 0.015 in V, ± 0.011 in $B-V$ and ± 0.016 in $U-B$.

PART VIII

CHOICE OF STANDARD STARS

REMARKS ON THE PHOTOMETRIC CRITERIA OF
CHOICE OF THE STANDARD STARS

M. GOLAY

Observatoire de Genève, Switzerland

Abstract. We propose, in several tables, a number of stars selected to be observed both by photometrists, spectroscopists, astrometrists. Properties of selected stars are discussed.

1. Definition of the 'HR Standard' Star

In astronomy we are familiar with several kinds of 'standard' stars, for instance: standard stars for spectral classification, those destined to define the scales of magnitudes and colours, stars selected to define an absolute energy distribution, etc. Practical considerations have often determined the choice of standard stars: uniform distribution in the sky, easy identification, necessity to cover a certain interval of magnitudes or colours. We propose here, in several tables, a number of stars selected to be observed by photometrists, spectroscopists and astrometrists. This list is of course not exhaustive. We will henceforth name the stars of our list 'HR standard' stars. Indeed, the object is to gather a sufficient amount of observations on each one of these stars to allow the precise fitting of stellar models liable to give us the absolute magnitude M_v, the gravity $\log g$, the effective temperature T_{eff}, the chemical composition χ. It would be ideal to dispose of these informations for typical stars distributed throughout a three-dimensional HR diagram (M_v, T_{eff}, χ). The three physical parameters M_v, T_{eff}, χ are generally obtained by means of indirect methods, very often photometrically. One must thus bear in mind that the photometric quantities such as magnitudes, indices, combinations of indices, depend in no negligible manner on multiplicity, rotational velocity of the star observed, on the quantity of interstellar matter and on the extinction law. Consequently, care must be given to avoid the use as 'HR standard' of stars having particularities liable to introduce errors in the photometric quantities.

2. Photometric Effects of Binarity, Rotation, Gravity and Chemical Composition

In the following diagrams, Figure 1a, b, Figure 2, Figure 3, Figure 4a, b, c, Figure 5a, b, we attempt to illustrate the deviations in colour to be expected for two stars slightly different in spectral class, or in luminosity class, or in chemical composition, or in rotational velocity, or due to the fact that one of them can be a binary. The conclusion derived from an inspection of these diagrams is that a difference in colour (between 3500 Å and 6500 Å) of ± 0.02 can be caused by:

a difference of 0.05 in spectral class;

a difference of 200° to 300° in T_{eff};

B. Hauck and B. E. Westerlund (eds.), Problems of Calibration of Absolute Magnitudes and Temperature of Stars, 275–297.
All Rights Reserved. Copyright © 1973 by the IAU.

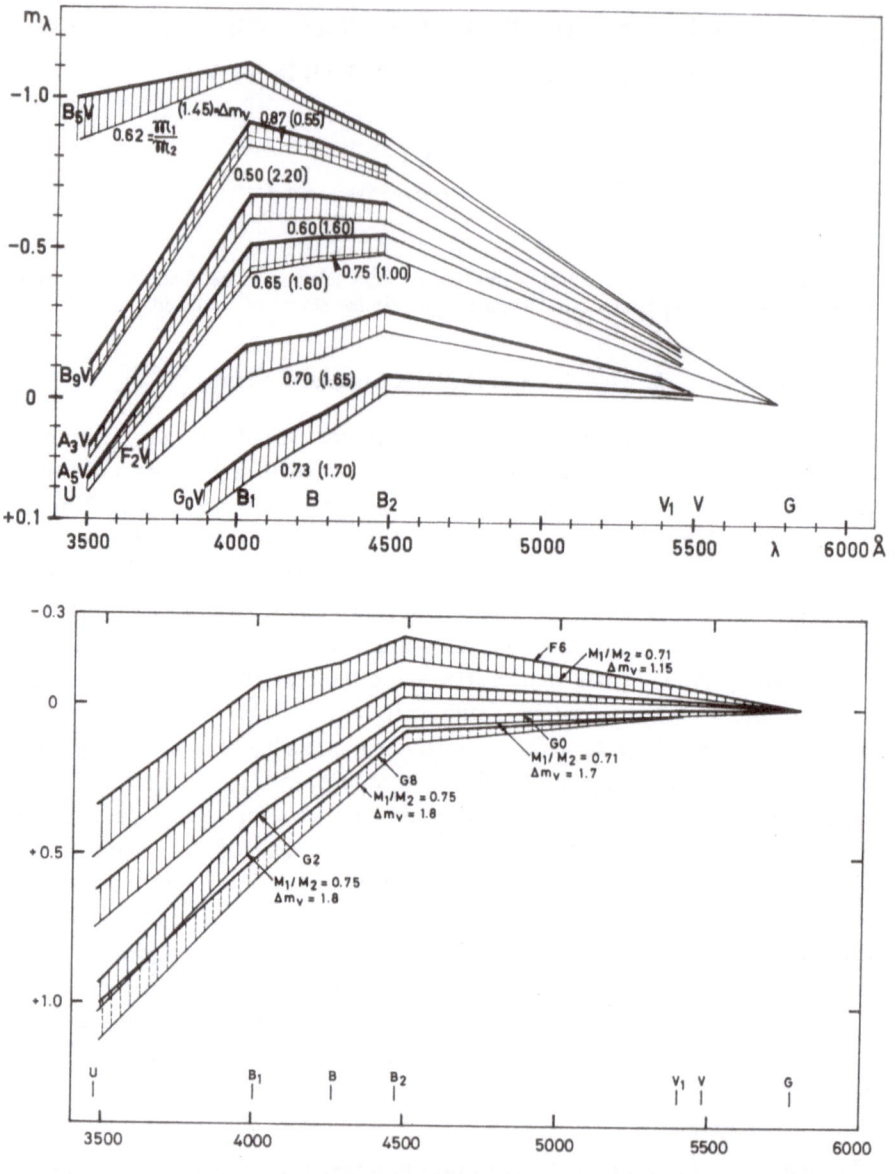

Fig. 1a, b. Effect of binarity for some spectral types. The thick solid line refers to a unit mass ratio. The thin solid line is for a mass ratio giving a pseudo-continuum of greatest deviation from the unit mass ratio.

a difference of 0.3 in $\log g$;

a difference of 0.1 in [Fe/H];

a difference of 150 km s^{-1} in rotational velocity;

a companion weaker by 5 mag.

These figures are only orders of magnitude, and can vary considerably individually according to the type of star (cold star, hot star, dwarf or supergiant).

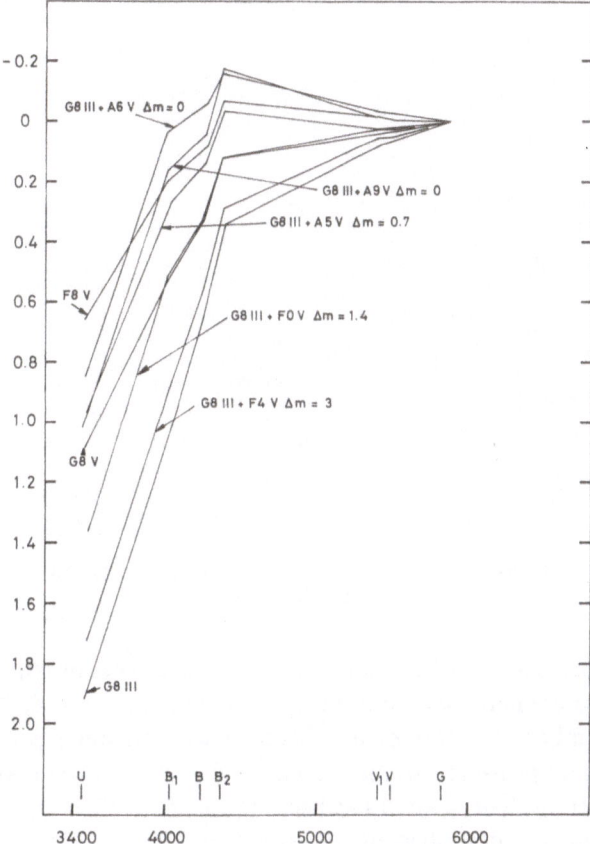

Fig. 2.   Effect of binarity for some G8 III stars.

3. Choice of the 'HR Standard' Stars

The 'HR standard' stars are to be used for the calibration of mean relations, such as:
M_v, colour indices;
θ_{eff}, colour indices or combinations of colour indices;
Spectrum, colour indices;
[Fe/H], colour indices or combinations of colour indices;
χ, colour indices or combinations of colour indices;
etc.

The colours can be those obtained with wide bandpasses (example $U\,B\,V\,R\,I$), narrow bandpasses (example $u\,v\,b\,y$), intermediary bandpasses (example $U\,B_1\,B_2\,V_1\,G$).

The determination of θ_{eff}, $\log g$, χ requires the fitting of stellar models to observed energy distributions or, at least, to the known colours of the star considered. This involves the necessity to know and to preserve the response functions of the photometric system with an accuracy equal to that of the intensity measurements. Thus,

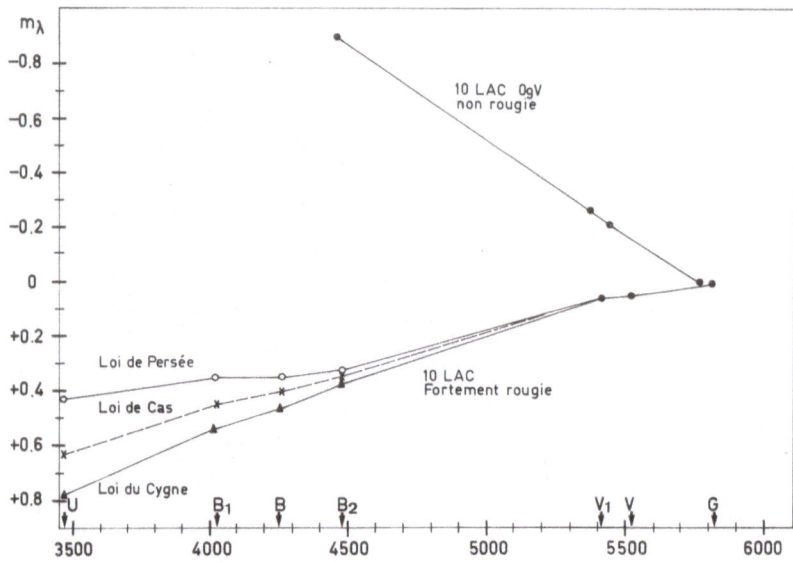

Fig. 3.   Deformation of the pseudo-continuum for an O star resulting
from three different laws.

these remarks imply that accurate photometric measurements of the stars proposed as 'HR standard' are available in a photometric system with well defined bandpasses. Moreover, the number of separate measurements of each star has to be sufficient to be able to detect any possible variability. The $U B V B_1 B_2 V_1 G$ photometric system meets these conditions. This system has been in use at the Geneva Observatory since 1959. Its properties are described by Golay (1969, 1971, 1972). The last catalogue published by Rufener (1971) contains more than 1500 stars, and will very soon reach 2500. This publication contains the response functions of the seven filters as well as a discussion of the accuracy of the measurements. Here, we have only considered stars with a weight $p \geqslant 3$, which corresponds to the following standard deviations expressed in thousandths of a magnitude (Table I). In Table I we give the standard deviations and the mean wavelengths of the bandpasses (whose equivalent rectangular bandpasses are between 300 Å and 600 Å).

TABLE I

Mean wavelengths and standard deviations

	U	B	V	B_1	B_2	V_1	G
λ_0 [Å]	3456	4245	5500	4024	4480	5405	5805
σ	6,1	3,2	3,9	3,2	3,3	3,7	4,4

The stars we propose as 'HR standards' all fulfill the following conditions:

(1) Are not binaries, (except the interesting case with well separated components) or the two components are practically identical ($\Delta m \leqslant 0.1$ and same spectral type) or the difference in magnitude Δm is $\geqslant 5$.

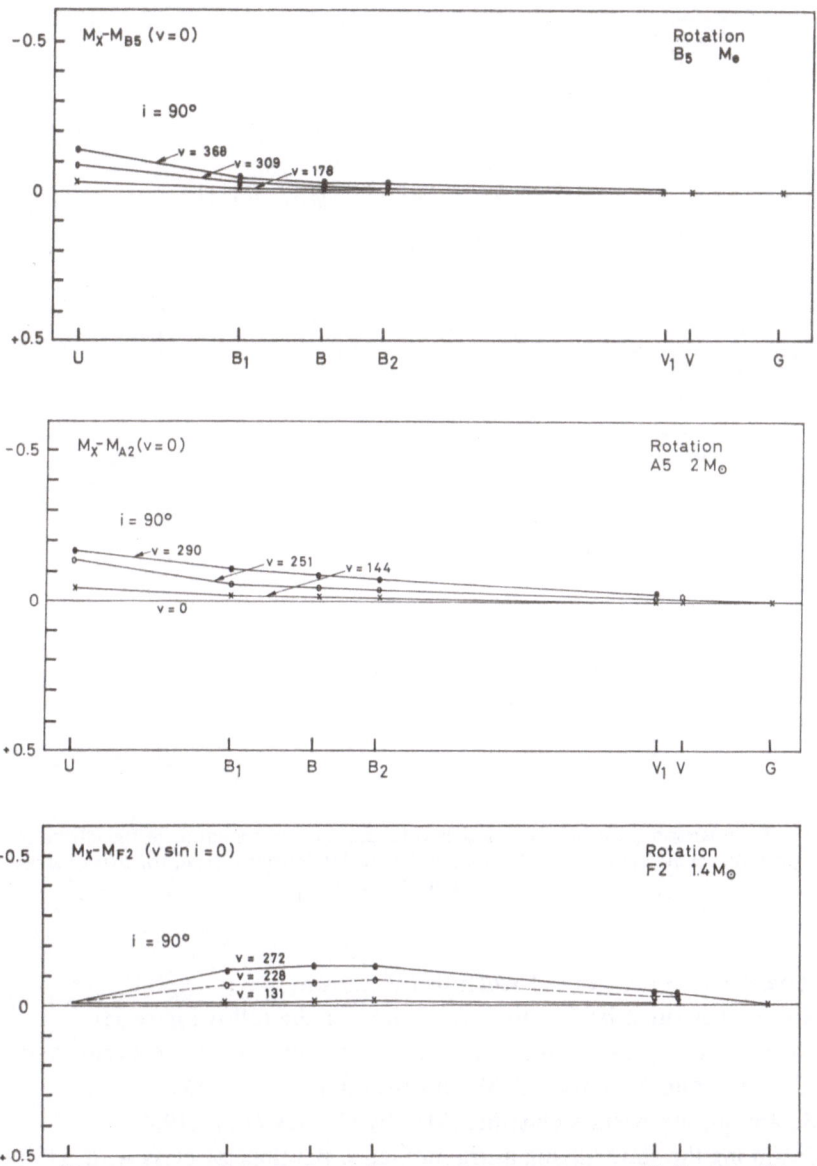

Fig. 4a, b, c.   Deformation of the pseudo-continuum caused by stellar rotation (colours computed by Maeder with models of Maeder and Peytremann 1970, 1972).

(2) Are not, or are but slightly reddened by interstellar matter.

(3) Are not suspected to be variable. We have retained as HR standard, stars suspected of variability in Literature, but not confirmed as such in our photometry. We identify with asterisks those stars not known to be variable, but which our measurements lead us to suspect of being variable.

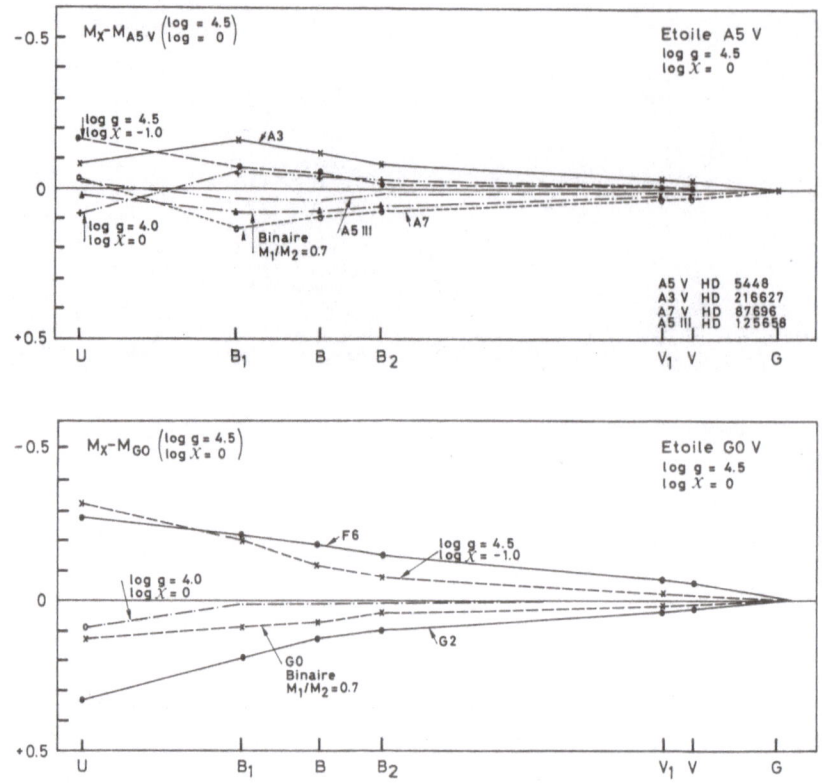

Fig. 5a, b. Comparison with help of models (Maeder and Peytremann) of the effects of binarity
gravity, chemical composition ($\log\chi \simeq 0.6$ [Fe/H]), spectral type, for two normal
stars A5 V, G0 V.

(4) Have $V \sin i > 150$ km s^{-1} when this is known.

Moreover, they must belong to one or more of the following series:

(a) They must be among the MK standards published by Morgan and Roman (1950), Johnson and Morgan (1953), and Morgan *et al.* (1953).

(b) Be among the A stars classified MK by Cowley *et al.* (1969).

(c) Be among the stars having a trigonometric parallax of class A, B, C in Gliese's catalogue (1969).

(d) Be among stars which have once been the object of a study of abundances by means of spectroscopic observations. The stars have been selected from the lists of Cayrel and Cayrel de Strobel (1966) and Powell (1970).

(e) Be probable members of open clusters of known distance modulus. Our list is limited for the moment to the Hydaes, Praesepe and Coma Berenices clusters.

In the following pages, we give a serie of tables. Tables 1 to 7 contain all the stars we propose as HR standard. Each table contains stars belonging to the same original list.

Table 1 contains the stars of series a
Table 2 contains the stars of series b
Table 3 contains the stars of series c
Table 4 contains the stars of series d
Table 5 members of the Hyades
Table 6 members of Praesepe
Table 7 members of Coma Berenices

TABLE 1

MK Standards

HD or BD	HR or other	Sp	m_v	$B_2 - V_1$	Variab.	$V \sin i$	Remarks
14633		O8	7.459	−0.285	***	126	
38771	2004	B0.5 Ia	(2.040)	−0.242	**	81	βy
163506	6685	F2 Ia	5.419	0.146	***	23	βy
212593	8541	B9 Iab	(4.530)	−0.038		29	
91316	4133	B1 Ib	3.858	−0.217		69	βy
87737	3975	A0 Ib	3.522	−0.142		18	1, 2
							βy
46300	2385	A0 Ib	(4.480)	−0.116		17	1, 2, βy
20902	1017	F5 Ib	1.811	0.241	**	18	1, 4, βy, DG
164136	6707	F2 II	4.402	0.177		27	1, 4, βy
571	27	F2 II	5.030	0.199		47	βy
35468	1790	B2 III	1.634	−0.297		64	1, 8, βy
30836	1552	B2 III	3.667	−0.248		42	βy
22928	1122	B5 III	3.012	−0.240		271	1, 8, βy
23302	1142	B6 III	3.704	−0.215		227	
123299	5291	A0 III	3.659	−0.180	***	12	1, 8, βy
89025	4031	F0 III	(3.430)	0.105		82	βy
13174	623	F2 III	(5.010)	0.127		154	βy
17584	840	F2 Ill	(4.220)	0.141		149	βy
21770	1069	F4 III	5.301	0.191		29	βy
27022	1327	G5 III	5.269	0.524		<19	βy, DG
28305	1409	K0 III	3.548	0.681		≤ 8	1, 5, DG
1013	45	M2 III	4.837	1.188	***		1, 8
47105	2421	A0 IV	1.939	−0.151		37	1, 2, βy
211336	8494	F0 IV	4.185	0.072		86	
17094	813	F0 IV	4.268	0.097		54	βy
89449	4054	F6 IV	4.783	0.228		16	βy
82328	3775	F6 IV	3.183	0.253	***	13	1, 4, βy
11443	544	F6 IV	(3.530)	0.254		95	1, 4, βy
216385	8697	F7 IV	(5.220)	0.261		0	βy
220657	8905	F8 IV	(4.510)	0.350		79	βy
121370	5235	G0 IV	2.705	0.328			βy
23249	1136	K0 IV	3.548	0.601	***	<17	1, 3, 4, 8, DG
36512	1855	B0 V	4.592	−0.320		17	βy
3360	153	B2 V	3.646	−0.276		22	βy
74280	3454	B3 V	4.291	−0.274		132	βy
32630	1641	B3 V	3.155	−0.270		139	βy
83754	3849	B5 V	(5.060)	−0.250	***	185	βy
23338	1145	B6 V	4.302	−0.218		134	βy

M. GOLAY

Table 1 (Continued)

HD or BD	HR or other	Sp	m_v	$B_2 - V_1$	Variab.	$V \sin i$	Remarks
87901	3982	B7 V	1.375	−0.219		354	1, 8, βy
214923	8634	B8 V	3.416	−0.215		196	βy
135742	5685	B8 V	(2.610)	−0.209	*	230	1, 8, βy
222173	8965	B8 V	4.280	−0.209	*	84	βy
103287	4554	A0 V	2.452	−0.154		163	1, 2, 8, βy
71155	3314	A0 V	3.902	−0.159		122	1, 2, 8, βy
139006	5793	A0 V	(2.230)	−0.148		132	βy
97633	4359	A2 V	3.320	−0.154		14	1, 2, 8, βy
1280	63	A2 V	4.608	−0.113	*	107	1, 2, βy
106591	4660	A3 V	(3.313)	−0.089	**	177	1, 2, 3, βy
11636	553	A5 V	2.664	−0.061		73	1, 8, βy
8538	403	A5 V	2.653	−0.045	**	116	1, 8, βy
87696	3974	A7 V	4.484	−0.000		157	1, 2, βy
58946	2852	F0 V	4.169	0.119		63	1, 3, βy
110379	4825	F0 V	2.766	0.141	*	27	1, 3, 8, βy
91480	4141	F1 V	5.156	0.137		79	βy
128167	5447	F2 V	4.474	0.168	*	0	βy
134083	5634	F5 V	4.930	0.210	*	44	1, 3, βy
210027	8430	F5 V	3.770	0.216	***	7	1, 4, βy
30652	1543	F6 V	3.199	0.230		16	1, 3, 4, βy
173667	7061	F6 V	4.194	0.249		14	1, 8, βy
142860	5933	F6 V	3.880	0.258	**	7	1, 3, 4, 8, βy
120136	5185	F7 V	4.498	0.255		14	βy
16895	799	F7 V	4.110	0.261	***	6	1, 4, βy
126660	5404	F7 V	4.051	0.264	*	31	1, 3, βy
184960	7451	F8 V	5.719	0.252		⩽ 6	βy
90839	4112	F8 V	4.838	0.293		0	1, 3, 4, βy
9826	458	F8 V	(4.080)	0.294		8	1, 3, 4, βy, DG
102870	4540	F8 V	3.636	0.301		0	1, 3, 4, 8, βy
114710	4983	G0 V	4.252	0.332		6	1, 3, 4, βy, DG
19373	937	G0 V	4.059	0.335		⩽10	1, 3, 4, βy, DG
13974	660	G0 V	4.873	0.351		⩽10	1, 3, 4, βy, DG
109358	4785	G0 V	4.276	0.350		⩽ 3	1, 3, 4, βy, DG
115043		G2 V	6.811	0.356		<50	1, 4, βy
10307	483	G2 V	4.966	0.362		⩽ 3	1, 3, 4, βy, DG
20630	996	G5 V	(4.820)	0.407		<17	1, 4, βy, DG
6582	321	G5 Vp	5.166	0.431	**	<17	βy, DG
117176	5072	G5 V	4.967	0.441		⩽10	βy, DG
10700	509	G8 Vp	3.481	0.424		<17	1, 3, βy, DG
101501	4496	G8 V	5.317	0.449	*	<17	1,3, DG
154345		G8 V	(6.770)	0.457	**		1, 3, βy, DG
103095	4550	G8 VI	6.434	0.482			1, 3, 4, βy, DG
10780	511	K0 V	5.622	0.492			1, 3, DG
124752		K0 V	8.517	0.519	*		
3651	166	K0 V	5.900	0.526	***		1, 3, βy, DG
166620	6806	K2 V	6.393	0.552	***		βy, DG
109011		K2 V	8.103	0.599	*		DG
128165		K3 V	7.255	0.616	***		
219134	8832	K3 V	(5.570)	0.628			1, 3, 4, βy, DG
151288		K7 V	8.091	0.919	***		1,3

TABLE 2

Stars classified by $C^2 J^2$, Cowley *et al.* (1969)

HD or BD	HR or other	Sp	m_v	B_2-V_1	Variab.	$V \sin i$	Remarks
6457	311	A0 Vn	5.549	−0.194		267	βy
85504	3906	A0 Vs	6.018	−0.185	***		βy
92728	4187	A0 Vs	5.794	−0.185		40	
118214	5109	A0 V	5.599	−0.167	*	197	βy
71155	3314	A0 V	3.902	−0.159		127	1, 2, 8, βy
23441	1152	A0 Vn	6.439	−0.154	**	267	
103287	4554	A0 V	2.452	−0.154		169	1, 2, 8, βy
6456	310	A1 Vn	5.329	−0.150	***	247	βy
2888	128	A1 Vn	6.697	−0.148		265	βy
14055	664	A1 Vnn	(4.080)	−0.147		225	βy
77327	3594	A1 Vn	3.584	−0.147		247	2, 8, βy
30739	1544	A1 V	4.353	−0.139		250	βy
25490	1251	A1 V	(3.900)	−0.127		110	2, 8, βy
111397	4865	A1 V	5.707	−0.125			βy
97633	4359	A2 V	3.320	−0.155		2	1, 2, 8, βy
50973	2585	A2 Vn	(4.890)	−0.129	*	210	βy
125642	5373	A2 V	6.308	−0.115			βy
12471	599	A2 V	5.505	−0.114		100	βy
1280	63	A2 V	4.608	−0.113	*	116	1, 2, βy
28978	1448	A2 Vs	5.685	−0.103			βy
146738	6074	A3 V	5.797	−0.108	***	80	βy
23848	1177	A3 V	(5.100)	−0.101		117	βy
216627	8709	A3 V	(3.290)	−0.096	***	96	βy
27820	1381	A3 V	5.112	−0.089		90	βy
106591	4660	A3 V	(3.313)	−0.088	**	179	1, 2, 3, βy
18331	875	A3 V	(5.170)	−0.084		300	βy
141003	5867	A3 V	3.674	−0.081		200	βy
56537	2763	A3 V	3.585	−0.076		157	βy
14417	684	A3 V	6.480	−0.069			βy
118098	5107	A3 V	3.373	−0.068		195	βy
119024	5142	A3 Vn	5.467	−0.068		215	βy
108382	4738	A4 V	4.986	−0.075		94	2, 8
105805	4633	A4 Vn	6.014	−0.055	**	172	2, 7
38091	1969	A4 Vn	5.935	−0.027	***	260	βy
5448	269	A5 V	(3.930)	−0.052		77	βy
116842	5062	A5 V	4.016	−0.020		250	βy
220061	8880	A5 V	4.589	−0.005		175	βy
79439	3662	A5 V	4.814	0.007		178	βy
32301	1620	A7 V	(4.640)	−0.036		131	2, 5, βy
177196	7215	A7 V	(5.000)	−0.009		140	βy
87696	3974	A7 V	4.484	0.000		168	1, 2, βy
27084	1330	A7 V	4.454	0.040		97	
95608	4300	A1 m:	4.423	−0.122		24	βy
17581	839	A1 m:	(6.380)	−0.072			βy
195479	7839	A1 m	6.202	−0.058			βy
78209	3619	A1 m	4.463	0.069		38	βy
36484	1850	A2 m	6.487	−0.091			βy
12869	613	A2 m:	5.031	−0.064		0	βy
223461	9025	A2 m	5.968	−0.020			βy
140232	5845	A2 m	5.795	0.003			βy

Table 2 (Continued)

HD or BD	HR or other	Sp	m_v	B_2-V_1	Variab.	$V \sin i$	Remarks
72037	3354	A2 m	5.452	0.003		38	βy
166926	6811	A2 m	(5.820)	0.018			
33254	1672	A2 m	5.432	0.031		25	
99945	4429	A2 m	6.117	0.048	*		
221675	8944	A2 m	5.886	0.081			βy
18769	905	A3 m	(5.860)	−0.049	***	50	βy
79193	3655	A3 m	6.104	0.033			βy
141675	5887	A3 m	5.866	0.033		82	βy
195217	7833	A3 m	(6.330)	0.037			βy
27045	1329	A3 m	4.944	0.049		60	βy
102660	4535	A3 m	6.045	0.063			βy
27628	1368	A3 m	(5.720)	0.104		15	2, 5
33641	1689	A4 m:	(4.740)	−0.001		84	βy
32428	1627	A4 m	6.602	0.070			βy
159560	6555	A4 m	4.867	0.070		47	βy
76756	3572	A5 m	4.266	−0.046	*	74	βy
15385	723	A5 m	6.190	−0.038		60	βy
6116	290	A5 m:	5.955	−0.024			βy
24141	1192	A5 m:	5.795	−0.023			βy
60652	2914	A5 m	5.913	0.108			βy
111421	4866	A6 m:	6.250	−0.012		40	βy
3883	178	A7 m	6.059	0.046	*		βy
107168	4685	A8 m:	6.250	−0.024		≤12	
90569	4101	A0 p	6.013	−0.207		90	βy
10221	478	A0 p	5.569	−0.205		30	βy
32549	1638	A0 p	4.666	−0.205		29	βy
111133	4854	A0 p	(6.350)	−0.203			βy
4778	234	A0 p	(6.140)	−0.173			βy
38104	1971	A0 p	5.467	−0.135		40	βy
74521	3465	A1 p	5.655	−0.242			βy
72968	3398	A1 p	5.735	−0.188			βy
151199	6226	A2 p	6.178	−0.106		110	βy
65339	3109	A2 p	6.035	−0.057			βy
81009	3724	A5 p	6.518	0.001	**		βy
87737	3975	A0 Ib	3.522	−0.142		29	1, 2, βy
46300	2385	A0 Ib	(4.480)	−0.115		17	1, 2, βy
196821	7903	A0 III	(5.910)	−0.185	***		βy
221756	8947	A1 III	(5.580)	−0.089	**	145	βy
50019	2540	A3 III	3.601	−0.063		140	βy
109307	4780	A5 III	(6.230)	−0.066		8	2, 7
173880	7069	A5 III	(4.300)	−0.048		81	βy
125658	5374	A5 III	6.450	−0.042	*		βy
47105	2421	A1 IV	1.939	−0.151		26	1, 2, βy
89021	4033	A2 IV	3.434	−0.115		43	βy
107966	4717	A3 IV	5.163	−0.085		50	2, 7
28527	1427	A6 IV	(4.780)	−0.019		65	
138341	5760	A4 lV	6.455	0.017			βy
27934	1387	A7 IV–V	(4.220)	−0.056		77	
107131	4684	A7 IV–V	6.423	−0.022		175	2, 7
203280	8162	A7 IV–V	2.461	0.019	*	260	2, 3, βy
21551	1051	B8 Vnn	5.819	−0.163		380	

Table 2 (Continued)

HD or BD	HR or other	Sp	m_v	B_2-V_1	Variab.	$V \sin i$	Remarks
79469	3665	B9.5 V	(3.880)	−0.208		91	2, 8, βy
47964	2461	B8 III	5.793	−0.208		95	
182691	7381	B9 III	6.493	−0.192			
205551	8259	B9 III	(5.940)	−0.122		200	
35600	1804	B9 Ib	5.709	0.041			
133029	5597	B9 p	(6.380)	−0.256			βy
205087	8240	B9 p	6.701	−0.230	*		βy
68351	3215	B9 p	5.618	−0.216		0	βy
207857	8349	B9 p	(6.160)	−0.195	***		βy
184961	7452	B9 p?	6.328	−0.191	**	50	βy
145389	6023	B9 p	4.237	−0.185		0	βy
219749	8861	B9 p	(6.270)	−0.163		70	βy
148112	6117	B9 p	(4.560)	−0.156		44	2, 8, βy
173650	7058	B9 p	(6.500)	−0.137			βy
27176	1331	F0 V	(5.650)	0.073		105	2, 5
107326	4694	F0 IV	(6.080)	0.099		125	
108283	4733	F0 IIInp	4.922	0.085		227	
118295	5116	F0 III	6.841	0.008			βy
126661	5405	F0 m	5.408	0.031		50	βy
176232	7167	F0 p	5.915	0.038		103	βy

TABLE 3

Stars with trigonometric parallax *A, B, C*

HD or DB	HR or other	Sp	m_v	B_2-V_1	Variab.	$V \sin i$	Remarks
63°0137		K7 V	8.983	0.873	**		8.29 DG
166	8	K0 V	6.085	0.462	*		5.3 3, DG
3651	166	K0 V	5.900	0.526	***		5.75 βy, DG, 1
4628	222	K2 V	(5.760)	0.549			6.55 3, 8, DG
9826	458	F8 V	(4.080)	0.294		8	3.06 1, 3, 4, βy, DG
10307	483	G2 V	4.966	0.362		⩽ 3	4.66 1, 3, 4, βy, DG
10700	509	G8 Vp	3.481	0.424		<17	5.72 1, 3, βy, DG
10780	511	K0 V	5.622	0.492			5.91 1, 3, DG
13974	660	G0 V	4.873	0.351		⩽10	4.80 1, 3, 4, βy, DG
17925	857	K0 V	6.057	0.542	**		6.57 DG
19373	937	G0 V	4.059	0.335		⩽10	3.72 1, 3, 4, βy, DG

M. GOLAY

Table 3 (Continued)

HD or BD	HR or other	Sp	m_v	B_2-V_1	Variab.	$V \sin i$	Remarks
23249		K0 IV	3.548	0.601	***	<17	3.77
							1, 3, 4, 8, DG
25329		K1 Vsd	8.495	0.561	**		7.10
							3, 4, βy, DG
25680	1262	G5 V	5.900	0.368			5.09
30652	1543	F6 V	3.199	0.230		16	3.76
							1, 3, 4, βy
34411	1729	G0 V	4.706	0.360		⩽ 3	3.84
							3, 4, βy, DG
37394	1925	K1 V	6.224	0.524			6.07
							DG
58946	2852	F0 V	4.169	0.119		63	2.84
							1, 3, βy
65583		G8 V	(7.000)	0.451			5.82
							βy, DG
72905	3391	G0 V	5.629	0.362		4	4.67
							βy, DG
76644	3569	A7 V	3.127	0.015		138	2.24
							βy
84035		K5 V	8.139	0.712	***		7.31
84737	3881	G1 V	5.089	0.362		⩽10	4.20
							βy
88230		K7 V	6.597	0.931	***		8.32
							βy, DG
89125	4039	DF3	5.820	0.279		⩽ 6	4.7
							βy
90839	4112	F8 V	4.838	0.293		0	4.44
							1, 3, 4, βy
95128	4277	G0 V	5.043	0.364		⩽ 3	4.4
							βy, DG
101501	4496	G8 V	5.317	0.449	*	<17	5.55
							1, 3, DG
102870	4540	F8 V	3.636	0.301		0	3.60
							1, 3, 4, 8, βy
103095	4550	G8 VI	6.434	0.482			6.71
							1, 3, 4, βy, DG
106591	4660	A3 V	(3.313)	−0.089	**	177	1.9
							1, 2, 3, βy
109358	4785	G0 V	4.276	0.350		⩽ 3	4.46
							1, 3, 4, βy, DG
110379	4825	F0 V	2.766	0.141	*	27	3.46
							1, 3, 8, βy
110833		K3 V	7.017	0.585			6.0
							DG
110897	4845	G0 V	5.961	0.329		⩽ 6	5.0
							3, 4, βy, DG
114710		G0 V	4.252	0.332		6	4.66
							1, 3, 4, βy, DG
126660	5404	F7 V	4.051	0.264	*	31	3.22
							1, 3, βy
131156	5544	G8 V	4.555	0.483		<16	A: 5.53 B: 7.69

Table 3 (Continued)

HD or BD	HR or other	Sp	m_v	B_2-V_1	Variab.	$V \sin i$	Remarks
131511	5553	K2 V	6.017	0.528	*		5.66 DG
134083	5634	F5 V	4.930	0.210	*	44	3.9 1, 3, βy
139323		K3 V	7.654	0.581	*		6.2
142860	5933	F6 V	3.880	0.258	**	7	3.4 1, 3, 4, 8, βy
145417		K0 V	7.521	0.514			6.64
145675		K0 V	6.611	0.542			5.6 DG
151288		K7 V	8.091	0.919	***		8.19 1, 3
154345		G8 V	(6.770)	0.457	**		5.73 1, 3, βy, DG
157214	6458	G8 V	5.388	0.381	*	0	4.71 βy, DG
160346		K3 V	6.529	0.606			6.06 DG
182488	7368	K0 V	(6.380)	0.504			5.14 DG
190406	7672	G IV	5.795	0.363	***	4	4.6 βy, DG
193664	7783	G5 V	5.905	0.345			5.07
203280	8162	A7 V–IV	2.461	0.018	*	240	1.5 2, 3, βy
217987		M2 V	7.356	1.131			9.59
219134	8832	K3 V	(5.570)	0.628			6.41 1, 3, 4, βy, DG

TABLE 4.1

Stars with abundances determinations, Cayrel and Cayrel de Strobel

HD or BD	HR or others	Sp	m_v	B_2-V_1	Variab.	$V \sin i$	Remarks
3627	165	K3 III	(3.210)	0.898		<17	0.00 DG
5015	244	F8 IV–V	(4.860)	0.306		6	−0.15 βy
5544		K0 III	(7.710)	0.783			0.33
10307	483	G2 V	4.966	0.362		⩽ 3	0.20 1, 3, 4, βy DG
10380	489	K3 III	4.466	0.995		<19	0.00 DG
11443	544	F6 IV	(3.530)	0.254		95	−0.12 1, 4, βy
37°00432		K2 III	(9.010)	0.680	***		−0.25 DG, βy

Table 4.1 (Continued)

HD or BD	HR or others	Sp	m_v	$B_2 - V_1$	Variab.	$V \sin i$	Remarks
13974	660	G0 V	4.873	0.351		≤10	−0.43
							−0.18
							−0.51
							1, 3, 4, βy, DG
16895	799	F7 V	4.110	0.261	***	6	+0.07
							−0.02
							1, 4, βy
18474	885	G4p	5.484	0.594			0.15
							DG
19373	937	G0 V	4.059	0.335		≤10	0.14
							0.26
							1, 3, 4, βy, DG
19445		F7 VI	8.041	0.269	***		−0.77
							−1.75
							−1.75
							βy
20630	996	G5 V	(4.820)	0.407		<17	0.38
							1, 4, βy, DG
20902	1017	F5 lb	1.811	0.241	**	18	−0.45
							1, 4, βy, DG
22484	1101	F8 V	4.288	0.335		0	0.37
22879		F9 V	6.673	0.316			−0.57
							βy
23230	1135	F5 II	(3.770)	0.209		44	−0.19
							βy, DG
23249	1136	K0 IV	3.548	0.601	***	<17	0.00
							−0.09
							1, 3, 4, 8, DG
25329		K1 V	8.495	0.561	**		−2.30
							3, 4, βy, DG
30455		G2 V	6.954	0.374	**		−0.09
							−0.26
							DG
30649		G1 V–VI	6.959	0.355			−0.32
							−0.20
							βy, DG
30652	1543	F6 V	3.199	0.230			−0.40
							1, 3, 4, βy
31398	1577	K3 ll	2.707	1.149		<17	0.00
							βy, DG
34411	1729	G0 IV	4.706	0.360		≤ 3	0.14
							0.22
							3, 4, βy, DG
55575	2721	G0 V	5.554	0.342		≤ 6	−0.21
							βy, DG
64491	3083	Ap	6.224	0.082		70	−0.70
							βy
72324	3369	G9 III	6.353	0.688			0.32
							DG
73665	3427	K0 III	6.396	0.648	*	0	−0.04
							DG, 4, 6, βy
73710	3428	K0 III	6.418	0.671		<45	−0.17

Table 4.1 (Continued)

HD or BD	HR or others	Sp	m_v	B_2-V_1	Variab.	$V \sin i$	Remarks
82328	3775	F6 IV	3.183	0.253	***	13	DG, βy / −0.44
86728	3951	G4 V	5.390	0.398		≤10	1, 4, βy / 0.34 / 0.34
90508	4098	G1 V	6.435	0.366		≤10	βy, DG / −0.23
90839	4112	F8 V	4.838	0.293		0	βy, DG / 0.23
102870	4540	F8 V	3.636	0.301		0	1, 3, 4, βy / 0.33
103095	4550	G8 VI	6.434	0.482			1, 3, 4, 8, βy / −1.50
106516	4657	F6 V	(6.110)	0.234		8	1, 3, 4, βy, DG / (+0.05) / (−0.86)
109358	4785	G0 V	4.276	0.350		≤ 3	βy / 0.02
109995		A0 V	7.589	−0.098		30	1, 3, 4, βy, DG / −1.20
110897	4845	G8 V	5.961	0.329		≤ 6	βy / −0.32
114710	4983	Go V	4.252	0.332		6	3, 4, βy, DG / 0.19 / 0.05 / 0.08
114762		F9 V	7.302	0.311	***		1, 3, 4, βy, DG / −0.59
115043		G0 V	6.811	0.356		<50	−0.06 / −0.14
122563	5270	Pop II gi.	6.177	0.642	*		1, 4, βy / −2.90 / −2.65
142267	5911	G2 V	6.087	0.359	**		βy, DG / −0.28
142860	5933	F6 IV	3.880	0.258	**	7	βy / −0.40 / −0.36
152792		G0 V	6.822	0.388			1, 3, 4, 8 / −0.45
157089		G0 V	(6.960)	0.333			βy, DG / −0.57
160693		G0 V	8.381	0.336	***		βy, DG / −0.69
161817		A2 VI	6.963	−0.008	***		DG / (−0.41)
164136	6707	F2 III	4.402	0.177		27	βy / 0.08
170153	6927	F7 V	3.546	0.279	**	11	1, 4, βy / −0.64
185657	7477	G6 V	(6.350)	0.677			4, 8, βy / −0.51

Table 4.1 (Continued)

HD or BD	HR or others	Sp	m_v	$B_2 - V_1$	Variab.	$V \sin i$	Remarks
187923	7569	G2 V	(6.154)	0.389	**	≤ 10	0.12 0.00 βy, DG
190404		K2 V	7.288	0.502	*		−0.20 DG
191046		K0 III	(7.200)	0.821			−0.42 DG
193370	7770	F5 Ib	(5.220)	0.401		13	−0.19 βy
197461	7928	A7 III	4.443	0.091	**	41	0.50 βy
198149	7957	K0 IV	(3.430)	0.605		<17	0.00 βy, DG
201626		K0 IIIp	8.121	0.823	***		−1.45 βy, DG
210027	8430	F5 V	3.770	0.216	***	7	−0.10 1, 4, βy
215648	8665	F7 V	(4.190)	0.265		7	−0.05 βy
218804	8825	F5 IV	(5.950)	0.224		18	−0.21, βy
219134	8832	K3 V	(5.570)	0.628			0.00 1, 3, 4, βy, DG
221170		Pop II gi.	7.677	0.803	**		−2.70 βy, DG
221345	8930	G8 III	(5.220)	0.705	*	<19	−0.20 DG
224930	9088	G3 V	(5.760)	0.420	*	≤ 6	−0.59 −0.70 −0.55 −0.60, βy, DG

TABLE 4.2

Stars with abundances determinations, Powell

HD or BD	HR or other	Sp	m_v	$B_2 - V_1$	Variab.	$V \sin i$	Remarks
9826	458	F8 V	(4.080)	0.294		8	−0.11 1, 3, 4, βy, DG
19373	937	G0 V	4.059	0.335		≤ 10	0.05 βy, DG
30652	1543	F6 V	3.199	0.230			0.18 βy
82328	3775	F6 IV	3.183	0.253	***	13	−0.03 βy
102870	4540	F8 V	3.636	0.301		0	0.15 1, 3, 4, 8, βy
109358	4785	G0 V	4.276	0.350		≤ 3	−0.23 βy, DG

Table 4.2 (Continued)

HD or BD	HR or other	Sp	m_v	$B_2 - V_1$	Variab.	$V \sin i$	Remarks
136202	5694	F8 IV–V	(5.060)	0.299		0	−0.17 βy
142373	5914	F9 V	4.614	0.340		0	−0.35 βy
142860	5933	F6 IV	3.880	0.258	**	7	−0.11 1, 3, 4, 8, βy
222368	8969	F7 V	4.130	0.269	**	6	0.09 βy

TABLE 5

Stars members of the Hyades

HD or BD	HR or other	Sp	m_v	$B_2 - V_1$	Variab.	$V \sin i$	Remarks
27176	1331	A8 V	(5.650)	0.073		97	2, 5, βy
27397	1351	F3 V	(5.590)	0.068		109	βy
27459	1356	F0 V	5.280	0.025		65	βy
27524		(F8)	(6.800)	0.203		94	βy
27628	1368	Am	(5.720)	0.104		15	2, 5, βy
27946	1388	A7 V	(5.280)	0.048		153	βy
28294	1408	(F0)	5.916	0.119		102	βy
28305	1409	K0 III	3.548	0.680		⩽ 8	1, 5, βy, DG
28406		(F8)	6.899	0.231		20	βy
28527	1427	A7 V	(4.780)	−0.019		69	βy
28546	1428	Am	5.499	0.043		23	βy
28556	1430	(F1)	(5.400)	0.048		95	βy
28568		(F2)	(6.510)	0.209		53	βy
30780	1547	A5	(5.100)	0.022		141	βy
32301	1620	A7 V	(4.640)	−0.036		127	2, 5, βy

TABLE 6

Stars members of Praesepe

HD or BD	HR or other	Sp	m_v	$B_2 - V_1$	Variab.	$V \sin i$	Remarks
	Prae 23	G		0.428	***		
19°02050	Prae 34	F2 V	9.446	0.193		<45	βy
73174	Prae 40	Am	7.759	−0.010		29	βy
19°02052	Prae 47	F4 V	9.812	0.239	**		βy
73345	Prae 114	F0 V	8.152	0.000		96	βy
	Prae 127	G2	10.818	0.340	***		βy
73430	Prae 143	A9 V	8.311	0.021	**	73	βy
20°02145	Prae 155	F6	9.393	0.192	**		βy
73598	Prae 212	K0 III	6.593	0.632	***	<45	βy, DG
73616	Prae 226	F2 V	8.899	0.111		131	βy
73641	Prae 227	F2 V	9.475	0.195	*	15	βy
73617	Prae 232	F5 V	9.221	0.156		127	βy

Table 6 (Continued)

HD or BD	HR or other	Sp	m_v	B_2-V_1	Variab.	$V \sin i$	Remarks
73640	Prae 239	F4 V	9.661	0.215		32	βy
20°02157	Prae 250	F6 V	9.775	0.230		120	βy
73665	3427 Prae 253	K0 III	6.396	0.648	*	0	4, 6, βy, DG
20°02161	Prae 271	F2 V	8.779	0.087	***	86	βy
73730	Prae 286	Am	8.006	−0.011	**	30	βy
	Prae 293	–	9.836	0.236			βy
20°02170	Prae 295	(F6)	9.347	0.189		95	βy
73746	Prae 318	F0 V	8.644	0.079		95	βy
73763	Prae 323	A9 V	7.814	0.018		130	βy
73798	Prae 340	F0 Vn	8.467	0.062		166	βy
73819	Prae 348	A6 Vn	6.766	−0.023		140	βy
73854	Prae 370	F5 V	9.014	0.133		116	βy
20°02180	Prae 396	F4 V	9.815	0.234			βy
73937	Prae 411	F4 V	9.321	0.170		49	βy
73974	Prae 428	K0 III	6.910	0.653		<45	βy, DG
73993	Prae 429	F0 V		0.093	**	195	βy
74028	Prae 445	A7 V	7.962	0.001		180	βy
20°02190	Prae 454	–	9.881	0.230	*		βy
74058	Prae 459	F2 V	9.204	0.164		130	βy
20°02192	Prae 472	–	9.765	0.213			βy
20°02193	Prae 478	F4 V	9.674	0.218		<45	βy
72779	3387 PraeVL 133	G8 III	6.584	0.429		95	βy
72846	PraeVL 166	A5 V	7.494	−0.039		140	βy

TABLE 7

Stars members of Coma Berenices

HD or BD	HR or other	Sp	m_v	B_2-V_1	Variab.	$V \sin i$	Remarks
105805	4633	A4 V	6.014	−0.055	**	172	2, 7, βy
106103		F5 V		0.190	***	<12	βy
106691		F2 V	8.090	0.188	**	30	βy
106946		F2 V	7.836	0.158	***	50	βy
107067		F8 V	8.696	0.282		≤12	βy
107132		G0 V	8.787	0.281		12	βy
107131	4684	A5 V	6.423	−0.022		175	2, 7, βy
107168	4685	Am	6.250	−0.024		≤12	βy
107276		Am	6.625	−0.011		95	βy
107399		G0 V	9.016	0.326			βy
107611		(F7)	8.503	0.237		15	βy
107685		(F7)	8.529	0.237	**	≤12	βy
107877		F5 V	8.358	0.219	***	20	βy
107966	4717	A4 p	5.163	−0.085		54	2, 7, βy
108154		(F8)	8.559	0.251		≤12	
108226		F6 V	8.337	0.221	**	≤12	βy
108486		Am	6.676	−0.012	*	30	
109307	4780	Am		−0.066		8	2,8

TABLE 8.1

Stars with known spectral energy distribution, Willstrop

HD or BD	HR or other	Sp	m_v	$B_2 - V_1$	Variab.	$V \sin i$	Remarks
1013	45	M2 III	4.837	1.188	***		1, 8
3196	142	F8 V		0.315		18	DG
4628	222	K2 V		0.549			3, 8, DG
9270	437	G8 III	3.654	0.664		<19	βy, DG
11636	553	A5 V	2.664	−0.061		73	1, 8, βy
14386	681	(gMbe)	4.684	1.500	***		
20630	996	G5 V		0.407		<17	βy, DG
23249	1136	K0 IV	3.548	0.601	***	<17	1, 3, 4, 8, DG
36673	1865	F0 Ib	2.607	0.018	***	13	βy
129247	5478	A3 III	3.769	−0.119		156	
142860	5933	F6 V	3.880	0.258	**	7	1, 3, 4, 8, βy
166197	6788	B1 V	6.117	−0.222			βy
173667	7061	F6 V	4.194	0.249		14	1, 8, βy
203504	8173	K1 III	4.104	0.758		<17	βy, DG
209747	8413	K4 III		1.041	**	<17	DG
217987		M2 V	7.356	1.131			

All these stars have also to satisfy conditions 1 to 4 (except complementary information given in the column 'remarks'). For each star we give the colour indice $B_2 - V_1$. The 7 heterochromatic colours normalised to B can be found in (1971); the 7 monochromatic colours (deduced from the 7 heterochromatic colours) normalised to G and the 7 effective wavelengths of the monochromatic colours can be obtained at the Geneva Observatory (Monochromatic colours Catalogue). Let us point out that the λ_{eff} of filter U may not always be very accurate for stars having a large Balmer discontinuity.

Many stars are common to several tables; the numbers of the tables are given in the 'remarks' column. The magnitudes given are the V magnitudes established by Rufener and Maeder (1972), the magnitudes in brackets are taken from Literature. The spectral classifications given in tables 3 to 7 are taken from the catalogues of Jaschek et al. (1964) and from its extension by Kennedy. In the remarks column we also indicate by βy the stars also measured in the $u v b y \beta$ system, the general catalogue of which is being prepared by Lindemann and Hauck (1973); and by DG the cool stars measured in the Copenhagen system (1970).

The selection presented here gathers together stars which may serve to calibrate criteria for M_v, Sp, [Fe/H], etc. But these are also stars to which we believe it is important to give some attention and to attempt to determine the fundamental parameters M_v, θ_{eff}, $\log g$, χ. The proposed selection should be enriched with hot stars of well known absolute magnitude and reddening. For this purpose, one must measure young clusters of well determined distance modulus.

TABLE 8.2

Stars with known spectralenergy distribution, Terechtchenko and Kharitonov

HD or BD	HR or other	Sp	m_v	$B_2 - V_1$	Variab.	$V \sin i$	Remarks
3196	142	F8 V		0.315		18	DG
4727	226	B5 V	4.525	−0.242		75	βy
8538	403	A5 V	2.653	−0.045	**	116	1, 8, βy
11636	553	A5 V	2.664	−0.061		73	1, 8, βy
16970	804	A2 V		−0.087		183	
20320	984	Am	4.797	0.023	*	68	βy
22928	1122	B5 III	3.012	−0.240		271	1, 8, βy
25490	1251	A1 V		−0.127		71	2, 8, βy
35468	1790	B2 III	1.634	−0.297		64	1, 8, βy
58715	2845	B8 V		−0.208		270	βy
71155	3314	A0 V	3.902	−0.159		122	1, 2, 8, βy
77327	3594	B9n	3.584	−0.147		219	2, 8, βy
79469	3665	B9.5 V	–	−0.208		86	2, 8, βy
87901	3982	B7 V	1.375	−0.219		354	1, 8, βy
90089	4084	F5 IV	5.259	0.180	*	107	βy
97633	4359	A2 V	3.320	−0.156		14	1, 2, 8, βy
102870	4540	F8 V	3.636	0.301		0	1, 3, 4, 8, βy
103287	4554	A0 V	2.452	−0.154		163	1, 2, 8, βy
106112	4646	Am	5.145	0.109		69	βy
108382	4738	A2	4.986	−0.074		89	2, 8
110379	4825	F0 V	2.766	0.141	*	27	1, 3, 8, βy
120315	5191	B3 V	1.856	−0.258		216	βy
123299	5291	A0 III	3.659	−0.180	***	12	1, 8
129247	5478	A3 III	3.769	−0.119		156	
135742	5685	B8 V		−0.209	*	230	1, 8, βy
148112	6117	A1p		−0.156		28	βy
170153	6927	F7 V	3.546	0.279	**	11	4, 8, βy
202444	8130	F0 IV		0.185		94	
207098	8322	Am		0.095	*	104	βy
212061	8518	B9 III		−0.193		82	βy
218658	8819	G2 III	4.401	0.510		22	DG
224617	9072	F4 IV		0.206	*	34	βy

4. Stars With Known Spectral Energy Distribution

In Table 8 we give the 7 colours of the stars whose spectral energy distribution is given by Willstrop (1965) and by Terechtchenko and Kharitonov (1972). The selected stars do not satisfy the 4 conditions given in paragraph 3. They have a weight $p \geqslant 3$ in the $U \, V \, B \, B_1 \, B_2 \, V_1 \, G$ photometry, but can be binaries, reddened, and have a high rotational velocity. Nevertheless, they are useful for the photometrist who can use them to check the quality of the determination of his system's bandpasses.

References

Cayrel, R. and Cayrel de Strobel, G.: 1966, 'Abundance Determinations from Stellar Spectra', in *Ann. Rev. Astron. Astrophys.* **4**.

Cowley, A. Cowley, Ch., Jaschek, M., and Jaschek, C.: 1969, *Astron. J.* **74**, 375.

Dickow, R., Gyldenkerne, K., Hansen, L., Jacobsen, P. U., Kjaergaard, K., and Olsen, E. H.: 1970, *Astron. Astrophys. Suppl. Ser.* **2**, 1.

Gliese, W.: 1969, 'Catalogue of Nearby Stars', *Veröffentl. Astron. Rechen-Inst. Heidelberg* No. 22.

Golay, M.: 1972, *Vistas in Astronomy* **14**, in press.

Golay, M.: 1969, in O. Gingerich (ed.), *Theory and Observations of Normal Stellar Atmospheres*, M.I.T. Press, Cambridge.

Golay, M.: 1971, in Ch. Fehrenbach and B. E. Westerlund (eds.), 'Spectral Classification and Multicolour Photometry', *IAU Symp.* **50**, 145.

Jaschek, C., Condé, H., and Sierra, A.: 1969, *Catalogue of Stellar Spectra Classified, in the Morgan-Keenan System*, La Plata.

Johnson, H. L. and Morgan, W. W.: 1953, *Astrophys. J.* **117**, 313.

Kennedy, P. M.: 1970, *Extension du Catalogue de Jaschek et al.*, not published.

Lindemann, E. and Hauck, B.: 1973, *Astron. Astrophys., Suppl.* **11**, 119.

Maeder, A. and Peytremann, E.: 1970, *Astron. Astrophys.* **7**, 120.

Maeder, A. and Peytremann, E.: 1972, *Astron. Astrophys.* **21**, 279.

Morgan, W. W. and Roman, N. G.: 1950, *Astrophys. J.* **112**, 362.

Morgan, W. W., Harris, D. L., and Johnson, H. L.: 1953, *Astrophys. J.* **118**, 92.

Peytremann, E.: 1970, Thesis, Obs. de Genève.

Powell, A. L. T.: 1970, *Monthly Notices Roy. Astron. Soc.* **148**, 477.

Rufener, F. G.: 1971, *Astron. Astrophys. Suppl. Ser.* **3**. 181.

Rufener, F. and Maeder, A.: 1972, this volume, p. 156.

Terechtchenko, V. M. and Kharitonov, A. V.: 1972, *Trudy Astrofiz. Inst. Alma-Ata*, 21.

Willstrop, R. V.: 1965, *Men. Roy. Astron. Soc.* **69**, part 3.

DISCUSSION

Pecker: When trying to apply to observed stars the very fine rotating models built by Maeder, for example, we should keep in mind that the measured '$V \sin i$' plotted in the tables of Golay are not necessary measured *rotations*, they are just measurements of broadening, and stellar lines can be broadened by other causes, such as macro-velocity fields. Therefore, the 'rotating models' are not necessarily adequate to fit the observations.

Kodaira: I am of the same opinion as Dr Pecker, but as the zeroth approximation, one can regard the broadening of lines as the result of the rotation. It is worthwhile to try to establish the possible rotation effect observationally. But Dr Golay limits the value of $V \sin i$ to below 150 km s^{-1}, in order to avoid the influence of the stellar rotation. In doing so, however, you would bias your data, by mixing the intrinsic slow rotators with stars of high or moderate rotational velocity whose axises have small or moderate inclination (near 'pole-on'). My suggestion is to extend the standard stars as far as possible to a higher value of $V \sin i$ (~ 350), so far as no emission lines are observed.

Maeder: In connexion with Prof. Pecker's remark, I will firstly add that the fact that we do not know the law of rotation inside the stars is a black point in the theory of rotating models. In that respect, we may perhaps hope that the comparisons between observations and models bring some information on this subject. Secondly, so-called *observable* quantities like $V \sin i$ in fact are estimated by means of *theoretical* models.

Garrison: This may seem to be a comical remark, but I will make it semi-seriously. Perhaps we should choose stars with large $V \sin i$ as standards. At least then we know they are not pole-on and it should be a more homogeneous group.

FitzGerald: To add to Dr Garrison's remark that perhaps the fast rotation should be used as standards, it should be remarked that there was a paper at Athens suggesting that Am stars were slow rotators, in which element differentiation could occur, whereas in the moderate rotation mixing occurs preventing element differentiation. Thus the 'normal' stars suitable for standards should be the moderate rotators.

Hanbury-Brown: The angular diameters of 32 stars have been measured at Narrabri Observatory ranging in spectral type from O5 to F8. I will send a list of these stars to Prof. Golay for consideration as possible members of his list of standard stars.

Pecker: When one shifts from a group of stars of a given brightness to a less bright, methods of observation are changed. For example, (1) very bright stars are studied with very large dispersion; abundances etc. are determined. (2) bright stars are studied with lower spectrographic resolution and are more numerous. (3) still more numerous, less bright stars are available for 7-colour photometry (for example) etc.... The stars to be used as standards for group (n) have to be taken in groups (1) (2)... ($n-1$), but the smaller the n is, the smaller is the number of standard stars that are not monsters, that are 'safe'. By going to larger and larger groups, one improves the sampling, one deteriorates the quality of the 'standard' physical characteristics of this sampling. The whole thing is essentially a matter of successive approximations, and of the choice of 'proper equilibrium' between the completeness of the sampling and the physical meaning of this sampling.

Lamla: Which effects have faint lines in your 7-colour system? You compare your integrated brightness values with model intensities which give, I guess, the continuum without lines.

Golay: No, we use the models of Peytremann (Thesis, Geneva (1970)) which include the opacities of metallic lines.

Garrison: It is perhaps useful to clarify the situation regarding 'standard' stars in the MK system. For this purpose we can talk about fundamental standards, which are those stars to which the MK system is anchored; about primary standards, which are those stars that have been studied extremely carefully, such as the ones listed in the Johnson-Morgan paper; about secondary standards, which are stars within a given set of plates which fit the above standards and look very normal; and about the non-standard types which are stars that are observed carefully and classified carefully using the MK system. These latter stars should not be used with the same weight, in general, as the fundamental or primary standards as representatives of the MK system. In Prof. Golay's paper, for example, he has referred to the paper by the Cowley's and the Jaschek's as standard. I think that Prof. Jaschek would agree that his paper is in the last category.

Hack: Because of the uncertainty of oscillator strengths and because of the difficulty to fit stellar atmospherical models and observations, probably the safest way to derive atmospherical chemical composition is the method of the differential curve of growth. Hence we need standard stars for which we have good data on spectral energy distribution, Balmer discontinuity etc. so as to have reliable data on T_{eff} and g, and $V \sin i < 20$ km s^{-1}, and to correlate all these standards to the Sun by a 'grid' of standards covering the whole HR diagram, that is by comparing a G0 V to the Sun, a F8 V to the G0 V and so on.

Keenan: In connection with your list of standard stars for observation in as many ways as possible I would recommend first the inclusion of all the stars in the lists that Morgan and I will include in our review article that is coming out in *Annual Reviews of Astronomy and Astrophysics* next year. These lists are not ready yet, but in the meantime I shall send you soon a short list of some stars, mostly fainter than $V = 6.5$, that can be considered as type stars of some of the special population groups. If you do observe these in your colour system it will be very interesting to see how you distinguish them from the ordinary stars of types G, K and M. The types that I give in the list represent my most recent ones.

Murray: I would like to put in a plea for more late type dwarfs among the standards; there is only one in Golay's lists. There are obvious difficulties due to faintness and probable duplicity of many of these stars, but accurate calibration of their luminosities are very important. Fortunately many of these stars have good trigonometric parallaxes, and soon there will be more parallaxes available for Vyssotsky's stars which are on the observing lists of several parallax observers.

Gliese: The distance determinations of Vyssotsky's stars (which are dK8 to dM2) is not sufficient since the maximum of the luminosity function in the solar neighbourhood is supposed to be in the region of the dM8 stars. Probably the most promising way is by the parallax program of the USNO in combination with R, I measurements which allow the calibration of the $(M_v, R-I)$ relation of these very red dwarfs. This relation should be used for late-type red dwarfs, for instance, near the galactic poles.

Jones: I have observed 700 M stars with 30 Å interference filters:

    7460 Window
    7100 TiO
    6830 Ca H
    6076 Window.

As was first shown by Ohman Ca H is luminosity sensitive. On the present system TiO: CaH is a dwarf/giant discriminator with amplitude ~ 0.4 mag. 7460–6076 correlates well with $R-I$ and photometric parallaxes can be determined from the m_{7460} vs $m_{7460} - m_{6076}$ plot. The TiO strength correlates well with that observed by Wing. At the Cordoba Symposium he proposed a new spectral type scale for dwarfs, based on giants with the same TiO strength. I suggest the same standards be adopted here. In particular I confirm Wing's conclusion that Proxima Centauri is ~ .04 of a spectral type earlier than Wolf 359. I have observed the 100 M stars discovered by McCarthy and his co-workers in the South Galactic Cap. About 25 are dwarfs with median proper motion ~ 0 ″.2. Only a couple are closer than 20 pars. Of course, the volume of space searched at M6 V is much smaller than at M0 V.

Pecker: Theoreticians want to get from observers L, M, R, – to induce results on evolutionary properties, etc. It seems to me that from our point of view, interesting stars are necessarily falling in the region of the conceptual (M_v, apparent Radius, apparent binary separation) diagram where data at present are sufficiently well known. 'Standards' are there in *small* numbers; but the use of photometry, using this small number of objects as primary standards, may extend the number of stars that could be used as secondary standards.

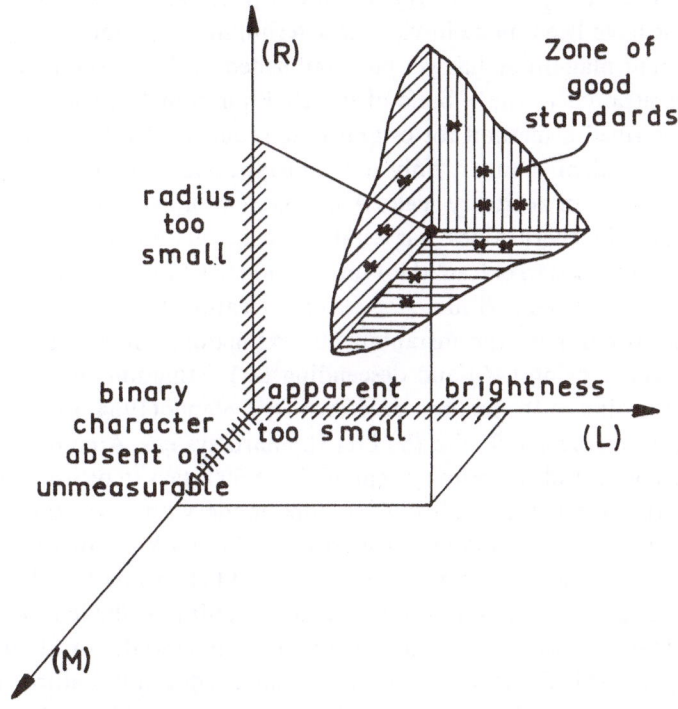

Fig. 6

DETERMINATION OF THE APPARENT V MAGNITUDES

F. RUFENER and A. MAEDER

Geneva Observatory, Switzerland

Abstract. V photoelectric magnitudes are determined on an independent basis. Correlation with V sequence of the UBV system confirms the coincidence of the two definitions. Comparisons with the *Catalogue of Bright Stars* revealed that more than 40% of BS stars have an apparent V magnitude with a probable error of $0^{\text{m}}1$.

1. Introduction

The Geneva Observatory Photometry has seven pass-bands. The observations and their reductions have been made in view of determining six colour indices with accuracy. No apparent magnitude has yet been published in this photometry. One of the pass-bands corresponds to the V band of the UBV photometry. Rufener and Maeder (1970) have described a method for establishing sequences of V magnitudes. This is done in an independent manner, only one constant has to be chosen to adjust the sequence obtained to an arbitrary scale. A first system of 240 standard stars covering the Northern and Southern Skies was established in this manner. Following this, a procedure was developed to extract the V magnitude from all the observations already made in the Geneva System. A first step was the treatment of those nights for which it was possible to compute the instantaneous extinction. These are more than 100 nights during which rising (M) and descending (D) extinction stars were observed. A reduction according to the method described by Rufener (1964) gave accurate magnitudes which were correlated with the first standard system. A compilation of these results produced a second standard system of about 500 stars to be used for the second step. This was the treatment of all observations made with a constant air mass and without measurements of extinction. This part of the work is not yet completed. A procedure that we will describe later on allows a weighted correlation of these observations. An intermediary compilation of all the magnitudes available at present has allowed us to obtain at least two measurements of V magnitude for about 1400 stars. We have compared this list with two sets of V magnitudes in literature. The object is to check whether the definition of V by Geneva is comparable to that of the UBV system. This question was already partly answered when Rufener and Maeder (1970) determined sequences of V magnitudes for the Pleiades and Praesepe clusters which were in very good agreement with those already published. The inner accuracy of the Geneva V magnitudes has not yet been analysed in detail; nevertheless, an average standard deviation of $0^{\text{m}}01$ is characteristic for the average values of three measurements of stable stars that are easy to measure.

2. Comparison with the BS Magnitudes

In the BS catalogue (*Bright Stars Catalogue*, Hoffleit 1964), one finds an estimate of the visual magnitude belonging to one of the three following groups:

B. Hauck and B. Westerlund (eds.), Problems of Calibration of Absolute Magnitudes and Temperature of Stars, 298–301.

(a) The photoelectric V magnitude of the UBV system (57% of stars).

(b) The visual magnitude of the first edition of the catalogue, corrected by E. Rybka in view of getting a better correspondence with the V system. This magnitude is accompanied by the suffix R (26% of stars).

(c) The visual magnitude of the first edition, with the suffix H (17% of stars).

For the stars of each of the above-mentioned groups, we have computed correlations of the type: $V^* = a_1[V] + a_0$, $[V]$ being here the Geneva V magnitude. Two standard deviations have been computed: σ_1, the standard deviation of the differences $\Delta_1 = [V] - V$ and σ_2, the standard deviation of the differences after correlation: $\Delta_2 = V^* - V$.

A. CASE OF THE PHOTOELECTRIC MAGNITUDES

Table I gives a summary of the correlation parameters obtained for a few selections made in group A. The supergiants, which are often variable, have been excluded from the first three samples.

TABLE I

Correlation parameters for photoelectric magnitudes

Selection criteria	Nb. of stars	σ_1	σ_2	a_0	a_1
Stars with indice $(B_2 - V_1)$ < −0.05 (blue stars)	92	0^m017	0^m017	−0.010	1.002
Stars of spectral type earlier than K	337	0^m020	0^m020	−0.023	1.005
K and M stars	134	0^m030	0^m025	−0.061	1.009
No selection	508	0^m024	0^m023	−0.031	1.006

The mean value of the correlated magnitudes being close to 5, one notices that the coincidence of the zeros of the sequences is good. The fact that the slopes are slightly larger than unity does not seem to signify an error in linearity, but would rather be related to the spectral type of the stars. Indeed, the relative frequency of the deviations $\Delta_1 = [V] - V$ exceeding 0^m04 is distinctly greater for K and M type stars. Let us point out that the consultation of the USNO (1970) UBV catalogue shows that there are several dispersed determinations of V for almost all stars having a large Δ_1. The one figuring in the BS is often extreme, and the choice of a favorable determination would allow the reduction of the majority of larger Δ_1 to a normal value.

B. CASE OF MAGNITUDES WITH SUFFIX R

The stars of this group are distributed over the whole Northern Hemisphere and are of mixed spectral types. Their V magnitudes are little dispersed since they have values between 5^m2 and 6^m7 with an average value close to 6^m0. The computed correlation parameters are in Table II. We deduce that the magnitudes with suffix R are too low, the mean error is

$$\bar{\Delta}_R = 0\overset{m}{.}10 \pm 0\overset{m}{.}08 .$$

C. CASE OF MAGNITUDES WITH SUFFIX H

The stars of this group are mainly distributed in the Southern Hemisphere. Like in group B, the range of their V magnitudes is small ($5\overset{m}{.}3$ to $6\overset{m}{.}6$) and has a mean value close to $6\overset{m}{.}0$. After having computed the correlations (see Table II), we find that these magnitudes are on an average too large. The mean error is

$$\bar{\Delta}_H = - 0\overset{m}{.}08 \pm 0\overset{m}{.}09 .$$

TABLE II

Correlation parameters for R and H magnitudes

Group	Nb. of stars	σ_1	σ_2	a_0	a_1
Suffix R	99	$0\overset{m}{.}133$	$0\overset{m}{.}082$	0.424	0.913
Suffix H	29	$0\overset{m}{.}138$	$0\overset{m}{.}091$	0.433	0.942

3. Comparison With the 'Gliese' Magnitudes

V magnitudes of the UBV photometry are given in the *Catalogue of Nearby Stars* published by Gliese (1969). We have our own estimate of V for 161 stars of magnitudes 1 to 10 of this list. The parameters of the computed correlation are

$$\sigma_1 = 0\overset{m}{.}025 \qquad \sigma_2 = 0\overset{m}{.}024 \qquad a_0 = - 0.016 \qquad a_1 = 1.003 .$$

The average magnitude of this sample being close to $5\overset{m}{.}5$, we observe again a good adjustment of scale. It is difficult to explain the calculated value of the coefficient $a_1 = 1.003$. The suppression of the stars having magnitudes smaller than $4\overset{m}{.}0$ does not reduce this small difference to unity, thus it does not arise from a lack of linearity due to saturation of the Geneva System. As for the large deviations $\Delta_1 = [V] - V$, these cannot be reduced with a colour equation.

4. Conclusions

At the present we retain the conformity of our V magnitudes with those of the UBV system. The zeros of both scales coincide. The correlations established here give a slope slightly larger than 1.0. The one computed by Rufener and Maeder (1970) for Praesepe gave 1.0, whereas the Pleiades gave a value slightly smaller than 1.0. Because of the very small improvement observed between σ_1 and σ_2, one must not attribute too great an importance to these small deviations. The characteristic standard deviation observed during the comparisons made above ($\sigma_1 = 0.020$ to 0.025) is of the same order as the one obtained by Johnson *et al.* (1966) when they compared the V observations made at Catalina with the original V observations; the photoelectric equipment (filter and cell) being the same. The only case where a standard deviation of $0^m.01$ was observed for correlations of our V magnitudes with other determinations occurred during comparisons with the V sequences of superior quality given in literature for Praesepe and the Pleiades.

More than 40% of the BS (edition 1964) stars have a visual magnitude with a probable error of a tenth of a magnitude.

References

Blanco, V. M., Demers, S., Douglass, G. G., and FitzGerald, M. P.: 1970, *Publ. U.S. Naval Obs.* **21.**
Gliese, W.: 1969, *Veröffentl. Astron. Rechen-Inst. Heidelberg* **22.**
Hoffleit, D.: 1964, *Catalogue of Bright Stars,* Yale University Observatory, New Haven, Connecticut.
Johnson, H. L., Mitchell, R. I., Iriarte, B., and Wisniewski, W. Z.: 1966, *Commun. Lunar Planetary Lab.* No. 63, **4,** 99.
Rufener, F.: 1964, *Publ. Obs. Genève, Série A,* Fasc. 66.
Rufener, F. and Maeder, A.: 1970, *Publ. Obs. Genève, Série A,* Fasc. 78.

DISTRIBUTION DE L'ÉNERGIE DANS LE
SPECTRE ULTRAVIOLET DE QUELQUES ÉTOILES

D. CHALONGE and L. DIVAN

Institut d'Astrophysique, Paris, France

and

C. T. HUA

Laboratoire d'Astronomie Spatiale, Marseille, France

Résumé. On donne les résultats préliminaires des comparaisons directes entre quelques étoiles, O, B (et également de α Lyr) et une source terrestre préalablement rapportée au corps noir. Il semble que les gradients ultra-violets utilisés jusqu'ici doivent être corrigés de $+0,10$ en moyenne.

Abstract. Preliminary results of recent direct comparisons between some early type O, B standards (and also α Lyr) and a terrestrial source previously compared to a black body are given for the ultra-violet range. It appears that the previously-used gradients ϕ_{uv} must be corrected on the average by $+0.10$.

Les recherches de spectrophotométrie stellaire en cours depuis plusieurs années à l'Institut d'Astrophysique ont fourni des données caractérisant la distribution de l'énergie dans le spectre continu de plusieurs centaines d'étoiles des premiers types spectraux. Ces données sont constituées par les valeurs d'un certain nombre de paramètres: les gradients ϕ_b et ϕ_{uv} définissant respectivement la température de couleur du spectre visible (région bleue-violette) et celle du spectre ultraviolet (3700–3150 Å) et les quantités D et λ_1 caractérisant la grandeur et la position dans le spectre de la discontinuité de Balmer. Elles sont déduites de la comparaison des étoiles étudiées à un certain nombre d'étoiles standard.

Les valeurs des paramètres de ces étoiles standard ont été déterminées à partir des comparaisons directes entre quelques étoiles et une source terrestre, comparaisons effectuées entre 1936 et 1938 par Barbier et Chalonge (1941).

Ces anciennes déterminations n'ont certainement pas la précision que l'on pourrait atteindre à partir des sources terrestres de comparaison dont on dispose actuellement et avec les méthodes photoélectriques de mesures modernes. Aussi de nouvelles comparaisons directes ont-elles été entreprises entre quelques étoiles brillantes des premiers types et une source terrestre aussi bien définie que possible.

Les mesures, couvrant l'intervalle spectral 6100–3150 Å, ont été effectuées en Octobre 1971 à 'Observatoire de Haute-Provence, en utilisant, pour les observations stellaires, le petit spectrographe à optique de quartz décrit par Baillet et Chalonge (1973) fixé au foyer Cassegrain d'un petit télescope de 25 cm de diamètre. Quelques étoiles O et B (et également α Lyr) ont été comparées à une étoile artificielle distante de 73 m et constituée par la source luminescente décrite par Chalonge et Servigne (1952), source dont la répartition d'énergie est très analogue à celle d'une étoile O. Ainsi les spectres stellaires et ceux de la source luminescente, pris sur la même plaque

B. Hauck and B. E. Westerlund (eds.), Problems of Calibration of Absolute Magnitudes and Temperature of Star, 302–304.

et avec le même temps de pose, présentent des noircissements toujours voisins, faciles à comparer avec précision par les méthodes de photométrie photographique. D'autre part, ces méthodes spectrophotométriques permettent de déterminer de façon précise et correcte (Chalonge et Divan, 1972; Chalonge et Hua, 1972) l'absorption atmosphérique pendant chaque nuit d'observation, pour tout l'intervalle spectral considéré et en tenant compte de l'absorption par l'ozone, variable d'une nuit à l'autre.

La comparaison, très délicate, entre la source luminescente de luminance faible et relativement riche en ultraviolet, à une lampe à ruban de tungstène, très brillante dans le visible et pauvre en ultraviolet, était effectuée au Laboratoire d'Astronomie Spatiale, au moyen d'un monochromateur double de Zeiss et d'un appareillage photoélectrique de grande précision. La lampe à ruban elle-même a été comparée au corps noir de l'Observatoire de Heidelberg.

Résultats

Les résultats donnés ici ont un caractère très préliminaire. En effet, pour la plupart des nuits d'observation, les clichés ont révélé une absorption atmosphérique très variable, notamment dans le domaine visible par suite de la présence de brume et les déterminations les plus précises obtenues jusqu'ici (le travail de dépouillement n'est pas achevé) se rapportent à la région ultraviolette.

Le Tableau I (colonne 2) donne les valeurs de ϕ_{uv} obtenues en utilisant seulement les observations faites au cours des deux meilleures nuits (15 et 17 Octobre). La colonne 3 donne les valeurs correspondant à des mesures relatives entre étoiles, mesures partiellement publiées (Divan, 1966) et dont le zéro de l'échelle est défini par les comparaisons absolues de Barbier et Chalonge (1941).

TABLEAU I

Comparaison des ϕ_{uv} actuels et des ϕ_{uv} anciens

Étoile	ϕ_{uv} actuels	ϕ_{uv} anciens	Terme correctif
S Mon	0.66	0.52	+0.12
10 Lac	0.79	0.62	+0,15
γ Ori	0,73	0,67	+0,06
γ Peg	0,74	0,69	+0,05
ε Per	0,75	0,67	+0,08
α Lyr	1,48	1,36	+0,12
		Moyenne	+0,10

On voit que les ϕ_{uv} anciens devraient être majorés de 0.10 en moyenne mais ce résultat devra être confirmé par de nouvelles observations faites dans des conditions meilleures permettant de déterminer également les ϕ_b et les D.

Remerciements

Nous remercions le Professeur D. Labs de l'Observatoire de Heidelberg d'avoir bien voulu mettre à notre disposition son laboratoire de calibration absolue pour la comparaison de la lampe à ruban au corps noir.

Bibliographie

Baillet, A. et Chalonge, D.: 1973, *Nouvelle Revue d'Optique Appliquée*, à paraître.
Barbier, D. et Chalonge, D.: 1941, *Ann. Astrophys.* **4**, 30.
Chalonge, D. et Servigne, M.: 1952, *Ann. Astrophys.* **15**, 151.
Chalonge, D. et Divan, L.: 1972, à paraître.
Chalonge, D. et Hua, C. T.: 1972, à paraître.
Divan, L.: 1966, dans K. Lodén, L. O. Lodén, and U. Sinnerstad (eds.), 'Spectral Classification and Multicolour Photometry', *IAU Symp.* **24**, 311.